TEUBNER-TEXTE zur Physik   Band 30

M. Bordag (Hrsg.)

Quantum Field Theory Under
the Influence of External Conditions

TEUBNER-TEXTE zur Physik

Herausgegeben von
Prof. Dr. Werner Ebeling, Berlin
Prof. Dr. Manfred Pilkuhn, Stuttgart
Prof. Dr. Bernd Wilhelmi, Jena

This regular series includes the presentation of recent research developments of strong interest as well as comprehensive treatments of important selected topics of physics. One of the aims is to make new results of research available to graduate students and younger scientists, and moreover to all people who like to widen their scope and inform themselves about new developments and trends.

A larger part of physics and applications of physics and also its application in neighbouring sciences such as chemistry, biology and technology is covered. Examples for typical topics are: Statistical physics, physics of condensed matter, interaction of light with matter, mesoscopic physics, physics of surfaces and interfaces, laser physics, nonlinear processes and selforganization, ultrafast dynamics, chemical and biological physics, quantum measuring devices with ultimately high resolution and sensitivity, and finally applications of physics in interdisciplinary fields.

Quantum Field Theory Under the Influence of External Conditions

Edited by

Dr. Michael Bordag

University of Leipzig

 B. G. Teubner Verlagsgesellschaft
Stuttgart · Leipzig 1996

Dr. rer. nat. habil. Michael Bordag

Born in 1952 in Dresden. Studied physics in Leningrad State University from 1972 to 1977. He received his first doctoral degree (Dr. rer. nat.) in 1980 from the University of Leipzig for research on Light Cone Expansion. He worked from 1983 to 1987 at JINR Dubna, Laboratory for Theoretical Physics, after he returned to Leipzig. In 1987 he received his second doctoral degree (in 1992 transformed into Dr. rer. nat. habil.) for work on Quantum Electrodynamics with boundary conditions. Teaching at the University of Leipzig, mainly Quantum Mechanics and special lectures. Research on Quantum Field Theory, vacuum structure, quantization with boundary conditions, Casimir effect.

Gedruckt auf chlorfrei gebleichtem Papier.

Die Deutsche Bibliothek – CIP-Einheitsaufnahme

Quantum field theory under the influence of external conditions /
ed. by Michael Bordag. -
Stuttgart ; Leipzig : Teubner, 1996
  (Teubner-Texte zur Physik ; Bd. 30)
    ISBN 978-3-663-01205-4   ISBN 978-3-663-01204-7 (eBook)
    DOI 10.1007/978-3-663-01204-7

NE: Bordag, Michael [Hrsg.]; GT

Bindung: Buchbinderei Bettina Mönch, Leipzig
Umschlaggestaltung: E. Kretschmer, Leipzig

Preface

This volume contains papers based on talks delivered at the

Third Workshop on Quantum Field Theory under the Influence of External Conditions

held at the University of Leipzig from September 18th to September 22nd, 1995. This series of workshops is organized by the quantum field theory group at the University of Leipzig. The first and the second Workshop in the row have been held in 1989 and 1992. The present meeting has been joined by more than 70 participants. This fact underlines the vivid interest in the field covered by the Workshop ranging from the Casimir effect to semiclassical gravity.

The material of this volume is organized along five subjects. The first part is devoted to the Casimir effect itself. The actual interest is mostly in non-stationary boundaries and a possible explanation of the phenomenon of sonoluminescence – a still open and challenging question.

The second part deals with QED and QCD in external fields. The main interest was focused on the effective action approach, on some aspects of the bag model and on quantum effects in a conical background.

The ground state in external fields is studied in the contributions of the third part. They demonstrate that the zeta function regularization together with the heat kernel expansion is the most powerful method at present. New calculations of functional determinants, of finite temperature effects and an interesting method applicable in the presence of smooth background fields are presented.

The fourth and the fifth part are devoted to quantum effects involving gravitational fields. Quantum corrections in the background of black holes – mainly corrections to the entropy – are discussed on a high standard. Questions of quantum gravity, geometrical quantization and bosonic strings as well as a few topics in Quantum Optics are also considered.

The daily program of the Workshop included 4 lectures in the morning and 8 shorter contributions in the afternoon. This led to a very demanding schedule and the participants listened with admirable attention. The comprehensive program created the basis for intensive work and fruitful discussions. People came together to look for new collaborations or to intensify the existing ones.

In this volume the morning lectures are represented almost completely, both in length and number. Due to the limited space available the afternoon talks are represented up to some exceptions by enlarged abstracts only. For some authors this was a quite strong restriction I would like to apologize for. The complete list of addresses of the participants is added so that the interested reader can contact the authors for further details. In contributions with several authors the name of the speaker is underlined.

I hope that the collection of papers presented here provides an instructive survey of the current problems of interest in the field and stimulates further work.

On behalf of all participants I have to thank the Deutsche Forschungsgemeinschaft, the Sächsisches Staatsministerium für Wissenschaft und Kunst and the Center for Higher Studies of the University of Leipzig for substantial financial support.

I am indebted to all people having contributed to the success of the Workshop. This includes all participants, the speakers and the session leaders. Especially I have to thank my colleagues from the Institute for Theoretical Physics of the University of Leipzig for assistance and my co-workers Klaus Kirsten and Joachim Lindig who did most of the administrative work.

Leipzig, January 1996   M. Bordag

Contents

III Ground State in External Fields   109

IV Quantum Fields in Black Hole Background   173

Part I

Casimir Effect

Casimir Energy for a Spherical Cavity in a Dielectric: Toward a Model for Sonoluminescence?

Kimball A. Milton

Abstract

In the final few years of his life, Julian Schwinger proposed that the "dynamical Casimir effect" might provide the driving force behind the puzzling phenomenon of sonoluminescence. Motivated by that exciting suggestion, I have computed the static Casimir energy of a spherical cavity in an otherwise uniform material with dielectric constant ϵ and permeability μ. As expected the result is divergent; yet a plausible finite answer is extracted, in the leading uniform asymptotic approximation. That result gives far too small an energy to account for the large burst of photons seen in sonoluminescence. If the divergent result is retained (which is different from that guessed by Schwinger), it is of the wrong sign to drive the effect. Dispersion does not resolve this contradiction. However, dynamical effects are not yet included.

1 Introduction

In a series of papers in the last three years of his life, Julian Schwinger proposed [1] that the dynamical Casimir effect could provide the energy that drives the copious production of photons in the puzzling phenomenon of sonoluminescence [2, 3]. In fact, however, he guessed an approximate (static) formula for the Casimir energy of a spherical bubble in water, based on a general, but incomplete, analysis [4]. He apparently was unaware that I had, at the time I left UCLA, completed the analysis of the Casimir force for a dielectric ball [5]. It is my purpose here to carry out the very straightforward calculation for the complementary situation, for a cavity in an infinite dielectric medium. In fact, I will consider the general case of spherical region, of radius a, having permittivity ϵ' and permeability μ', surrounded by an infinite medium of permittivity ϵ and permeability μ.

Of course, this calculation is not directly relevant to sonoluminescence, which is anything but static. It is offered as only a first step, but it should give an idea of the orders of magnitude of the energies involved. It is an improvement over the crude estimation used in [1]. Attempts at dynamical calculations exist [6, 7]; but they are subject to possibly serious methodological objections. Sonoluminescence aside, this calculation is of interest for its own sake, as one of a relatively few nontrivial Casimir calculations with nonplanar boundaries [8, 9, 10, 11, 12, 13, 14, 20]. It represents a significant generalization on the calculation of Brevik and Kolbenstvedt [16], who consider the same geometry with $\mu\epsilon = \mu'\epsilon' = 1$, a special case, possibly relevant to hadronic physics, in which the result is unambiguously finite.

In the next section we review the Green's dyadic formalism we shall employ, and compute the Green's functions in this case for the TE and TM modes. Then, in Section 3, we compute the force on the shell from the discontinuity of the stress tensor. The energy is computed similarly in Section 4, and the expected relation between stress and energy is found. Estimates in Section 5 show that the result so constructed, even with physically required subtractions, and including both interior and exterior contributions, is divergent, but that if one supplies a plausible contact term, a finite result (at least in leading approximation) follows. (Physically, we expect that the divergence is regulated by including dispersion.) Numerical estimates of both the divergent and finite terms are given in the conclusion, and comparison is made with the calculations of Schwinger.

2   Green's Dyadic Formulation

I follow closely the formulation given in [10, 5]. We start with Maxwell's equations in rationalized units, with a polarization source $\mathbf{P}$: (in the following we set $c = \hbar = 1$)

$$\boldsymbol{\nabla} \times \mathbf{H} = \frac{\partial}{\partial t}\mathbf{D} + \frac{\partial}{\partial t}\mathbf{P} \ , \qquad \boldsymbol{\nabla} \cdot \mathbf{D} = -\boldsymbol{\nabla} \cdot \mathbf{P},$$

$$-\boldsymbol{\nabla} \times \mathbf{E} = \frac{\partial}{\partial t}\mathbf{B} \ , \qquad \boldsymbol{\nabla} \cdot \mathbf{B} = 0, \tag{1}$$

where, for an homogeneous, isotropic, nondispersive medium

$$\mathbf{D} = \epsilon\mathbf{E}, \qquad \mathbf{B} = \mu\mathbf{H}. \tag{2}$$

We define a Green's dyadic $\boldsymbol{\Gamma}$ by

$$\mathbf{E}(\mathbf{r}, t) = \int (d\mathbf{r}')\, dt'\, \boldsymbol{\Gamma}(\mathbf{r}, t; \mathbf{r}', t') \cdot \mathbf{P}(\mathbf{r}', t') \tag{3}$$

and introduce a Fourier transform in time

$$\boldsymbol{\Gamma}(\mathbf{r}, t; \mathbf{r}', t') = \int_{-\infty}^{\infty} \frac{d\omega}{2\pi} e^{-i\omega(t-t')}\boldsymbol{\Gamma}(\mathbf{r}, \mathbf{r}'; \omega), \tag{4}$$

where in the following the ω argument will be suppressed. Maxwell's equations then become (which define $\boldsymbol{\Phi}$)

$$\boldsymbol{\nabla} \times \boldsymbol{\Gamma} = i\omega\boldsymbol{\Phi} \ , \qquad \boldsymbol{\nabla} \cdot \boldsymbol{\Phi} = 0,$$

$$\frac{1}{\mu}\boldsymbol{\nabla} \times \boldsymbol{\Phi} = -i\omega\epsilon\boldsymbol{\Gamma}' \ , \qquad \boldsymbol{\nabla} \cdot \boldsymbol{\Gamma}' = 0, \tag{5}$$

in which $\mathbf{\Gamma}' = \mathbf{\Gamma} + \mathbf{1}/\epsilon$, where $\mathbf{1}$ includes a spatial delta function. The two solenoidal Green's dyadics given here satisfy the following second-order equations:

$$(\nabla^2 + \omega^2 \epsilon\mu)\mathbf{\Gamma}' = -\frac{1}{\epsilon}\nabla \times (\nabla \times \mathbf{1}), \tag{6}$$

$$(\nabla^2 + \omega^2 \epsilon\mu)\mathbf{\Phi} = i\omega\mu\nabla \times \mathbf{1}. \tag{7}$$

They can be expanded in vector spherical harmonics [17, 18] defined by

$$\mathbf{X}_{lm} = \frac{1}{\sqrt{l(l+1)}}\mathbf{L}Y_{lm}, \tag{8}$$

as follows:

$$\mathbf{\Gamma}'(\mathbf{r},\mathbf{r}') = \sum_{lm}\left(f_l(r,\mathbf{r}')\mathbf{X}_{lm}(\Omega) + \frac{i}{\omega\epsilon\mu}\nabla \times g_l(r,\mathbf{r}')\mathbf{X}_{lm}(\Omega)\right), \tag{9}$$

$$\mathbf{\Phi}(\mathbf{r},\mathbf{r}') = \sum_{lm}\left(\tilde{g}_l(r,\mathbf{r}')\mathbf{X}_{lm}(\Omega) - \frac{i}{\omega}\nabla \times \tilde{f}_l(r,\mathbf{r}')\mathbf{X}_{lm}(\Omega)\right). \tag{10}$$

When these are substituted in Maxwell's equations (5) we obtain, first,

$$g_l = \tilde{g}_l, \qquad f_l = \tilde{f}_l + \frac{1}{\epsilon}\frac{1}{r^2}\delta(r - r')\mathbf{X}_{lm}(\Omega'), \tag{11}$$

and then the second-order equations

$$(D_l + \omega^2\mu\epsilon)g_l(r,\mathbf{r}') = i\omega\mu\int d\Omega''\,\mathbf{X}_{lm}^*(\Omega'') \cdot \nabla'' \times \mathbf{1}, \tag{12}$$

$$(D_l + \omega^2\mu\epsilon)f_l(r,\mathbf{r}') = -\frac{1}{\epsilon}\int d\Omega''\,\mathbf{X}_{lm}^*(\Omega'') \cdot \nabla'' \times (\nabla'' \times \mathbf{1}),$$

$$= \frac{1}{\epsilon}D_l\frac{1}{r^2}\delta(r - r')\mathbf{X}_{lm}^*(\Omega'), \tag{13}$$

where

$$D_l = \frac{\partial^2}{\partial r^2} + \frac{2}{r}\frac{\partial}{\partial r} - \frac{l(l+1)}{r^2}. \tag{14}$$

These equations can be solved in terms of Green's functions satisfying

$$(D_l + \omega^2\epsilon\mu)F_l(r,r') = -\frac{1}{r^2}\delta(r - r'), \tag{15}$$

which have the form

$$F_l(r,r') = \begin{cases} ik'j_l(k'r_<)[h_l(k'r_>) - Aj_l(k'r_>)], & r,r' < a, \\ ikh_l(kr_>)[j_l(kr_<) - Bh_l(kr_<)], & r,r' > a, \end{cases} \tag{16}$$

where

$$k = |\omega|\sqrt{\mu\epsilon}, \qquad k' = |\omega|\sqrt{\mu'\epsilon'}, \tag{17}$$

and $h_l = h_l^{(1)}$ is the spherical Hankel function of the first kind. Specifically, we have

$$\tilde{f}_l(r, \mathbf{r}') = \omega^2 \mu F_l(r, r') \mathbf{X}_{lm}^*(\Omega'), \tag{18}$$

$$g_l(r, \mathbf{r}') = -i\omega\mu \boldsymbol{\nabla}' \times G_l(r, r') \mathbf{X}_{lm}^*(\Omega'), \tag{19}$$

where F_l and G_l are Green's functions of the form (16) with the constants A and B determined by the boundary conditions given below. Given F_l, G_l, the fundamental Green's dyadic is given by

$$\begin{aligned}
\boldsymbol{\Gamma}'(\mathbf{r}, \mathbf{r}') = \sum_{lm} \Big\{ &\omega^2 \mu F_l(r, r') \mathbf{X}_{lm}(\Omega) \mathbf{X}_{lm}^*(\Omega') \\
&- \frac{1}{\epsilon} \boldsymbol{\nabla} \times G_l(r, r') \mathbf{X}_{lm}(\Omega) \mathbf{X}_{lm}^*(\Omega') \times \overleftarrow{\boldsymbol{\nabla}}' \\
&+ \frac{1}{\epsilon} \frac{1}{r^2} \delta(r - r') \sum_{lm} \mathbf{X}_{lm}(\Omega) \mathbf{X}_{lm}^*(\Omega') \Big\}.
\end{aligned} \tag{20}$$

Now we consider a sphere of radius a centered at the origin, with properties ϵ', μ' in the interior and ϵ, μ outside. Because of the boundary conditions that

$$\mathbf{E}_\perp, \quad \epsilon E_r, \quad B_r, \quad \frac{1}{\mu} \mathbf{B}_\perp \tag{21}$$

be continuous at $r = a$, we find for the constants A and B in the two Green's functions in (20)

$$A_F = \frac{\sqrt{\epsilon\mu'}\,\tilde{e}_l(x')\tilde{e}_l'(x) - \sqrt{\epsilon'\mu}\,\tilde{e}_l(x)\tilde{e}_l'(x')}{\Delta_l}, \tag{22}$$

$$B_F = \frac{\sqrt{\epsilon\mu'}\,\tilde{s}_l(x')\tilde{s}_l'(x) - \sqrt{\epsilon'\mu}\,\tilde{s}_l(x)\tilde{s}_l'(x')}{\Delta_l}, \tag{23}$$

$$A_G = \frac{\sqrt{\epsilon'\mu}\,\tilde{e}_l(x')\tilde{e}_l'(x) - \sqrt{\epsilon\mu'}\,\tilde{e}_l(x)\tilde{e}_l'(x')}{\tilde{\Delta}_l}, \tag{24}$$

$$B_G = \frac{\sqrt{\epsilon'\mu}\,\tilde{s}_l(x')\tilde{s}_l'(x) - \sqrt{\epsilon\mu'}\,\tilde{s}_l(x)\tilde{s}_l'(x')}{\tilde{\Delta}_l}. \tag{25}$$

Here we have introduced $x = ka$, $x' = k'a$, the Riccati-Bessel functions

$$\tilde{e}_l(x) = x h_l(x), \qquad \tilde{s}_l(x) = x j_l(x), \tag{26}$$

and the denominators

$$\begin{aligned}
\Delta_l &= \sqrt{\epsilon\mu'}\,\tilde{s}_l(x')\tilde{e}_l'(x) - \sqrt{\epsilon'\mu}\,\tilde{s}_l'(x')\tilde{e}_l(x), \\
\tilde{\Delta}_l &= \sqrt{\epsilon'\mu}\,\tilde{s}_l(x')\tilde{e}_l'(x) - \sqrt{\epsilon\mu'}\,\tilde{s}_l'(x')\tilde{e}_l(x),
\end{aligned} \tag{27}$$

and have denoted differentiation with respect to the argument by a prime.

3 Stress on the sphere

We can calculate the stress (force per unit area) on the sphere by computing the discontinuity of the (radial-radial component) of the stress tensor:

$$\mathcal{F} = T_{rr}(a-) - T_{rr}(a+), \tag{28}$$

where

$$T_{rr} = \frac{1}{2}\langle[\epsilon(E_\perp^2 - E_r^2) + \mu(H_\perp^2 - H_r^2)]\rangle. \tag{29}$$

The vacuum expectation values of the product of field strengths are given directly by the Green's dyadics computed in Section 2:

$$i\langle\mathbf{E}(\mathbf{r})\mathbf{E}(\mathbf{r}')\rangle = \boldsymbol{\Gamma}(\mathbf{r},\mathbf{r}'), \tag{30}$$

$$i\langle\mathbf{B}(\mathbf{r})\mathbf{B}(\mathbf{r}')\rangle = -\frac{1}{\omega^2}\boldsymbol{\nabla}\times\boldsymbol{\Gamma}(\mathbf{r},\mathbf{r}')\times\overleftarrow{\boldsymbol{\nabla}}', \tag{31}$$

where here and in the following we ignore δ functions because we are interested in the *limit* as $\mathbf{r}' \to \mathbf{r}$. It is then rather immediate to find for the stress on the sphere (the *limit* $t' \to t$ is assumed)

$$\mathcal{F} = \frac{1}{2ia^2}\int_{-\infty}^{\infty}\frac{d\omega}{2\pi}e^{-i\omega(t-t')}\sum_{l=1}^{\infty}\frac{2l+1}{4\pi}$$

$$\times\Bigg\{(\epsilon'-\epsilon)\left[\frac{k^2}{\epsilon}a^2 F_l(a+,a+) + \left(\frac{l(l+1)}{\epsilon'} + \frac{1}{\epsilon}\frac{\partial}{\partial r}r\frac{\partial}{\partial r'}r'\right)G_l(r,r')\bigg|_{r=r'=a+}\right]$$

$$+(\mu'-\mu)\left[\frac{k^2}{\mu}a^2 G_l(a+,a+) + \left(\frac{l(l+1)}{\mu'} + \frac{1}{\mu}\frac{\partial}{\partial r}r\frac{\partial}{\partial r'}r'\right)F_l(r,r')\bigg|_{r=r'=a+}\right]\Bigg\}$$

$$= \frac{i}{2a^4}\int_{-\infty}^{\infty}\frac{dy}{2\pi}e^{-iy\delta}\sum_{l=1}^{\infty}\frac{2l+1}{4\pi}x\frac{d}{dx}\ln\Delta_l\tilde{\Delta}_l, \tag{32}$$

where $y = \omega a$, $\delta = (t-t')/a$, and

$$\ln\Delta_l\tilde{\Delta}_l = \ln\left[(\tilde{s}_l(x')\tilde{e}_l'(x) - \tilde{s}_l'(x')\tilde{e}_l(x))^2 - \xi^2(\tilde{s}_l(x')\tilde{e}_l'(x) + \tilde{s}_l'(x')\tilde{e}_l(x))^2\right] + \text{constant}. \tag{33}$$

Here the parameter ξ is

$$\xi = \frac{\sqrt{\frac{\epsilon'}{\epsilon}\frac{\mu}{\mu'}} - 1}{\sqrt{\frac{\epsilon'}{\epsilon}\frac{\mu}{\mu'}} + 1}. \tag{34}$$

This is not yet the answer. We must remove the term which would be present if either medium filled all space (the same was done in the case of parallel dielectrics [19]). The corresponding Green's function is

$$F_l^{(0)} = \begin{cases} ik'j_l(k'r_<)h_l(k'r_>), & r,r' < a \\ ikj_l(kr_<)h_l(kr_>), & r,r' > a \end{cases} \tag{35}$$

The resulting stress is

$$\mathcal{F}^{(0)} = \frac{1}{a^3} \int_{-\infty}^{\infty} \frac{d\omega}{2\pi} e^{-i\omega\tau} \sum_{l=1}^{\infty} \frac{2l+1}{4\pi} \Big\{ x' [\tilde{s}_l'(x')\tilde{e}_l'(x') - \tilde{e}_l(x')\tilde{s}_l''(x')]$$
$$- x[\tilde{s}_l'(x)\tilde{e}_l'(x) - \tilde{e}_l(x)\tilde{s}_l''(x)] \Big\}. \tag{36}$$

The final formula for the stress is obtained by subtracting (36) from (32):

$$\mathcal{F} = -\frac{1}{2a^4} \int_{-\infty}^{\infty} \frac{dy}{2\pi} e^{iy\delta} \sum_{l=1}^{\infty} \frac{2l+1}{4\pi} \{ x\frac{d}{dx} \ln \Delta_l \tilde{\Delta}_l$$
$$+ 2x'[s_l'(x')e_l'(x') - e_l(x')s_l''(x')] - 2x[s_l'(x)e_l'(x) - e_l(x)s_l''(x)]\}, \tag{37}$$

where we have now performed a Euclidean rotation,

$$y \rightarrow iy, \quad x \rightarrow ix, \quad \tau = t - t' \rightarrow i(x_4 - x_4') \quad [\delta = (x_4 - x_4')/a],$$
$$\tilde{s}_l(x) \rightarrow s_l(x) = \sqrt{\frac{\pi x}{2}} I_{l+1/2}(x), \quad \tilde{e}_l(x) \rightarrow e_l(x) = \frac{2}{\pi}\sqrt{\frac{\pi x}{2}} K_{l+1/2}(x). \tag{38}$$

4 Total energy

In a similar way we can directly calculate the Casimir energy of the configuration, starting from the energy density

$$U = \frac{\epsilon E^2 + \mu H^2}{2}. \tag{39}$$

In terms of the Green's dyadic, the total energy is

$$E = \int (d\mathbf{r})\, U$$
$$= \frac{1}{2i} \int r^2 dr\, d\Omega \left[\epsilon \mathrm{Tr}\, \mathbf{\Gamma}(\mathbf{r},\mathbf{r}) - \frac{1}{\omega^2\mu} \mathrm{Tr}\, \mathbf{\nabla} \times \mathbf{\Gamma}(\mathbf{r},\mathbf{r}) \times \overleftarrow{\mathbf{\nabla}} \right] \tag{40}$$
$$= \frac{1}{2i} \int_{-\infty}^{\infty} \frac{d\omega}{2\pi} e^{-i\omega(t-t')} \sum_{l=1}^{\infty} (2l+1) \int_0^{\infty} r^2 dr$$
$$\times \left\{ 2k^2[F_l(r,r) + G_l(r,r)] + \frac{1}{r^2}\frac{\partial}{\partial r} r \frac{\partial}{\partial r'} r' [F_l + G_l](r,r')|_{r'=r} \right\}, \tag{41}$$

where there is no explicit appearance of ϵ or μ. (However, the value of k depends on which medium we are in.) As in [10] we can easily show that the total derivative term integrates to zero. We are left with

$$E = \frac{1}{2i} \int_{-\infty}^{\infty} \frac{d\omega}{2\pi} e^{-i\omega\tau} \sum_{l=1}^{\infty} (2l+1) \int_0^{\infty} r^2 dr\, 2k^2[F_l(r,r) + G_l(r,r)]. \tag{42}$$

However, again we should subtract off that contribution which the formalism would give if either medium filled all space. That means we should replace F_l and G_l by

$$\tilde{F}_l, \tilde{G}_l = \begin{cases} -ik' A_{F,G} j_l(k'r) j_l(k'r'), & r, r' < a \\ -ik B_{F,G} h_l(kr) h_l(kr'), & r, r' > a \end{cases} \tag{43}$$

so then (42) says

$$\begin{aligned} E = & -\sum_{l=1}^{\infty}(2l+1)\int_{-\infty}^{\infty}\frac{d\omega}{2\pi}e^{-i\omega\tau}\Big\{\int_0^a r^2 dr\, k'^3(A_F + A_G)j_l^2(k'r) \\ & +\int_a^{\infty} r^2 dr\, k^3(B_F + B_G)h_l^2(kr)\Big\}. \end{aligned} \tag{44}$$

The radial integrals may be done by using the following indefinite integral for any spherical Bessel function j_l:

$$\int dx\, x^2 j_l^2(x) = \frac{x}{2}[((xj_l)')^2 - j_l(xj_l)' - xj_l(xj_l)'']. \tag{45}$$

But we must remember to add the contribution of the total derivative term in (41) which no longer vanishes when the replacement (43) is made. The result is precisely that expected from the stress (37),

$$E = 4\pi a^3 \mathcal{F}, \qquad \mathcal{F} = \frac{1}{4\pi a^2}\left(-\frac{\partial}{\partial a}\right)E, \tag{46}$$

where the derivative is the naive one, that is, the cutoff δ has no effect on the derivative.

5 Asymptotic analysis and numerical results

The result for the stress (37) is an immediate generalization of that given in [5], and therefore, the asymptotic analysis given there can be applied nearly unchanged. The result for the energy is new, and seems not to have been recognized earlier.

We first remark on the special case $\sqrt{\epsilon\mu} = \sqrt{\epsilon'\mu'}$. Then $x = x'$ and the energy reduces to

$$E = -\frac{1}{4\pi a}\int_{-\infty}^{\infty} dy\, e^{iy\delta}\sum_{l=1}^{\infty}(2l+1)x\frac{d}{dx}\ln[1 - \xi^2((s_l e_l)')^2], \tag{47}$$

where

$$\xi = \frac{\mu - \mu'}{\mu + \mu'}. \tag{48}$$

If $\xi = 1$ we recover the case of a perfectly conducting spherical shell, treated in [10], for which E is finite. In fact (48) is finite for all ξ, and if we use the leading uniform asymptotic approximation for the Bessel functions we obtain

$$U \sim \frac{3}{64a}\xi^2. \tag{49}$$

K.A. Milton

Further analysis of this special case is given by Brevik and Kolbenstvedt [16].

In general, using the uniform asymptotic behavior, with $x = \nu z$, $\nu = l + 1/2$, and, for simplicity looking at the large z behavior, we have

$$E \sim -\frac{1}{2\pi a}\frac{1}{\sqrt{\epsilon\mu}}\sum_{l}\nu^2\int_{-\infty}^{\infty} dz\, e^{iz\nu\delta/\sqrt{\epsilon\mu}} z\frac{d}{dz}\ln\left[1+\frac{1}{16z^4}\left(\frac{\epsilon\mu}{\epsilon'\mu'}-1\right)^2(1-\xi^2)\right],$$

$$(50)$$

which exhibits a cubic divergence as $\delta \to 0$. To be more explicit, let us content ourselves with with the case when $\epsilon - 1$, $\epsilon' - 1$ are both small and $\mu = \mu' = 1$. Then, the leading ν term is

$$
\begin{aligned}
E &\sim -\frac{(\epsilon'-\epsilon)^2}{16\pi a}\sum_{l=1}^{\infty}\nu^2\frac{1}{2}\int_{-\infty}^{\infty} dz\, e^{i\nu z\delta}z\frac{d}{dz}\frac{1}{(1+z^2)^2}\\
&= -\frac{(\epsilon'-\epsilon)^2}{64a}\left(\frac{16}{\delta^3}+\frac{1}{4}\right) \to -\frac{(\epsilon'-\epsilon)^2}{256a}.
\end{aligned}
$$

$$(51)$$

Here, the last arguable step is made plausible by noting that since $\delta = \tau/a$ the divergent term represents a contribution to the surface tension on the bubble, which should be cancelled by a suitably chosen counter term (contact term). This argument is given somewhat more weight by the discussion in [20]. In essence, justification is provided there for the use of zeta-function regularization, which directly gives the finite part here:

$$E \sim \frac{(\epsilon'-\epsilon)^2}{32\pi a}\sum_{l=1}^{\infty}\nu^2\frac{\pi}{2} = \frac{(\epsilon'-\epsilon)^2}{64a}\left(-\frac{1}{4}\right),$$

$$(52)$$

because $\sum_{l=0}^{\infty}\nu^s = (2^{-s}-1)\zeta(-s)$ vanishes at $s = 2$.

Alternatively, one could argue that dispersion should be included [21, 22, 23], crudely modelled by

$$\epsilon(\omega) - 1 = \frac{\epsilon_0 - 1}{1 - \omega^2/\omega_0^2}.$$

$$(53)$$

If this rendered the expression for the stress finite [we consider the stress, not the energy, for it is not necessary to consider the dispersive factor $d(\omega\epsilon(\omega))/d\omega$ there], we could drop the cutoff δ and the sign of the force would be positive: (at last, we set $\epsilon' = 1$)

$$\mathcal{F} \sim +\frac{(\epsilon_0-1)^2}{128\pi^2 a^4}\sum_{l=1}^{\infty}\nu^2\int_{-\infty}^{\infty} dz\,\frac{1}{(1+z^2)^2}\frac{1}{(1+z^2/z_0^2)^2},$$

$$(54)$$

where $z_0 = \omega_0 a/\nu$. As $\nu \to \infty$, $z_0 \to 0$, and the integral here approaches $\pi z_0/2$, and so

$$\mathcal{F} \sim \frac{(\epsilon_0-1)^2}{256\pi a^3}\omega_0\sum_{l=1}^{\nu_c}\nu \sim \frac{(\epsilon_0-1)^2}{512\pi a}\omega_0^3,$$

$$(55)$$

if we take as the cutoff[1] of the angular momentum sum $\nu_c \sim \omega_0 a$. The corresponding energy is obtained by integrating $-4\pi a^2 \mathcal{F}$,

$$E \sim -\frac{(\epsilon_0 - 1)^2}{256} \omega_0^3 a^2,\tag{56}$$

which is of the form of (51) with $1/\delta \to \omega_0/4$.

6 Conclusions

So finally, what can we say about sonoluminescence? To calibrate our remarks, let us recall (a simplified version of) the argument of Schwinger [1]. On the basis of a provocative but incomplete analysis he argued that a bubble ($\epsilon' = 1$) in water ($\epsilon \approx (4/3)^2$) possessed a positive Casimir energy[2]

$$E_c \sim \frac{4\pi a^3}{3} \int \frac{(d\mathbf{k})}{(2\pi)^3} \frac{1}{2} k \left(1 - \frac{1}{\sqrt{\epsilon}}\right) \sim \frac{a^3 K^4}{12\pi} \left(1 - \frac{1}{\sqrt{\epsilon}}\right),\tag{57}$$

where K is a wavenumber cutoff. Putting in his estimate, $a \sim 4 \times 10^{-3}$cm, $K \sim 2 \times 10^5$cm^{-1} (in the UV), we find a large Casimir energy, $E_c \sim 13$MeV, and something like 3 million photons would be liberated if the bubble collapsed.

What does our full (albeit static) calculation say? If we believe the subtracted result, the last form in (51), and say that the bubble collapses from an initial radius $a_i = 4 \times 10^{-3}$cm to a final radius $a_f = 4 \times 10^{-4}$cm, we find that the change in the Casimir energy is $\Delta U \sim +10^{-4}$eV. This is far to small to account for the observed emission.

On the other hand, perhaps we should retain the divergent result, and put in reasonable cutoffs. If we do so, we have

$$E = -\frac{(\epsilon - 1)^2}{4} a^2 K^3 \sim -4 \times 10^5 \text{eV},\tag{58}$$

perhaps of acceptable magnitude, but of the *wrong* sign. The same conclusion follows if one uses dispersion, as (56) shows.

So we are unable to see how the Casimir effect could possibly supply energy relevant to the copious emission of light seen in sonoluminescence. Of course, dynamical effects could change this conclusion, but elementary arguments suggest that this is impossible unless ultrarelativistic velocities are achieved. (See also [7].) Yet the subject of vacuum energy is sufficiently subtle that surprises could be in store. A more complete analysis will be provided elsewhere.

---

[1] Inconsistently, for then $z_0 \sim 1$. If $z_0 = 1$ in (54), however, the same angular momentum cutoff gives 5/12 of the value in (55).

[2] Note, for small $\epsilon - 1$, Schwinger's result goes like $(\epsilon - 1)$, indicating that he had not removed the "vacuum" contribution corresponding to (36). This is the essential physical reason for the discrepancy between his results and mine.

Acknowledgements

I thank the US Department of Energy for partial financial support of this research. I am grateful to M. Bordag and the other organizers for inviting me to participate in this very interesting Workshop and for providing such gracious hospitality. I am happy to acknowledge useful conversations with J. Ng, I. Brevik, and C. Eberlein. I dedicate this paper to the memory of Julian Schwinger, who taught me so much, and first stirred my interest in this subject and then revived that interest, with his remarkable suggestion that the Casimir effect is behind sonoluminescence. Would that it be so!

References

[1] J. Schwinger, Proc. Natl. Acad. Sci. USA **90**, 958, 2105, 4505, 7285 (1993); **91**, 6473 (1994).

[2] H. Frenzel and H. Schultes, Z. Phys. Chem., Abt. B **27**, 421 (1934).

[3] D.F. Gaitan, L.A. Crum, C.C. Church, and R.A. Roy, J. Acoust. Soc. Am. **91**, 3166 (1992); B. P. Barber and S.J. Putterman, Nature **352**, 318 (1991); B.P. Barber, R. Hiller, K. Arisaka, H. Fetterman, and S.J. Putterman, J. Acoust. Soc. Am. **91**, 3061 (1992); R.G. Holt, D.F. Gaitan, A.A. Atchley, and J. Holzfuss, Phys. Rev. Lett. **72**, 1376 (1994); R. Hiller, S.J. Putterman, and B.P. Barber, Phys. Rev. Lett. **69**, 1182 (1992); B.P. Barber and S.J. Putterman, Phys. Rev. Lett. **69**, 3839 (1992).

[4] J. Schwinger, Lett. Math. Phys. **24**, 59, 227 (1992); Proc. Natl. Acad. Sci. USA **89**, 4091, 11118 (1992).

[5] K.A. Milton, Ann. Phys. (N.Y.) **127**, 49 (1980).

[6] E. Sassaroli, Y.N. Srivastava, and A. Widom, Phys. Rev. D **50**, 1027 (1993).

[7] C. Eberlein, Illinois preprints P-95-05-037 (quant-ph/9506023), P-95-06-039 (quant-ph/9506024).

[8] T.H. Boyer, Phys. Rev. **174**, 1764 (1968).

[9] R. Balian and B. Duplantier, Ann. Phys. (N.Y.) **112**, 165 (1978).

[10] K.A. Milton, L.L. DeRaad, Jr., and J. Schwinger, Ann. Phys. (N.Y.) **115**, 388 (1978).

[11] L.L. DeRaad, Jr. and K.A. Milton, Ann. Phys. (N.Y.) **136**, 229 (1981).

[12] I. Brevik and G.H. Nyland, Ann. Phys. (N.Y.) **230**, 321 (1994).

[13] K.A. Milton, Ann. Phys. (N.Y.) **150**, 432 (1983).

[14] K.A. Milton and Y.J. Ng, Phys. Rev. D **46**, 842 (1992).

[15] C.M. Bender and K.A. Milton, Phys. Rev. D **50**, 6547 (1994).

[16] I. Brevik and H. Kolbenstvedt, Phys. Rev. D **25**, 1731 (1982); Phys. Rev. D **26**, 1490 (1983) (E); Ann. Phys. (N.Y.) **143**, 179 (1982); **149**, 237 (1983).

[17] J.D. Jackson, *Classical Electrodynamics*, 2nd ed. (Wiley, New York, 1975), p. 746 ff.

[18] J.A. Stratton, *Electromagnetic Theory* (McGraw-Hill, New York, 1941), Chapter 7.

[19] J. Schwinger, L.L. DeRaad, Jr, and K.A. Milton, Ann. Phys. (N.Y.) **115**, 1 (1978).

[20] K.A. Milton, Phys. Rev. D **27**, 439 (1983).

[21] I. Brevik and G. Einevoll, Phys. Rev. D **37**, 2977 (1988); I. Brevik and I. Clausen, Phys. Rev. D **39**, 603 (1989).

[22] P. Candelas, Ann. Phys. (N.Y.) **143**, 241 (1982); **167**, 257 (1986).

[23] I. Brevik, H. Skurdal, and R. Sollie, J. Phys. A **27**, 6853 (1994).

Conformal Invariance with Boundaries

Hugh Osborn

Quantum field theories become relevant to statistical physics for systems at or near their critical points. Under such circumstances physical observables obey simple scaling relations and the parameters or critical exponents describing the power law relations are essentially independent of the microscopic details of the interactions and depend only on general features of the system such as the overall symmetry invariance. This crucial property of universality allows critical exponents to be calculated by using continuum quantum field theories with the same symmetry content. The couplings of the quantum field theory must be tuned to a critical point, or surface, when the theory becomes scale invariant and correlation functions of various operators have a power law dependence on their spatial separations determined by the dimensions of the operators which are directly related to the statistical mechanical critical exponents. Thus the Ising model, describing simple magnetic systems on a lattice, is related to the field theory with a single scalar field ϕ. For the renormalisable ϕ^4 theory there is one coupling g and at the critical point, determined by the vanishing of the β function, $\beta(g_*) = 0$, the theory becomes scale invariant. The important critical exponents are then given by the dimensions of the operators ϕ, ϕ^2 which can be written in dimension d respectively as $\frac{1}{2}d - 1 + \gamma_\phi(g_*)$, $d - 2 + \gamma_{\phi^2}(g_*)$, where $\gamma_\phi(g), \gamma_{\phi^2}(g)$ are the anomalous scale dimensions, reflecting the non trivial renormalisation of ϕ, ϕ^2.

For most quantum field theories scale invariance also implies conformal invariance since we may write for the trace of the energy momentum tensor

$$T_{\mu\mu} = \sum_i \beta^i(g)\mathcal{O}_i \,, \tag{1}$$

where $\mathcal{O}_i$ are a basis of scalar operators in the theory which may appear in the interaction with corresponding couplings g^i. At a critical point the corresponding β functions all vanish so that $T_{\mu\mu} = 0$ (more generally there may be derivative operators appearing in the trace which may be removed by a suitable redefinition of the energy momentum tensor, this has been analysed by Polchinski [1]). This condition implies that the theory is invariant under the full conformal group, in d Euclidean dimensions this is $O(d+1,1)$ if $d > 2$. The full conformal group is generated by combining the usual Euclidean group of translations and rotations, $x_\mu \longrightarrow R_{\mu\nu}x_\nu + a_\nu$, with inversions,

$$x_\mu \longrightarrow \frac{x_\mu}{x^2} \,. \tag{2}$$

Conformal invariance is much stronger than simple scale invariance and is a significant constraint on the correlation functions of the theory. The functional dependence of both two and three point functions is essentially determined while the four point function is expressible in terms of a function of two conformal invariants. However since measurements are really only possible for two point functions conformal invariance has little significance from the experimental point of view, although employing explicit conformal invariance has proved essential in carrying out calculations of critical exponents in the $1/N$ expansion for the $O(N)$ sigma model which contains N component scalar fields [2].

For systems with boundaries the situation changes. Experimentally at a critical point there are additional exponents which describe the scaling behaviour of the system in the neighbourhood of the boundary [3]. For any bulk critical point there are in general more than one possible sets of exponents relevant to the behaviour near a boundary and which correspond to different possible boundary conditions on the fields compatible with scale invariance. If we consider a plane boundary and choose coordinates $x_\mu = (y, \mathbf{x})$, where y is the perpendicular distance from the boundary, with $y \geq 0$, and $\mathbf{x} \in \mathbb{R}^{d-1}$ is the position coordinate on the boundary itself, then the two point function for a scalar operator $\mathcal{O}$ may be taken to have the limiting form

$$\langle \mathcal{O}(y, \mathbf{x}) \mathcal{O}(y', \mathbf{x}') \rangle \propto \begin{cases} \frac{1}{y^{\eta_\perp}}, & \text{as } y \to \infty; \\ \frac{1}{|\mathbf{s}|^{\eta_\parallel}}, & \text{as } |\mathbf{s}| \to \infty; \quad \mathbf{s} = \mathbf{x} - \mathbf{x}', \end{cases} \tag{3}$$

which defines exponents $\eta_\perp, \eta_\parallel$ related to the boundary conditions. Only if both $y, y' \to \infty$ is the behaviour of the two point function independent of the boundary.

In the presence of a boundary conformal invariance may still have non trivial consequences, as was first suggested by Cardy [4], if the conformal group is restricted to transformations leaving the boundary invariant. The appropriate subgroup for a plane $d - 1$ dimensional boundary is generated by translations and rotations on $\mathbb{R}^{d-1}$ and also inversions as in eq.(2), since this also leaves $y = 0$ unchanged. In this case the restricted conformal group is then $O(d, 1)$ (other conformal transformations map the plane boundary $\mathbb{R}^{d-1}$ into the sphere S^{d-1}). For two points on $\mathbb{R}^d_+$, corresponding to $y \geq 0$, then the general transformations take the form

$$(x - x')^2 \to \frac{(x - x')^2}{\Omega(x)\Omega(x')}, \quad y \to \frac{y}{\Omega(x)}, \quad y' \to \frac{y'}{\Omega(x')}, \tag{4}$$

where $\Omega(x)$ represents the local rescaling of lengths under conformal transformations. It is easy to see from (4) that an invariant may be constructed by defining

$$\xi = \frac{(x - x')^2}{4yy'}, \quad 0 \leq \xi < \infty. \tag{5}$$

Hence the general form for the two point function of a scalar operator assuming invariance under the restricted conformal group is

$$\langle \mathcal{O}(y, \mathbf{x}) \mathcal{O}(y', \mathbf{x}') \rangle = \frac{1}{(4yy')^{\eta_O}} F(\xi), \tag{6}$$

where $\eta_\mathcal{O}$ is the dimension of the operator $\mathcal{O}$. This result is clearly more restrictive than assuming just scale invariance, together with translational and rotational invariance on $\mathbb{R}^{d-1}$, since in that case F could depend on the two dimensionless scalars $|\mathbf{s}|/y$, $|\mathbf{s}|/y'$. For compatibility with eq.(3) we must assume that as $\xi \to \infty$ then $F(\xi) \propto \xi^{-\frac{1}{2}\eta_\parallel}$. Using the form in eq.(6) it is then easy to see that in eq.(3) we must take

$$\eta_\perp = \eta_\mathcal{O} + \tfrac{1}{2}\eta_\parallel \, . \tag{7}$$

Besides two point functions which have the form given by eq.(6) it is also possible for the one point function of an operator to be non zero in the neighbourhood of a boundary (except for the identity operator scale invariance implies that $\langle \mathcal{O} \rangle = 0$ away from the boundary). Simply from scale invariance at a critical point we may also assume

$$\langle \mathcal{O}(y, \mathbf{x}) \rangle = \frac{A_\mathcal{O}}{(2y)^{\eta_\mathcal{O}}} \, , \tag{8}$$

where conformal invariance further requires that $A_\mathcal{O} = 0$ unless $\mathcal{O}$ is a scalar operator.

For any quantum field theory with a boundary at a critical point, assuming conformal invariance also holds, a crucial task is to determine the single variable functions such as appear in eq.(6). The presence of a boundary violates translational invariance which simplifies the evaluation of Feynman graphs by allowing transformation to momentum space. However the simple form exhibited in eq.(6) enables some simplification [5, 6] by integrating the two point function over planes parallel to the boundary which gives

$$\int \mathrm{d}^{d-1}\mathbf{x} \, \langle \mathcal{O}(y, \mathbf{x})\mathcal{O}(y', \mathbf{x}') \rangle = \frac{1}{(4yy')^{\eta_\mathcal{O}-\lambda}} \hat{F}(\rho) \, , \quad \rho = \frac{(y - y')^2}{4yy'} \, , \quad \lambda = \tfrac{1}{2}(d-1) \, . \tag{9}$$

where

$$\hat{F}(\rho) = \frac{\pi^\lambda}{\Gamma(\lambda)} \int_0^\infty \mathrm{d}u \, u^{\lambda-1} F(u + \rho) \, . \tag{10}$$

It is crucial that the transformation $F \longrightarrow \hat{F}$ is invertible, for $d = 3$, or $\lambda = 1$, it is easy to see from eq.(10) that

$$F(\xi) = -\frac{1}{\pi} \hat{F}'(\xi) \, , \tag{11}$$

and for general d F is given by a simple integral (this transform and its corresponding inverse is very similar to the radon transform when a function on $\mathbb{R}^d$ has an invertible transformation to functions on planes, $f(x) \to \hat{f}(\hat{n}, \lambda)$, where the plane is defined by $x \cdot \hat{n} = \lambda$, $\hat{n}^2 = 1$, and $\hat{f}$ is obtained defined by integrating f over the plane [7]). What makes the transformation useful in this context is that $\hat{F}(\rho)$ is much easier to evaluate perturbatively than $F(\xi)$ directly since the integrals over the plane $\mathbb{R}^{d-1}$ are usually straightforward when calculating $\hat{F}(\rho)$. The essential integrals can be written in the form

$$\int_{\mathbb{R}^d_+} \mathrm{d}^d x \, \frac{1}{(2y)^d} \, F_1(\xi_1) F_2(\xi_2) = F(\xi_{12}) \,, \quad \xi_i = \frac{(x_i - x)^2}{4y_i y} \,, \quad \xi_{12} = \frac{(x_1 - x_2)^2}{4y_1 y_2} \,, \tag{12}$$

where conformal invariance has been used to restrict the integral to be a function of ξ_{12} only. Using the above transforms we may show that

$$\hat{F}(\sinh^2(\theta_1 - \theta_2)) = \int_{-\infty}^{\infty} \mathrm{d}\theta \, \hat{F}_1(\sinh^2(\theta - \theta_1)) \hat{F}_2(\sinh^2(\theta - \theta_2)) \,, \tag{13}$$

and the convolution may be further simplified using Fourier transforms. As an illustration of the applicability of these methods the following non trivial integrals, which are of the form which appear in perturbative calculations, can be evaluated in terms of standard hypergeometric functions [6]

$$(2y_1)^{\eta_1} (2y_2)^{\eta_2} \int_{\mathbb{R}^d_+} \mathrm{d}^d x \, \frac{1}{(2y)^b} \frac{1}{(s_1^2)^{\eta_1} (s_2^2)^{\eta_2}}$$

$$= \pi^{\frac{1}{2}d} \frac{\Gamma(\frac{1}{2}d - b)\Gamma(\frac{1}{2}d - \eta_1)\Gamma(\frac{1}{2}d - \eta_2)}{\Gamma(b)\Gamma(\eta_1)\Gamma(\eta_2)} \frac{1}{\xi_{12}^{\frac{1}{2}d-b}} F(\tfrac{1}{2}d - \eta_1, \tfrac{1}{2}d - \eta_2, 1 + b - \tfrac{1}{2}d; -\xi_{12})$$

$$+ \tfrac{1}{2} S_d \frac{\Gamma(1 - b)\Gamma(b - \frac{1}{2}d)}{\Gamma(1 - \frac{1}{2}d)} F(\eta_1, \eta_2; 1 + \tfrac{1}{2}d - b; -\xi_{12}) \,,$$

$$(2y_1)^{\eta_1} (2y_2)^{\eta_2} \int_{\mathbb{R}^d_+} \mathrm{d}^d x \, \frac{1}{(2y)^b} \frac{1}{(\bar{s}_1^2)^{\eta_1} (\bar{s}_2^2)^{\eta_2}}$$

$$= \tfrac{1}{2} S_d \frac{\Gamma(1 - b)\Gamma(\frac{1}{2}d)}{\Gamma(1 + \frac{1}{2}d - b)} F(\eta_1, \eta_2; 1 + \tfrac{1}{2}d - b; -\xi_{12}) \,,$$

$$(2y_1)^{\eta_1} (2y_2)^{\eta_2} \int_{\mathbb{R}^d_+} \mathrm{d}^d x \, \frac{1}{(2y)^b} \frac{1}{(s_1^2)^{\eta_1} (\bar{s}_2^2)^{\eta_2}}$$

$$= \tfrac{1}{2} S_d \frac{\Gamma(1 - b)\Gamma(\frac{1}{2}d - \eta_1)}{\Gamma(1 - \frac{1}{2}d + \eta_2)} F(\eta_1, \eta_2; \tfrac{1}{2}d; -\xi_{12}) \,, \tag{14}$$

where $s_i^2 = (\mathbf{x} - \mathbf{x}_i)^2 + (y - y_i)^2$, $\bar{s}_i^2 = (\mathbf{x} - \mathbf{x}_i)^2 + (y + y_i)^2$, $S_d = 2\pi^{\frac{1}{2}d}/\Gamma(\frac{1}{2}d)$ and it is necessary to take $\eta_1 + \eta_2 + b = d$ in order to ensure conformal invariance.

The expression given eq.(6) is appropriate for a scalar operator $\mathcal{O}$. For operators with spin the expression has of course to be extended to include a dependence on spin indices [5]. For a theory without boundary conformal invariance requires that the two point function is zero unless both operators have the same spin and scale dimension. Both these restrictions may be violated in the presence of a boundary. As an illustration of this it is important to note that, for two points x, x', it is possible to construct a vector $X_\mu(x)$ under conformal transformations at x by

$$X_\mu = \frac{y}{[\xi(1 + \xi)]^{\frac{1}{2}}} \partial_\mu \xi = \frac{1}{[\xi(1 + \xi)]^{\frac{1}{2}}} \frac{1}{4yy'} (y^2 - y'^2 - \mathbf{s}^2, 2y\mathbf{s}) \,, \tag{15}$$

where the normalisation coefficient has been chosen so that $X^2 = 1$. As an illustration of the use of these expressions we consider the two point function involving the traceless energy momentum tensor $T_{\mu\nu}$, which has dimension d, and a scalar operator $\mathcal{O}$. This has then the general form

$$\langle T_{\mu\nu}(y, \mathbf{x})\mathcal{O}(y', \mathbf{x}')\rangle = \frac{1}{(2y)^d (2y')^{\eta_{\mathcal{O}}}}\left(X_\mu X_\nu - \frac{1}{d}\delta_{\mu\nu}\right) R(\xi) . \tag{16}$$

Applying the conservation equation $\partial_\mu T_{\mu\nu} = 0$ determines the function $C(\xi)$ up to an overall constant in this case

$$R(\xi) = \frac{r}{[\xi(1 + \xi)]^{\frac{1}{2}d}} . \tag{17}$$

In any quantum field theory the energy momentum tensor $T_{\mu\nu}$ is fundamentally important and its two point function is of general significance. In the presence of a plane boundary even in the conformal limit there are three possible terms in the tensorial expansion of $\langle T_{\mu\nu}(y, \mathbf{x})T_{\sigma\rho}(y', \mathbf{x}')\rangle$ which require three functions A, B, C of the conformal invariant ξ. The three terms may be constructed using the conformal vector X_μ, its partner X'_σ which is defined similarly at x', and also the inversion tensor $I_{\mu\sigma}(s) = \delta_{\mu\sigma} - 2s_\mu s_\sigma/s^2$ which for $s = x - x'$ transforms appropriately as vectors at both x and x' ($I_{\mu\sigma}(s)$ is akin to a parallel transport matrix from x' to x). The conservation equation $\partial_\mu T_{\mu\nu} = 0$ in this case gives two coupled differential equations amongst A, B, C leaving in general one undetermined function [5]. If $d = 2$ there are only two independent terms in the tensorial basis so the equations have a unique solution, using the boundary condition $T_{1i}(0, \mathbf{x}) = 0$. With complex coordinates $z = x + iy, \bar{z} = x - iy$ this has the form [8]

$$\langle T_{zz}(x)T_{zz}(z')\rangle = \tfrac{1}{2}c\frac{1}{(z - z')^4} , \qquad \langle T_{zz}(z)T_{\bar{z}\bar{z}}(\bar{z}')\rangle = \tfrac{1}{2}c\frac{1}{(z - \bar{z}')^4} , \tag{18}$$

where c is the Virasoro central charge.

The functions $F(\xi)$ which appear in the two point function as in eq.(6) are dependent on the universality class for the conformal invariant fixed point. The limiting behaviour for small and large ξ is constrained by the detailed operator content of the theory. For $\xi \to 0$ we may make use of the standard operator product expansion. The contribution of a scalar operator $\mathcal{O}_1$ to the operator product of two scalar operators $\mathcal{O}$ takes the form

$$\mathcal{O}(x)\mathcal{O}(x') \sim \frac{C_{\mathcal{O}\mathcal{O}}{}^{\mathcal{O}_1}}{(s^2)^{\frac{1}{2}(2\eta_{\mathcal{O}} - \eta_{\mathcal{O}_1})}}\mathcal{O}_1(x') \quad \text{for} \quad s = x - x' \sim 0 , \tag{19}$$

neglecting less singular terms involving derivatives of $\mathcal{O}_1$. The coefficients $C_{\mathcal{O}\mathcal{O}}{}^{\mathcal{O}_1}$ are determined by the bulk theory and are independent of any specific boundary conditions. If we use eq.(8), where of course $A_{\mathcal{O}_1}$ does depend on the particular boundary behaviour, then it is easy to obtain

$$F(\xi) \sim C_{\mathcal{O}\mathcal{O}}{}^{\mathcal{O}_1} A_{\mathcal{O}_1} \xi^{-2\eta_\mathcal{O}+\eta_{\mathcal{O}_1}} \quad \text{as} \quad \xi \to 0 \ . \tag{20}$$

In general the operator product expansion for conformal invariant theories, for which the leading term may be expressed as in eq.(19), may be extended to an expansion involving all operators in the theory x compatible with the basic symmetries. This leads to a (hopefully) convergent expansion for $F(\xi)$ at $\xi = 0$ containing powers of ξ depending on the operator content. For this situation only scalar operators, for which as mentioned earlier $\langle \mathcal{O}_1 \rangle \neq 0$, are relevant. For two identical operators, as illustrated in eq.(6), the leading operator is expected to be the identity 1 whose scaling dimension is manifestly zero. More generally for operators of differing spins and dimensions the leading singular behaviour in their two point functions as $\xi \to 0$ is determined by operators with dimension $\eta > 0$. In the example given in eqs.(16,17) the relevant operator is $\mathcal{O}$ itself. In this case the coefficient $C_{T\mathcal{O}}{}^{\mathcal{O}}$ appearing in the operator product expansion is determined by Ward identities [9] and it is possible to obtain in eq.(17)

$$r = -\frac{\eta_\mathcal{O} d}{d-1} \frac{A_\mathcal{O}}{S_d} , \qquad S_d = \frac{2\pi^{\frac{1}{2}d}}{\Gamma(\frac{1}{2}d)} \ . \tag{21}$$

It is possible [5] to sum up explicitly all contributions from derivative operators in the operator product expansion since these are determined by the requirement of reproducing the bulk three point function, which has an essentially unique form by conformal invariance. This leads to contributions to $F(\xi)$ which are expressible as hypergeometric functions but in any theory there is also an infinite sum over non derivative quasi-primary operators (which are defined by having simple transformation properties under conformal transformations). For the two point function in eq.(16) the entire functional dependence on ξ given in eq.(17) is obtained from the contribution of derivatives of the operator $\mathcal{O}$ in the operator product expansion.

For the alternative limit $\xi \to \infty$ the behaviour of the function $F(\xi)$ is determined by the spectrum of boundary operators in the theory. The boundary operators are local operators defined on the boundary $y = 0$ and are in general independent of the bulk operators. If we consider a scalar boundary operators $\hat{\mathcal{O}}(\mathbf{x})$, with dimension $\hat{\eta}$, then the two point function for this operator on the boundary may be assumed to have the form

$$\langle \hat{\mathcal{O}}(\mathbf{x}) \hat{\mathcal{O}}(\mathbf{x}') \rangle = \frac{\hat{C}_{\hat{\mathcal{O}}}}{\mathbf{s}^{2\hat{\eta}}} , \qquad \mathbf{s} = \mathbf{x} - \mathbf{x}' \ . \tag{22}$$

The leading contribution of such an operator to the boundary operator expansion for an operator $\mathcal{O}(x)$ can be taken as

$$\mathcal{O}(y, \mathbf{x}) \sim \frac{B_\mathcal{O}{}^{\hat{\mathcal{O}}}}{(2y)^{\eta-\hat{\eta}}} \hat{\mathcal{O}}(\mathbf{x}) \ . \tag{23}$$

The corresponding behaviour of the function $F(\xi)$ is then

$$F(\xi) \sim \left(B_{\mathcal{O}}^{\hat{\mathcal{O}}}\right)^2 \hat{C}_{\hat{\mathcal{O}}}\, \xi^{-\hat{\eta}} \quad \text{as} \quad \xi \to \infty \, . \tag{24}$$

The result eq.(23) may be extended to an expansion at $\xi = \infty$ involving all boundary operators with the same quantum numbers as $\mathcal{O}$. This is a very non trivial extension of the ordinary Taylor expansion of a classical field $\mathcal{O}(y, \mathbf{x})$ about $y = 0$. The boundary operator expansion involves derivatives of boundary operators whose coefficients are determined by reproducing the mixed bulk/boundary two point function

$$\langle \mathcal{O}(y, \mathbf{x})\hat{\mathcal{O}}(\mathbf{x}')\rangle = \frac{B_{\mathcal{O}}^{\hat{\mathcal{O}}}\hat{C}_{\hat{\mathcal{O}}}}{(2y)^{\eta - \hat{\eta}}(y^2 + \mathbf{s}^2)^{\hat{\eta}}} \, , \tag{25}$$

where we have assumed a basis of boundary operators for which the two point functions as in eq.(22) are diagonal. The spectrum of boundary operators depends on the particular boundary conditions and in general of course includes other than scalar operators. For the simplest scalar field theory with a field $\phi(x)$ the leading boundary operator with Neumann boundary conditions may be taken as $\hat{\phi}(\mathbf{x})$ which has the same dimension, $\frac{1}{2}d - 1$, classically and for Dirichlet boundary conditions corresponding operator is $\hat{\phi}_n(\mathbf{x})$ with a classical dimension $\frac{1}{2}d$ when it is given by the normal derivative of ϕ on the boundary. However in a quantum field theory the renormalisation of boundary operators is very different from the bulk operators so their scaling dimensions are no longer simply related. The contribution of all derivatives of a given boundary operator may again be summed explicitly since their contributions to the boundary operator expansion is determined and again leads to contributions to $F(\xi)$ involving hypergeometric functions.

One boundary operator present in any theory may be defined as the limit of the perpendicular components of the energy momentum tensor, $\hat{T}(\mathbf{x}) = T_{11}(0, \mathbf{x})$ where the limit is well defined as a consequence of the Ward identities following from the conservation equations satisfied by $T_{\mu\nu}$. Hence the scalar operator $\hat{T}$ has dimension d and its normalisation is uniquely specified by $B_T^{\hat{T}} = 1$. In general it follows that

$$\langle T_{\mu\nu}(y, \mathbf{x})\hat{T}(\mathbf{x}')\rangle = \frac{d}{d-1}\frac{\hat{C}_{\hat{T}}}{(y^2 + \mathbf{s}^2)^d}\left(\hat{X}_\mu \hat{X}_\nu - \frac{1}{d}\delta_{\mu\nu}\right), \quad \hat{X}_\mu = \frac{1}{y^2 + \mathbf{s}^2}(y^2 - \mathbf{s}^2, 2y\mathbf{s}) \, , \tag{26}$$

where $\hat{X}_\mu$ is the vector given by the limit of X_μ, defined in eq.(15), as $y' \to 0$. $\hat{C}_{\hat{T}}$ is the normalisation factor for the two point function for the operator $\hat{T}$ on the boundary as in eq.(22). In two dimensions it is not difficult to see from the exact expression in eq.(18) that $\hat{C}_{\hat{T}} = 2c$. However, contrary to initial suppositions [10], there is no direct relation between $\hat{C}_{\hat{T}}$ and the normalisation coefficient for the energy momentum tensor two point function in the bulk theory for $d > 2$, in accord with specific calculations in particular models [11, 12, 5]. Using eq.(26) it may be shown that the contribution of the boundary operator $\hat{T}$, and its derivatives, in the boundary operator expansion of $\mathcal{O}$ gives the entire expression for the two point function given by eqs.(16,17) if, using eq.(21),

$$B_{\mathcal{O}}{}^{\hat{T}}\hat{C}_{\hat{T}} = -\eta_{\mathcal{O}}\,\frac{A_{\mathcal{O}}}{S_d}\,, \tag{27}$$

which was first obtained by Cardy [10] by applying both the operator product and boundary operator expansions. This case is a simple example of the mutual consistency of these expansions. In two dimensions, for minimal models, it is possible to formulate a general bootstrap consistency [4, 13]. This depends crucially on using the complete Virasoro algebra for classifying operators into conformal blocks and also the simplifying feature that with a one dimensional boundary there are only scalar boundary operators.

In particular models the universal properties at critical points may be calculated by standard techniques. For instance in renormalisable ϕ^4 field theory critical exponents and functions such as $F(\xi)$ may be calculated in the ε expansion since the critical coupling $g_* = \mathrm{O}(\varepsilon)$ [14, 12, 5]. Perturbative calculations for field theories with a boundary are of course much less straightforward than for the usual case on flat space, as appropriate for bulk theories, due to the fact that the usual momentum space methods are no longer valid. The standard ε expansion is restricted formally to d close to 4. An alternative approximation, valid for all d, $2 < d \leq 4$, is to consider the $O(N)$ sigma model in which exponents and other quantities at the critical point can be calculated as a series in $1/N$. This model is in the same universality class as ϕ^4 theory when it is extended to describe an N component field ϕ^i with $O(N)$ symmetric interactions. In the sigma model treatment there is also an auxiliary field λ (which up to a factor may be identified with ϕ^2). The basic two point functions can be written as

$$\langle \phi^i(x)\phi^j(x')\rangle = \delta^{ij}G_\phi(x, x')\,, \quad \langle \lambda(x)\lambda(x')\rangle = \langle \lambda(x)\rangle\langle \lambda(x')\rangle + G_\lambda(x, x')\,, \tag{28}$$

assuming manifest $O(N)$ invariance which requires $\langle \phi^i\rangle = 0$. To zeroth order in $1/N$ the basic equations for G_ϕ, G_λ are

$$(-\nabla^2 + \langle \lambda(x)\rangle)G_\phi(x, x') = \delta^d(x - x')\,, \tag{29}$$

and also

$$\int \mathrm{d}^d x\, G_\phi(x_1, x)^2\, G_\lambda(x, x_2) = -\frac{2}{N}\delta^d(x_1 - x_2)\,, \tag{30}$$

where the normalisation of the fields corresponds to an elementary interaction $\mathcal{L}_I = \frac{1}{2}\lambda\phi^2$ (the theory is exactly equivalent to renormaliseable ϕ^4 theory if an additional λ^2 interaction is introduced but for $d < 4$ this is an irrelevant operator and so does not contribute at the critical point). In the large N limit the scale dimensions of the fields at the critical point are determined to be

$$\eta_\phi = \tfrac{1}{2}d - 1 + \mathrm{O}(1/N)\,, \qquad \eta_\lambda = 2 + \mathrm{O}(1/N)\,, \tag{31}$$

32
H. Osborn

where explicit formulae to $O(N^{-2})$ and, for η_ϕ, $O(N^{-3})$ have been calculated making essential use of conformal invariance [2]. With a boundary, and using the leading order result for η_λ in eq.(31), we may write

$$\langle \lambda(y, \mathbf{x}) \rangle = \frac{A_\lambda}{4y^2} \ . \tag{32}$$

The basic equations defining the $O(N)$ model can be applied to the case with a boundary, although in eq.(30) the integration is restricted to $\mathbb{R}^d_+$, and it is necessary to impose boundary conditions. At the critical point the two point functions are determined by functions $F_\phi(\xi), F_\lambda(\xi)$. With Dirichlet boundary conditions on ϕ (this corresponds to the so called ordinary transition) the results are relatively straightforward to derive [5] (for earlier discussions see [15])

$$F_\phi(\xi) = \frac{C_\phi}{[\xi(1+\xi)]^{\frac{1}{2}d-1}} \ , \tag{33}$$

and, by using the method of integrating over planes parallel to the boundary described earlier,

$$F_\lambda(\xi) = 2C_\lambda \frac{Q^2_{d-3}(1+2\xi)}{\xi(1+\xi)} \ , \quad A_\lambda = -(4-d)(d-2) \ , \tag{34}$$

where Q^2_{d-3} is an associated Legendre function. The constants C_ϕ, C_λ are determined in terms of Gamma functions by solving eqs.(29,30). When $d = 3$, which is the case of presumed interest, $C_\phi = 1/4\pi$ and F_λ simplifies to

$$F_\lambda(\xi) = \frac{16}{\pi^2 N} \frac{1+2\xi}{\xi^2(1+\xi)^2} \ . \tag{35}$$

Using the calculational methods already described, which make essential use of conformal invariance, results to leading order in the large N limit have also been obtained for the two point function of the energy momentum tensor [5]. The results are rather complicated but reveal how the conservation equations are satisfied for general d. The results for Neumann boundary conditions on the fields are more involved although it is easy to obtain

$$F_\phi(\xi) = C_\phi \frac{1+2\xi}{[\xi(1+\xi)]^{\frac{1}{2}d-1}} \ , \qquad A_\lambda = (4-d)(6-d) \ . \tag{36}$$

The result for F_λ is expressed in terms of generalised $_3F_2$ hypergeometric functions. The results in both cases are in accord with the ε expansion as $d \to 4$.

Acknowledgement

This work was undertaken in collaboration with David McAvity, for whose stimulus I am most grateful.

References

[1] J. Polchinski, Nucl. Phys. **B303** (1988) 226.

[2] A.N. Vasil'ev, Yu.M. Pis'mak and Yu.R. Khonkenen, Theoretical and Mathematical Physics **46** (1981) 104; **47** (1981) 465; **50** (1982) 127.

[3] H.W. Diehl, in *Phase Transitions and Critical Phenomena*, vol 10, p. 75, (C. Domb and J.L. Lebowitz eds.) Academic Press, London (1986).

[4] J.L. Cardy, Nucl. Phys. **B240** [FS12] (1984) 514.

[5] D.M. McAvity and H. Osborn, Nucl. Phys. B, to be published, cond-mat/9505127.

[6] D.M. McAvity, J. Phys. A, to be published, hep-th/9507028.

[7] I.M. Gel'fand, M.I. Graev and N.Ya Vilenkin, "Generalised Functions", vol. 5, Academic Press, New York (1966).

[8] J. Cardy, *in* "Phase Transitions and Critical Phenomena", vol. 11, p. 55 (C. Domb and J.L. Lebowitz eds.) Academic Press, London (1987);

J. Cardy, *in* "Champs, Cordes et Phénomènes Critiques", (E. Brézin and J. Zinn-Justin eds.) North Holland, Amsterdam 1989.

[9] A. Petkou, Ann. Phys. (N.Y.), to be published, hep-th/9410093.

[10] J.L. Cardy, Phys. Rev. Lett., **65** (1990) 1443.

[11] E. Eisenriegler, M. Krech and S. Dietrich, Phys. Rev. Lett. **70** (1993) 619, E, 2051.

[12] D.M. McAvity and H. Osborn, Nucl. Phys. **B394** (1993) 728.

[13] E. Eisenriegler, "Polymers Near Surfaces", World Scientific, Singapore (1993).

[14] G. Gompper and H. Wagner, Zeit. f. Phys. B **59** (1985) 193.

[15] A.J. Bray and M.A. Moore, Phys. Rev. Lett. **38** (1977) 735; J. Phys. A, (1977) 1927. K. Ohno and Y. Okabe, Phys. Lett. A **95** (1983) 41; **99** (1983) 54; Prog. Theor. Phys. **70** (1983) 1226.

Hard, Semihard and Soft Boundary Conditions

<u>Alfred Actor</u> and Iring Bender[1]

1   Introduction

It is easy to introduce a broad range of static background spatial "objects" into bosonic quantum field theory (QFT) on flat space time. One replaces the wave equation $[-\Delta]\Phi_n = \omega_n^2\Phi_n(\vec{x})$ for the spatial modes k of the quantum field Φ by the Schrödinger equation

$$[-\Delta + V]\Phi_n = \omega_n^2\Phi_n(\vec{x}) \tag{1}$$

in which the static potential V is approximately chosen to represent the rigid spatial object or structures of interest. These objects are rather literally "immersed" in Φ, and each of the interacts with this field, in the way displayed mode for mode by eq. (1) (which of course must be solved to implement the entire procedure). This description is Dirichlet-like. In regions where is large the quantum field and all its modes are at least partially excluded or expelled. Where is infinite the quantum field and its modes have to vanish. A Dirichlet boundary on which vanishes is represented in this picture by an infinite step-function potential. Any kind of background spatial object immersed in will distort this field, forcing it to be different from the uniform translation-invariant field it would be in free (boundaryless, backgroundless) space. Compelled to be spatially nonuniform, of course exerts a back reaction on each object responsible for its distortion – the famous Casimir forces. If the region of field distortion caused by two or more objects significantly overlap, these objects experience mutual (true many-body) global Casimir forces from the distorted quantum field. Also, local Casimir forces act on and throughout individual objects. The idea underlying eq. (1) – quantization about a background field – is not new. However, refs. [1, 2] seem to be the first to suggest using this idea to model impenetrable boundaries with arbitrary surface texture.

A practical classification [1, 2] was introduced for confining boundaries. A "hard" boundary in one whose surface is precisely located in space. in eq. (1) makes an abrupt, infinite jump at such a surface. A "semihard" boundary is a hard boundary with surface texture added. This means the jump in is smoothed, which of course can be done in may ways. Finally, a "soft" boundary can be regarded as one having infinitely deep surface texture. A soft boundary is necessarily remote, with its surface texture extending into finite space and represented by a potential which is finite everywhere except at infinity. Besides confining boundaries, it is also interesting to consider finite objects in space. We continue to use the term "hard" for <u>any</u> boundary with precise spatial localization, where makes an abrupt (but not necessarily infinite) jump. A smoothed infinite (finite) jump is called semihard (soft). Two contributions of these

---

[1] Institut für Hochenergiephysik, Universität Heidelberg, 69120 Heidelberg, Germany

proceedings (from M. Bordag and J. Lindig) discuss, from much the same perspective, finite objects immersed in a quantum field.

Dirichlet-like background objects can be introduced into spinor field theory by Yukawa-coupling the spinor field to a background scalar potential which has dimension (mass). The in the Dirac mode equation, appears as an arbitrary potential. The parallel with eq. (1) is rather close. Such use of potential barriers to quantized Fermi fields has a fairly long history (see e.g. refs. [3]).

Practically all of existing Casimir theory has been done using hard boundaries. For several years the authors have performed nontraditional calculations employing semihard and soft boundaries, as well as hard ones. We generally use a scalar quantum field $\hat{\phi}(x)$ for simplicity, and always view the field's nonuniformity as the central phenomenon. We often measure $\hat{\phi}$'s nonuniformity using the canonical vacuum stress tensor $< \hat{T}_{\mu\nu}(x) >$. Calculations can routinely be done for both zero and nonzero mass and temperature. Many calculations begin with the formula

$$< \hat{\phi}(x)\hat{\phi}(y) >= \sum_n \frac{1}{2\omega_n} e^{-i\omega_n(x_0-y_0)} \phi_n(\vec{x})\bar{\phi}_n(\vec{y}) \tag{2}$$

which is readily obtained from standard equal-time quantization. The right side od eq. (2) can be differentiated as needed, the the limit $x - y \to 0$ taken. To evaluate the mode sums on the right it is usually most convenient to employ heat-kernel methods. In this report we discuss the results of some of our calculations, but cannot present the calculations themselves.

A very important aspect which also cannot be discussed in any detail is ultraviolet regularization. When the limit $x - y \to 0$ is taken, one finds in eq. (2) the usual (uniform) free-space divergence, and additional divergences which depend on $V(\vec{x})$. Spatial nonuniformity pervades the UV renormalization process, creating a situation very different from and yet comparable to UV renormalization in curved space. $V(\vec{x})$ is a local, changing property of space, as is curvature. Both make quantum fields nonuniform <u>before</u> renormalization. Inevitably, the renormalization process becomes nonuniform.

2 Some hard boundaries immersed in a scalar field

We consider now structures constructed from flat Dirichlet and Neumann boundaries joined together in various ways, interacting with a massless scalar field $\hat{\phi}(x)$. Space between boundaries is featureless, making UV renormalization uniform. Ref. [4] gives the vacuum stress tensors for all of these systems, from which we obtain the local Casimir forces on all surfaces.

For an isolated Dirichlet plane positioned at $x_1 = 0$ in flat d-dimensional space one finds $< \hat{T}_{\mu\nu} >= (const)|x_1|^{-d-1}(1, 0, -1, ..., -1)_{\mu\nu}$. The divergent factor $|x_1|^{-d-1}$ expresses the extreme distortion ($\hat{\phi} \to 0$) of $\hat{\phi}(x)$ as the boundary is approached. Nonetheless, due to planar geometry, $< T_{11} >= 0$ everywhere so the boundary plane experiences no Casimir force from either side. (The conservation of $< T_{\mu\nu} >$ enables one to predict this known result.)

To the full (vertical Dirichlet plane at $x_1 = 0$ we attach a (horizontal) perpendicular Dirichlet half plane extending from one side. We find a vanishing Casimir force on the half plane, and an <u>attractive</u> local Casimir force on the full plane

$$\frac{Force}{area} = \frac{\Gamma((d+1)/2)}{2(\sqrt{4\pi x})^{d+1}} \tag{3}$$

pointing into the space occupied by the half plane, with x being the perpendicular distance to the line of attachment of the half plane. For overall equilibrium there must be a counterbalancing mechanical force within the half plane. No global Casimir force is visible in this problem. But if the half plane were separated horizontally from the vertical full plane by distance L, a global Casimir force $F \sim L^{-L-1}$ would come into existence between these separated objects. This (hard to calculate) force would be, we believe, attractive. There would be a position-dependent local Casimir force acting on the vertical plane, and the force (3) above is its $L \to 0$ limit. By symmetry the global $L > 0$ Casimir force must lie within the horizontal half plane, and coincide with the mechanical force pushing this half plane toward the vertical plane. For $L \to 0$ this force becomes the counterbalancing mechanical force mentioned above.

Many additional examples of connected, apparently attracting, Dirichlet planar objects can be given. These include a second perpendicular Dirichlet half plane attached to the <u>other</u> side of the vertical full plane. The local Casimir force on this vertical plane continues to point straight outward, to the side from which the nearer half plane extends. One knows, of course, that parallel Dirichlet planes attract. It is tempting conclude that <u>distinct</u> Dirichlet objects generally attract. ¿From this statement we specifically exclude <u>enclosing</u> boundary configurations such as cavities and waveguides. For example, a spherical shell can be constructed by bringing together two hemispherical boundary surfaces. As long as the latter are well-separated they experience some "distinct-object" Casimir effect. But when the hemispheres are moved together and joined, this global Casimir effect disappears, and the cavity Casimir effect that remains may have very different character. For nonenclosing final structures we expect more predictability as the different parts are brought together.

Let us next attach to the vertical Dirichlet plane a horizontal <u>Neumann</u> half plane. ($\hat{\phi}$ satisfies the Neumann conditions $\partial_? \hat{\phi} = 0$ on any Neumann surface.) The vertical Dirichlet plane now experiences a <u>repulsive</u> force/area given by eq. (3), but now pointing <u>away</u> from the space occupied by the half plane. Other more complicated systems also suggest that Distinct (nonenclosing) Dirichlet and Neumann objects repel. In particular, a Dirichlet plane and a parallel Neumann plane mutually repel, as one can easily verify.

3 Effects of boundary semihardening

Boundary semihardening means, adding surface texture to a Dirichlet boundary which already exists. The core boundary remains impenetrable. This affects the lower modes, perhaps very strongly, while higher modes are less affected, and asymptotically the effects of semihardening become negligible. A standard topic within differential equation theory, associated with the names Sturm and Liouville, deals with this kind of spectral distortion. In general, when one or more boundaries in a hard boundary problem are made semihard, the mode spectrum is distorted, but not unrecognizable so. A one-to-one correspondence persists. As semihardening is removed, the entire spectrum of modes, smoothly becomes the original hard-boundary mode spectrum. Therefore the mathematics of semihard and hard boundaries are quite solidly related. This is important, because the physical consequences of boundary semihardening are striking and seemingly unpredictable.

A very explicit mathematical example of boundary semihardening is available [2]. The spatial potential in the mode equation (1) can be chosen to be $V(x_1) = (\alpha^2 - 1/2)/x_1^2$ plus an infinite step-function potential (from 0 to ∞) at $x_1 = L$. The latter represents an impenetrable Dirichlet wall at $x_1 = L$, while the potential $V(x_1)$ represents a semihard planar boundary at $x_1 = 0$. We call this semihard boundary a "Bessel wall", because the mode equation separates into a Bessel equation in the x_1 direction, plus plane-wave equations for $x_{2,3}$. The orthonormal x_1 mode factors are

$$\phi_n(x_1)0\frac{\sqrt{2x_1}}{L}\,|J_{\alpha+1}(j_{\alpha n})|^{-1}\,J_\alpha(k_{1n}x_1) \tag{4}$$

where $k_{1n} = j_{\alpha n}/L$, $n = 1, 2, 3, ...$ and $j_{\alpha n}$ is the spectrum of zeroes of $J_\alpha(z)$. In the limit $L \to \infty$ which isolates the Bessel wall, the x_1 modes become $\phi_{k_1}(x_1) = (k_1 x_1)^{1/2}J_\alpha(k_1 x_1)$ with $k_1 \geq 0$. Semihardening is removed by letting $\alpha \to 1/2$, and the Bessel wall smoothly (mode by mode) becomes a Dirichlet wall: $k_{1n} \to n\pi/L$ and $\phi_n(x_1) \to (2/L)^{1/2}\sin(n\pi x_1/L)$.

The effects of semihardening on the individual modes (4) is made very clear by plotting a few of them for increasing semihardening parameter $\alpha \geq 1/2$. The Bessel potential $V(x_1)$ strongly suppresses all the lower modes near $x_1 = 0$. Higher modes do, of course, penetrate the potential of the $x_1 = 0$ wall, or the region of mode suppression. The x_1 mode factors (4) are, of course, normalized. Thus if $\phi_n(x_1)$ is squeezed in the small-x_1 region, it must bulge out somewhere else. Since both walls are impenetrable, ϕ_n is enhanced towards the Dirichlet wall. In particle language, semihardening (nonuniformly) displaces somewhat the virtual particle sea—the entire quantum field actually—towards the Dirichlet wall. This displacement of $\hat{\phi}(x)$ increases without limit as α increases, and can be expected to have physical consequences.

To probe the effects of semihardening we have investigated the global Casimir effect of the parallel Bessel-Dirichlet wall system [2]. The results seemed to us, initially, surprising. Let $E_\alpha(L)$ be the total energy contained within a tube of length L and unit cross section extending from $x_1 = 0$ to $x_2 = L$ between the walls. For odd spatial dimension $d = 1, 3, \ldots$ ($d =$ even requires another formula) one easily verifies by ζ function

$$
\begin{aligned}
E_\alpha(L) \;=\; & -\frac{1}{2}\left(\frac{\sqrt{\pi}}{2L}\right)^d \Gamma\left(-\frac{d}{2}\right) \{c_n(\alpha) \\
& +\frac{1}{2}R_n(\alpha)\left[\gamma + \Psi(n + 1/2) + 2\ln\frac{\mu L}{\pi}\right]\}
\end{aligned}
\tag{5}
$$

where $n = (d+1)/2$, $\gamma \approx 0.5772$, $\Psi(z)$ is the digamma function and μ is the usual UV renormalization mass parameter. The calculation involves the ζ function [5] $\zeta_\alpha = \sum(j_{\alpha_n}/\pi)^{-s}$ constructed from the spectrum of zeros $\{j_{alpha_n}\}$. $\zeta_\alpha(s)$ has poles at $s = 1, -1, -3, -5, \ldots$,

$$
\zeta_\alpha(\epsilon + 1 - 2n) = \frac{1}{\epsilon}R_n(\alpha) + C_n(\alpha) + \mathcal{O}(\epsilon)
\tag{6}
$$

with residues R_n and finite parts C_n which contribute to the Casimir energy as indicated in eq. (5). The residues are known polynomials in α while the finite parts are not known in closed form. All complications in eq. (5) reside in the $C_n(\alpha)$. We have two very different formulae for these function of α: an integral formula obtained from the defining series for $\zeta_\alpha(s)$ by applying the Cauchy theorem [6], and an algebraic formula [7] obtained from the first eight terms of the MaMahon expansion of j_{α_n} in inverse powers of $[n + (2\alpha - 1)/4]$ [8]. These two ways of computing $C_n(\alpha)$ agree numerically very well out to $\alpha = 50$ and higher, giving us confidence this problem is under quantitative control.

Dimensions $d = 1, 3$ have been studied numerically. The global Casimir force defined by $F_\alpha/area 0 E'_\alpha(L)$ is given by eq. (5), modified by an overall factor $(-d/L)$ and an additional term $-2/d$ inside the square bracket. The Casimir energy $E_\alpha(L)$ gives us basically the same information so we continue to use it. For fixed L we find that increasing α above $\alpha = 1/2$ at first strengthens the attraction experienced by parallel Dirichlet planes ($E_{1/2}(L) = -|const|/L^d$ for any d). But then the attraction begins to weaken, and $E_\alpha(L)$ eventually passes through zero at some $\alpha = \alpha_0$ and becomes positive for $\alpha > \alpha_0$, indicating the Casimir effect becomes <u>repulsive</u>. Moreover, this repulsive Casimir effect becomes much stronger than the original attractive one. However, the system does not remain repulsive to arbitrarily large α. In $d = 3$ there is another turnaround, with $E_\alpha(L)$ again vanishing at some $\alpha'_0 \gg \alpha_0$ and becoming negative again, and eventually stronger than the repulsive stage was. In $d = 1$ we have not yet seen this second turnaround, perhaps because we have not yet gotten to large enough α. Dimension $d = 2$ has also been studied numerically, and behaviour much like $d = 3$ is found.

Global calculations such as those just described reveal a system's global behaviour better than they are able to explain it. Local calculations are needed to probe more deeply. The parallel Bessel-Dirichlet wall system is being studied locally. It now seems that the first reversal of the Casimir effect can be understood physically, at a rather general level. The size limitations of this report preclude a discussion of the mechanism here.

4   Boundary softening

Our results for soft boundaries seem undramatic in comparison with those for the Bessel wall. In ref. [1] we chose the potential in eq. (1) to be the general harmonic oscillator (HO) potential in $d = 3$ spatial dimensions,

$$V(\vec{x}) = \frac{1}{4}\left[\alpha_1^4 x_1^2 + \alpha_2^4 x_2^2 + \alpha_3^4 x_3^2\right] + M^2. \tag{7}$$

Then separation of coordinates yields three 1D HO equations, whose solution in terms of Hermite polynomials is given in every text on quantum mechanics. The spectrum of the Schrodinger operator $[-\Delta + V(\vec{x})]$ can be written down almost by inspection,

$$\omega_n^2 - M^2 = \alpha_1^2(n_1 + 1/2) + \alpha_2^2(n_2 + 1/2) + \alpha_3^2(n_3 + 1/2) \tag{8}$$

where each $n_i = 0, 1, 2, \ldots$ This spectrum is unrelated to any hard-boundary spectrum. When the potential is removed (i.e. $\alpha_i \to 0$) the result is free space: a free quantum field with mass M. Tuning the α_i up (down) strengthens (weakens) the potential—or moves the soft barriers inward (outward)—and consequently weakens (strengthens) $\hat{\phi}(x)$ in regions away from the origin. However, there is nowhere an absolute barrier to the quantum field. This is very unlike any hard or semihard boundary system.

In ref. [1] we chose $M = 0$ and looked at various choices of $\alpha_{1,2,3}$. All choices considered (we did not make an exhaustive search) led to an attractive Casimir effect— i.e. to a global Casimir energy which decreases as the soft boundaries are moved inward by tuning the α_i upward. Matters become less simple when Dirichlet planes are used to slice through the center of the HO cavity, leading in many cases to mixed HO-Dirichlet cavities with repulsive Casimir effects. These calculations have appeared [1], so rather than describing them in more detail we proceed to a brief comment about nonzero field mass.

Consider parallel HO boundaries ($\alpha_1 = \alpha$, $\alpha_{2,3} = 0$) in 3D space. The Casimir energy within an infinite tube of unit cross section perpendicular to the HO boundaries is

$$E(\alpha) = -\frac{\alpha^3}{6\pi}\zeta\left(-\frac{3}{2}, \frac{1}{2} + \frac{M^2}{\alpha^2}\right) \tag{9}$$

where $\zeta(s, a)$ is th Hurwitz ζ function. Because $\zeta(-3/2, 1/2) > 0$ the energy (9) is "attractive" for $M = 0$; increasing α decreases the energy. However, for sufficiently large $M > 0$ the sign of the Casimir energy (9) becomes positive. For example, if (numerically) $M^2 = \alpha^2/2$, then $\zeta(-3/2, 1) = \zeta(-3/2) \approx -0.0255$ shows that $E(\alpha)$

is positive. Nonetheless, the Casimir effect continues to be attractive. The reason is, attraction or repulsion is decided not by the sign of $E(\alpha)$, but rather by the sign of

$$E'(\alpha) = -\frac{\alpha^2}{2\pi}\zeta\left(-\frac{3}{2},\frac{1}{2}+\frac{M^2}{\alpha^2}\right) - \frac{M^2}{2\pi}\zeta\left(-\frac{1}{2},\frac{1}{2}+\frac{M^2}{\alpha^2}\right) \qquad (10)$$

which is negative for $M \geq 0$. The easiest way to see this for large M is to make the identification $1/2 + M^2/\alpha^2 = r + N$ where N is an integer, and use the property $\zeta(s, r + N) = \zeta(s, r) - [r^{-s} + (r + 1)^{-s} + ... + (r + N - 1)^{-s}]$.

5　Lattice Casimir theory

QFT on spatial or spacetime lattices is an important alternative to continuum QFT. The lattice provides UV regularization by preventing adjacent spacetime points from being closer together than the lattice spacing a. As $a \to 0$ the lattice theory should smoothly become the continuum theory. This requirement does not determine the lattice theory uniquely. Extensive use of lattice QFT is made in elementary particle physics (see e.g. ref. [9]) and in some other areas of quantum physics. It has not been used in Casimir theory to our knowledge.

Some time ago we set out to develop lattice Casimir theory, our motivation being twofold. First, to find out if lattice QFT closely reproduces the local and global features characteristic of continuum Casimir theory. While we have not checked everything, our experience up to this point is that indeed, lattice quantum fields in the presence of background spatial structures do closely resemble continuum quantum fields in the presence of the same structures.

Our second motivation is more ambitious. On a d-dimensional spatial lattice $\vec{x} = a(m_1, m_2, ..., m_d)$ with $m_i = 1, 2, ..., N$ one can, in principle, numerically compute everything—all modes $\phi_n(a(m_1, m_2, ..., m_d))$ and their frequencies ω_n in the lattice equivalent of the mode equation (1) for an arbitrary lattice potential $V(a(m_1, m_2, ..., m_d))$. At lattice points where V is large, the modes and the quantum field $\hat{\phi}(x)$ itself are diminished. At points where $V = \infty$ all modes and $\hat{\phi}(x)$ vanish. One can introduce arbitrarily-shaped hard, semihard and soft objects on the lattice and numerically compute the system's vacuum stress tensor, all back forces, the global Casimir energy, and anything else one may wish to know. Arrangements of hard boundaries whose Casimir problems cannot be solved analytically in the continuum theory can be dealt with numerically on a finite lattice. This will enable us to establish or disprove the conjecture in Sec. 2, that distinct Dirichlet objects always attract.

In continuous space it is difficult to semiharden the second Dirichlet boundary in the parallel Bessel-Dirichlet wall system discussed in Sec. 3. On a finite lattice this is straightforward. Moreover, semihardening with many potentials other than the Bessel potential is also straightforward. Once it is established that lattice Casimir theory works as one believes it should, one can use it to probe the semihardening idea quite deeply. This is a principle goal of our lattice Casimir theory investigation.

We have no space here to present numerical results. Instead, to give the flavor of lattice Casimir theory we discuss two simple systems, each consisting of a massless scalar field defined on a 1D lattice.

The "topological Casimir effect" for a scalar field $\hat{\phi}$ on a circle of circumference L was first studied by Ford [10], who found the continuum Casimir energy $E(L) = -\pi/6L$. The quantum fluctuations of continuum $\hat{\phi}$ try to collapse the spatial circle. This problem reformulated on a compact lattice with N sites (so that $L = aN$ with lattice constant a) yields the lattice Casimir energy $E(a, N) = (1/2)\sum_1^N (2/a)\sin(\pi k/N) = a^{-1}\cot(\pi/2N)$ which, although positive, also decreases with decreasing N. Thus quantum fluctuations of lattice $\hat{\phi}$ also try to collapse the spatial circle. Note that in the continuum limit $a \to 0$, $N \to \infty$ with $L = aN$ fixed, $E(a, N) \to 2N/a\pi - \pi/6(aN) + \mathcal{O}(1/aN^3)$. By discarding the UV-divergent first term we recover the continuum Casimir energy. Note also that finite-lattice QFT uses periodic boundary conditions (thereby preserving a discrete remnant of translation invariance) to model free, infinite space. What lattice theorists call "finite-size effects" are, from our perspective here, topological Casimir effects.

The 1D continuum Casimir energy for two Dirichlet points with separation L is $E(L) = (1/2)\sum_1^\infty (n\pi/L) = -\pi/24L$. This is also the Casimir energy of a compact line or circle of circumference L having one Dirichlet point (this system being just a reinterpretation of the other one). Either of these systems put on an $(N+1)$-site lattice with Dirichlet endpoints $n = 0, N$ identified in the compact version has Casimir energy $E(a, N) = (1/2)\sum_1^N (2/a)\sin(\pi k/2N) = (1/2a)[1 + \cot(\pi/4N)]$. In the continuum limit with $L = aN$ fixed, $E(a, N) \to [1 + 4N/\pi]/2a - \pi/24(aN) + \mathcal{O}(1/aN^3)$ and the continuum Casimir energy is recovered.

Finally, let us consider the same system locally. Its continuum vacuum stress tensor is constant, $< \hat{T}_{\mu\nu}(x) > = \delta_{\mu\nu}(\pi/24L^2)$, due to all dependence on x canceling away. Thus it is better to consider a particular mode sum in $< \hat{T}_{m\mu\nu} >$, say

$$\sum_n \omega_n |\phi_n|^2 \;=\; \frac{2}{L}\sum_{n=1}^\infty \left(\frac{n\pi}{L}\right)\left(\sin\frac{n\pi x}{L}\right)^2 \tag{11}$$

$$=\; \frac{1}{4\pi x^2} - \frac{pi}{L^2}\sum_{a=1}^\infty (-)^a \left(\frac{2\pi x}{L}\right)^{2a} \frac{1}{(2a)!}\zeta(-1 - 2a) \tag{12}$$

where $\zeta(s)$ is the Riemann ζ function. Here the unique term $1/4\pi x^2$ having the dimension $(\text{length})^{-2}$ of the mode sum expressed entirely in terms of x (not x and L) is the expected boundary divergence for a Dirichlet "surface" in one spatial dimension.

For the lattice circle with one Dirichlet point the vacuum stress tensor is not quite constant in n, but nonetheless is quite flat away from the endpoints. To probe the behaviour near $n = 0$ we compute the lattice version of eq. (12);

$$\sum_n \omega_n |\phi_n|^2 \;=\; \frac{1}{a}\sum_{k=1}^N \left(\frac{2}{a}\sin\frac{\pi k}{2N}\right)\left(\sqrt{\frac{2}{N}}\sin\frac{\pi k n}{N}\right)^2$$

$$\approx \quad \frac{1}{a^2 N}\left[1 + \cot\frac{\pi}{4N}\right] \tag{13}$$

$$+ \frac{1}{2Na^2}\left\{-\frac{1}{\sin\frac{\pi}{2N}\left(2n + \frac{1}{2}\right)} + \frac{1}{\sin\frac{\pi}{2N}\left(2n - \frac{1}{2}\right)}\right\}, N \gg 1,$$

$$\approx \quad \frac{1}{4\pi a^2}\frac{1}{n^2 - 1/16} + \ldots, 0 < n \ll N.$$

Here we clearly see the emergence of the divergent term in eq. (12).

From these and other explicit examples we take confidence that lattice quantum fields distorted into nonuniformity by background spatial objects do closely resemble their continuum counterparts, both locally and globally.

In conclusion, it is a pleasure to thank this workshop's organizers for an instructive and enjoyable meeting.

References

[1] A. Actor and I. Bender, *Phys. Rev.* **D52** (1995) 3581.

[2] A. Actor and I. Bender, *Casimir Effect with a Semihard Boundary*, preprint, 1995.

[3] R. Friedberg and T.D. Lee, *Phys. Rev.* **D16** (1977) 1096.
T.D. Lee, *Phys. Rev.* **D19** (1979) 1802.

[4] A. Actor and I. Bender, *Boundaries Immersed in a Scalar Quantum Field*, preprint, 1995.

[5] F. Steiner, *Fortschr. Phys.* **35** (1987) 87.

[6] S. Leseduarte and A. Romeo, *J. Phys. A: Math. Gen.* **27** (1994) 2483.

[7] A. Actor and I. Bender, *The Zeta Function Constructed from the Zeros of the Bessel Function*, in preparation.

[8] Royal Society Mathematical Tables, vol. 7, Bessel Functions, Part III. Zeros and associated values, edited by F.W.J. Olver (Cambridge University Press, Cambridge, England, 1960).

[9] H.J. Rothe, *Lattice Gauge Theories—an Introduction*, World Scientific Lecture Notes in Physics, vol. 43 (World Scientific, Singapore, 1992).

[10] L.H. Ford, *Phys. Rev.* **D11** (1975) 3370.

Quantum Radiation from Mirrors Moving Sideways

Gabriel Barton

We calculate the radiation (of scalar "photon" pairs) by a halfspace oscillating harmonically and rigidly with speed $v = v_0 \cos(\Omega t)$, where $v_0/c \ll 1$. As shown earlier [1], a treatment by standard canonical local quantum field theory is possible only if the material is taken to be imperfectly reflecting: here we consider a real nondispersive refractive index n. The perfect-mirror limit $n \to \infty$ can be taken (only) at the end. Then in one dimension (1D), where motion is necessarily in-and-out, the power fed into photons traveling both inward, one inward one outward, and both outward, becomes proportional to $1/n^2$, $1/n$, and 1, respectively, so that all inward emission stops completely.

For sideways oscillation (parallel to the surface) in 3D, where the surface remains undisplaced and the region occupied by the material never changes, one expects perfect conductors not to radiate outwards at all, because fields outside then decouple totally from the interior, while the boundary conditions remain unaffected by such motion. In the opposite extreme, ie to leading order [2] in $(1 - 1/n^2)$, the total power per unit surface area goes half in and half out, and reads

$$ W = \frac{(1 - 1/n^2)^2 \hbar v_0^2 \Omega^4}{5760 \pi^2 c^4}. $$

For an arbitrary n [3], one must distinguish carefully between three types of modes: outward-directed (O), inward-directed and totally-internall-reflected (T), and inward-directed but not TIR (N). Then, and especially as $n \to \infty$, integrations over phase space become very delicate, and there is a subtle interplay between divergent phase-space factors and vanishing matrix elements. Asymptotically, the total power $W(ij)$ fed into photon pairs of given type (ij) behaves as follows:

$$ W(OO) \sim 1/n^2, \qquad W(ON) \sim 1/n, \qquad W(OT) \sim (1/n) \log n; $$
$$ W(NN) \sim 1/n^2, \qquad W(NT) \sim (1/n^2) \log n; \qquad W(TT) \sim n^0 = 1. $$

Remarkably, the wholly-inward emission W(TT) becomes finite and nonzero in the perfect-reflector limit. This implies in particular (i) that the energy density inside increases indefinitely (since speed $\sim 1/n$); and (ii) by virtue of the fluctuation-dissipation theorem, that even a perfect mirror experiences fluctuating forces with nonzero mean-square deviation, due to the zero-point fields inside the material, in addition to the more familiar force fluctuation [4] due to the zero-point fields in the adjacent vacuum.

For in-and-out oscillations in 3D, one finds as $n \to \infty$ that $W(OO) \sim 1$, reproducing a long-known result [5], but that $W(TT) \sim 1$ likewise. Thus the paradoxes just described of inward emission into perfect reflectors are not peculiar to sideways motion, but generic in 3D.

References

[1] G. Barton and C. Eberlein (1993) Ann. Phys. (NY) **227** 222.

[2] G. Barton (1996) Ann. Phys. (NY) to appear.

[3] This work is being done jointly with Clive North.

[4] G. Barton (1991) J. Phys. A **24** 991 & 5533; and (1994) in "Cavity QED"; ed. P.R. Berman, Academic Press, New York.

[5] L.H. Ford and A. Vilenkin (1982) Phys. Rev. D **25** 2569.

Sonoluminescence as Quantum Vacuum Radiation

Claudia Eberlein

Sonoluminescence is the phenomenon of light emission by sound-driven gas bubbles in fluids. Ultra-sound makes bubbles collapse and rebound, and a flash of light shorter than 10ps is observed after each collapse. The emitted spectrum resembles the tail of a black-body spectrum of several tens of thousand Kelvin. So far no theory has been found that convincingly explains the radiation process.

This new theory explains sonoluminescence as the excitation of vacuum fluctuations by the moving interface between two media of different polarizability, as for instance water and air. It relates to the Unruh effect which predicts radiation by non-uniformly moving mirrors, i.e. interfaces of perfect conductors with the vacuum, but which has so far never been observed experimentally because the number of radiated photons is minute for common macroscopically achievable accelerations. However, the violent collapse of a sonoluminescent gas bubble is exceptionally non-uniform; the acceleration and higher moments of the interface motion reach huge values which lead to a vast number of photons created. The correlations of the vacuum fluctuations make this a highly coherent process, whence extremely short pulses are inherent in this theory. The apparent thermal distribution of the photons stems from the fact that the photons are created in pairs and that the measurement of the single-photon spectrum involves the tracing over the other photon in the pair. This is known to entail thermal-like properties even though no temperature is involved; it is purely an effect of correlated quantum fluctuations at zero temperature.

The theory is able to explain a number of previously unresolved issues and seems in good agreement with experimental observations. Experiments that discriminate this from other theories of sonoluminescence have been proposed. A detailed account of this theory will be published elsewhere.[1]

---

[1] C. Eberlein, Paper submitted to Phys. Rev. A

Dynamical Casimir Effect, "Particle Emission" and Squeezing

Carlos Villarreal, R. Jáuregui, and S. Hacyan

We studied the quantization of the electromagnetic field between moving boundaries in four dimensional spacetime. We considered the particular case of a field confined between two infinite conducting plates moving with constant relative velocity [1]. This motion produces a distortion of vacuum that can be interpreted as "particle production". The work necessary is done at the expense of the Casimir energy of the system [2]. We evaluated the energy momentum tensor and its spectrum from the field correlations. We have also found that a vacuum or coherent state between the plates becomes squeezed. Analytic formulas for the Bogolubov and the squeezing coefficients [3] have been obtained. We analyzed the interaction of the "emitted photons" with a realistic quantum detector [4] and we have come to the conclusion that there is no spontaneous excitation of the detector. However, the spontaneous decay ratio shows strong oscillations as compared with the free case.

References

[1] R. Jáuregui, C. Villarreal, and S. Hacyan, Mod. Phys. Lett. A **10**, 619 (1995), C. Villarreal, S. Hacyan, and R. Jáuregui, Phys. Rev. A **52**, 594 (1995).

[2] M. Bordag, F.M. Dittes, and D. Robaschik, Sov. J. Nucl. Phys. **43**, 1034 (1986).

[3] V.V. Dodonov, A.B. Klimov and V.I. Man'ko, Phys. Lett. A **149**, 225 (1990); J. Sov. Laser Research **12**,439 (1991).

[4] R. Jáuregui and C. Villarreal, Preprint IFUNAM (1995)

Photon Generation and Squeezing in a Cavity with Vibrating Walls

Victor V. Dodonov

The problem of the electromagnetic field inside cavities with moving walls attracted the attention of researchers beginning with early sixties. Exact solutions for uniformly moving boundaries were found in [1] for the classical field and in [5] for the quantized field. Approximate solutions valid for an arbitrary nonrelativistic law of motion of a boundary were proposed in [3] (quantized field) and [4] (classical field). They were used to investigate the case of an oscillating boundary (which seems the most interesting from the point of view of possible experimental realizations) in [5]. However, these solutions hold only in the adiabatic case, when the frequency of wall's vibrations is much smaller than the frequency of some unperturbed electromagnetic mode. Consequently, no photons can be created under such conditions. The evolution of the classical field in the case when both the frequencies were close to each other was investigated in [6]. Here the solution of the quantum problem in the resonance case is presented.

Consider first the field between two infinite ideal plates, one being at rest at $x = 0$, while another oscillates at twice the eigenfrequency of the principal unperturbed mode: $L(t) = L_0 \left[1 + \varepsilon \sin(2\omega_1 t)\right]$, $\omega_1 = \pi/L_0$ (we assume $c = \hbar = 1$). Confining ourselves to the electromagnetic modes whose vector potential is directed along z-axis, we can write down the field operator in the Heisenberg representation as follows,

$$\hat{A}(x,t) = 2 \sum_n \sum_k \sqrt{\frac{L_0}{nL(t)}} \hat{b}_n Q_k^{(n)}(t) \sin\left(\frac{\pi k x}{L(t)}\right) + \text{h.c.}, \tag{1}$$

where $\hat{b}_n$ means the usual annihilation photon operator, and $Q_k^{(n)}(t) = \delta_{kn} \exp\left(-i\omega_n t\right)$ at $t \leq 0$, when both the plates are assumed to be at rest ($\omega_n = \pi n/L_0$). At $t > 0$ the wave equation $A_{tt} - A_{xx} = 0$ leads to an infinite set of coupled differential equations

$$\ddot{Q}_k^{(n)} + \omega_k^2(t) Q_k^{(n)} = 2\gamma(t) \sum_j g_{kj} \dot{Q}_j^{(n)} + \dot{\gamma}(t) \sum_j g_{kj} Q_j^{(n)} + \gamma^2(t) \sum_{jl} g_{jk} g_{jl} Q_l^{(n)}, \tag{2}$$

$$\omega_k(t) = k\pi/L(t), \quad \gamma(t) = \dot{L}(t)/L(t), \quad g_{kj} = -g_{jk} = (-1)^{k-j} \frac{2kj}{j^2 - k^2}, \quad (j \neq k),$$

with the initial conditions $Q_k^{(n)}(0) = \delta_{kn}$, $\dot{Q}_k^{(n)}(0) = -i\omega_n \delta_{kn}$ (cf. [7]).

Assuming that $|\varepsilon| \ll 1$, it is natural to look for the solutions of Eq. (2) in the form $Q_k^{(n)}(t) = \xi_k^{(n)}(t) \exp(-i\omega_k t) + \eta_k^{(n)}(t) \exp(i\omega_k t)$, the coefficients $\xi_m^{(n)}$ and $\eta_m^{(n)}$ being *slowly varying functions of time.* In the *parametric resonance* case these coefficients depend on the "slow" time $\tau = \frac{1}{2}\varepsilon\omega_1 t$, according to the following system of equations:

$$\frac{\mathrm{d}}{\mathrm{d}\tau}\xi_1^{(n)} = -\eta_1^{(n)} + 3\xi_3^{(n)}, \qquad \frac{\mathrm{d}}{\mathrm{d}\tau}\xi_k^{(n)} = (k+2)\xi_{k+2}^{(n)} - (k-2)\xi_{k-2}^{(n)}, \quad k \geq 2, \quad (3)$$

$$\frac{\mathrm{d}}{\mathrm{d}\tau}\eta_1^{(n)} = -\xi_1^{(n)} + 3\eta_3^{(n)}, \qquad \frac{\mathrm{d}}{\mathrm{d}\tau}\eta_k^{(n)} = (k+2)\eta_{k+2}^{(n)} - (k-2)\eta_{k-2}^{(n)}, \quad k \geq 2. \quad (4)$$

Here we neglect the terms of the order of ε^2 and rapidly oscillating terms proportional to $\exp(\pm 2im\tau/\varepsilon)$, $m = \pm 1, \pm 2, \ldots$. The system (3)-(4) with the initial conditions $\xi_k^{(n)}(0) = \delta_{kn}$, $\eta_k^{(n)}(0) = 0$ can be solved *exactly* [8]. In particular,

$$\xi_1^{(1)} = \frac{2}{\pi}\frac{\mathbf{E}(\kappa) + \tilde{\kappa}\mathbf{K}(\kappa)}{1 + \tilde{\kappa}}, \qquad \eta_1^{(1)} = -\frac{2}{\pi}\frac{\mathbf{E}(\kappa) - \tilde{\kappa}\mathbf{K}(\kappa)}{1 - \tilde{\kappa}}, \tag{5}$$

where $\kappa = \sqrt{1 - e^{-8\tau}}$, $\tilde{\kappa} = \sqrt{1 - \kappa^2} = e^{-4\tau}$, and $\mathbf{K}(\kappa), \mathbf{E}(\kappa)$ are the complete elliptic integrals of the first and the second kind, respectively.

If the wall comes back to its initial position L_0 after some interval of time T, then one may introduce a new set of "physical" operators $\hat{a}_m$ in accordance with the decomposition

$$\hat{A}_f(x,t) = 2\sum_n \frac{1}{\sqrt{n}}\sin\frac{n\pi x}{L_0}\hat{a}_n\exp\left(-i\omega_n t\right) + \text{h.c.}. \tag{6}$$

"New" and "old" operators are related by means of the Bogoliubov transformation

$$\hat{a}_m = \sum_n \sqrt{\frac{m}{n}}\left(\hat{b}_n\xi_m^{(n)} + \hat{b}_n^\dagger\eta_m^{(n)}\right). \tag{7}$$

Assuming that initially the field was in the vacuum state $|0_b\rangle$, one can write the amount of photons created in the m-th mode as

$$\mathcal{P}_m \equiv \langle 0_b|\hat{a}_m^\dagger\hat{a}_m|0_b\rangle = m\sum_{n=1}^\infty \frac{1}{n}|\eta_m^{(n)}|^2. \tag{8}$$

The last sum can be calculated exactly. In particular, for the principal mode

$$\mathcal{P}_1(\kappa) = \frac{2}{\pi^2}\mathbf{E}(\kappa)\mathbf{K}(\kappa) - \frac{1}{2}. \tag{9}$$

The plot of this dependence versus time t is very simple: the initial parabola $\mathcal{P}_1(t) = (\varepsilon\omega_1 t/2)^2$ at $\varepsilon\omega_1 t \ll 1$ is smoothly transformed into the linear dependence $\mathcal{P}_1(t) \approx \frac{4}{\pi^2}\varepsilon\omega_1 t$ for $\varepsilon\omega_1 t \gg 1$. These asymptotics were found earlier in the framework of Moore's approach [3] in [9]. For the total amount of photons created in all the modes we found the expression

$$\mathcal{P} = \frac{1}{\pi^2}\left[\left(1 - \frac{1}{2}\kappa^2\right)\mathbf{K}^2(\kappa) - \mathbf{E}(\kappa)\mathbf{K}(\kappa)\right]. \tag{10}$$

Thus we have quadratic dependences on time of the total amount of photons both in the short-time and long-time limits: $\mathcal{P} = \tau^2$ at $\tau \ll 1$, and $\mathcal{P} \approx \frac{8}{\pi^2}\tau^2$ at $\tau \gg 1$.

The *three-dimensional* resonance case turns out more simple than the 1D model, because the 3D frequency spectrum *is not equidistant*. As a consequence, different modes do not interact with each other, and the problem can be reduced to that of a one-dimensional parametric oscillator. The average number of photons in the resonance mode (with the unperturbed eigenfrequency ω_0) equals [10] $\langle n \rangle = \sinh^2 \tau$, where $\tau = \frac{1}{2}\omega_0 \varepsilon t$. Moreover, the initial vacuum state of the field oscillator is transformed into a *highly squeezed vacuum state*, with the squeezing coefficient $s = e^{-2\tau}$. On the contrary, in the 1D cavity the squeezing coefficient of the quadrature component $\hat{x}_m = \left(\hat{a}_m + \hat{a}_m^\dagger\right)/\sqrt{2}$ tends to the constant (and rather moderate) value [9] $s_m = 1 - 2\left[1 - (-1)^m - \pi m \, \mathrm{si}(\pi m)\right]/(\pi^2 m)$, where $\mathrm{si}(x)$ means the integral sine function. This coefficient assumes the minimal value for the principal mode: $s_1 \approx 0.78$.

The maximal possible value of the dimensionless displacement of the boundary is [10] $\varepsilon_{max} = (v_s/2\pi c)\delta_{max} \sim 3 \cdot 10^{-8}$, where $v_s \sim 5 \cdot 10^5$ cm/s is the sound velocity inside the wall, and $\delta_{max} \sim 10^{-2}$ is the maximal possible relative deformation of the material which the wall is made from. For $\omega_1/2\pi = 10$ GHz and $t = 1$ s (i.e., for the quality factor $Q \sim 3 \cdot 10^{10}$) the amount of photons which can be generated in the principal mode in the 1D case reaches 600. In the 3D cavity under the same conditions one could create about $4 \cdot 10^4$ photons even for the amplitude $\varepsilon \sim 10^{-2}\varepsilon_{max}$.

References

[1] V.I. Kurilko, Zhurn. Tekhnicheskoi Fiziki, **30**, 504 (1960); N.L. Balazs, J. Math. Anal. Appl., **3**, 472 (1961); O.A.Stetsenko, Izvestiya VUZ – Radiotekhnika, **6**, 695, 701 (1963); R.I. Baranov and Yu.M. Shirokov, Zhurn. Eksper. Teor. Fiz., **53**, 2121 (1967); A.I.Vesnitskii, Izvestiya VUZ – Radiofizika, **12**, 935 (1969).

[2] M. Bordag, F.M. Dittes and D. Robaschik, Sov. J. Nucl. Phys., **43**, 1034 (1986); C. Villarreal, S. Hacyan and R. Jáuregui, Phys. Rev. **A52**, 594 (1995).

[3] G.T. Moore, J. Math. Phys., **11**, 2679 (1970).

[4] A.I. Vesnitskii, Izvestiya VUZ – Radiofizika, **14**, 1432 (1971).

[5] A.I. Vesnitskii, Izvestiya VUZ – Radiofizika, **14**, 1538 (1971); S. Sarkar, in *Photons and Quantum Fluctuations*, eds. E.R. Pike and H. Walther, Hilger, Bristol, 1988, p. 151.

[6] V.N. Krasil'nikov and A.M. Pankratov, in *Problems of Diffraction and Wave Propagation*, **8**, 59 (1968) (Leningrád State University Publ., in Russian).

[7] M. Razavy and J. Terning, Phys. Rev. D, **31**, 307 (1985); C.K. Law, Phys. Rev. A, **51**, 2537 (1995).

[8] V.V. Dodonov and A.B. Klimov, to be published

[9] V.V. Dodonov, A.B. Klimov, and V.I. Man'ko, Phys. Lett. A, **149**, 225 (1990); V.V. Dodonov and A.B. Klimov, Phys. Lett. A, **167**, 309 (1992); V.V. Dodonov, A.B. Klimov, and D.E. Nikonov, J. Math. Phys., **34**, 2742 (1993).

[10] V.V. Dodonov, Phys. Lett. A (in press).

Fluctuations of the Casimir Pressure at Finite Temperature

Dieter Robaschik

We consider special topics related to the Casimir effect [1]. We extend the considerations of the fluctuations of the Casimir pressure [2], [3] to finite temperature [4]. In the unphysical high-temperature limit the Green's functions become more singular. This explains why these Green's functions allow the determination of the Casimir pressure in this limit, but not of its fluctuations. We remark, that the boundary conditions for conductors in classical Electrodynamics are not sufficient for a unique definition of the Green's functions.

For ideal conductors, the boundary conditions $E_t = B_n = 0$ can be written in terms of the electromagnetic potentials A_μ by $\epsilon_{\mu\nu\rho\sigma} n^\rho \partial^\sigma A^\nu|_S = 0$, where n^ρ denotes the normal vector of the surface S. For the case of parallel plates perpendicular to the x_3-direction, we get the following general structure of all Green's functions [5], [3] in covariant gauge

$$< 0|T A_\mu(x) A_\nu(y)|0 > = \tag{1}$$

$$i(g_{\tilde\mu\tilde\nu} - \frac{\partial_{\tilde\mu}\partial_{\tilde\nu}}{\tilde\partial^2})\,{}^sD^c(\tilde x - \tilde y, x_3, y_3) + i \begin{pmatrix} \frac{\tilde\partial_\mu\tilde\partial_\nu}{\tilde\partial^2} & 0 \\ 0 & -1 \end{pmatrix} D^c(x - y)$$

$$- i n_\mu n_\nu \tilde D^c(\tilde x - \tilde y).$$

In the physical situation, we are considering, the third direction plays a special role. Therefore, we write $\tilde x = (x^0, x^1, x^2)$, and $\tilde\nu$ runs from 0 to 2. ${}^sD^c(x, y)$ denotes the scalar Feynman propagator satisfying the Dirichlet boundary condition at the position of the plates. It is interesting that the boundary conditions for conductors do not fix the physical problem uniquely, because the mode propagating parallel to the plates, not subject to the boundary conditions, can be treated as a plane wave extended over the full x_3-axis, or restricted to the "inner" region between the two plates. For "thin" plates, we consider the complete three-dimensional Euclidean space, whereas for "thick" plates, the space between the two plates only. Correspondingly for "thin" plates $D^c(x - y)$ denotes the standard free space Feynman propagator and $\tilde D \equiv 0$. For the case of "thick" plates, $D^c(x - y)$, is a free Feynman propagator satisfying periodicity conditions and $\tilde D^c$, a special Feynman propagator containing the contributions from the special mode. In standard manner we obtain Wightman functions as well as the necessary temperature dependent functions functions in the Real-Time-Formalism [6].

The Casimir pressure is defined as the difference of $< T_{33} >$ across the plates

$$p(x) = < T_{33}(x_3 = a + \epsilon) > - < T_{33}(x_3 = a - \epsilon) > . \tag{2}$$

In our case we choose $a = 0$. For the investigation of the fluctuations we consider T_{33} at a position of the inner side of the plate at $a = 0$. The essential information for the fluctuation of an observable is contained in the expectation value

$$W_{2,\beta}(x, x') \;\; = \;\; < T_{33}(x)T_{33}(x') >_\beta - < T_{33}(x) >_\beta < T_{33}(x') >_\beta . \tag{3}$$

Experimentally, local observables are measured over the finite space and finite time intervals $T = \int f(x)T_{33}(x)dx$ described by the function $f(x)$. We factorize the function $f(x_0, \vec{x}_\perp)$ according to $f(x_0, \vec{x}_\perp) = g(x_0)h(\vec{x}_\perp)$. As an example we choose [2], [3]

$$g(x_0) = \frac{\tau}{\pi} \frac{1}{x_0^2 + \tau^2}, \qquad h(\vec{x}_\perp) = \delta^{(2)}(\vec{x}_\perp).$$

This means, we perform a local measurement over a finite time characterized by the parameter τ. The function $g(x_0)$ is normalized to unity, moreover $g(x^0)$ has the property $\lim_{\tau \to 0} g(x^0) = \delta(x^0)$. With these functions, the fluctuation $(\Delta T)^2$ is expressed by means of the correlation function $W_{2,\beta}(\tilde{x}, \tilde{x}')$ as

$$(\Delta T_\tau)^2 = \int d\tilde{x} d\tilde{x}' f(\tilde{x}) f(\tilde{x}') W_{2,\beta}(\tilde{x}, \tilde{x}'), \quad \tilde{x} = (x_0, x_1, x_2). \tag{4}$$

We determine the fluctuations of the Casimir pressure at the inner side of the plates for large times τ and small temperature $T = 1/\beta$. We obtain for "thick" plates [4]

$$(\Delta T_\tau)^2_{\beta, \, \tau \gg d} = \frac{3}{2^{10}\pi^2} \frac{1}{d^2} \frac{1}{\tau^6} + \frac{3}{2^5 \pi^2} \frac{1}{\tau^3 d^2} \frac{1}{\beta^3} \zeta(3) + ... \tag{5}$$

and for "thin" plates

$$(\Delta T_\tau)^2_{\beta, \, \tau \gg d} = \frac{\pi^2}{4} \frac{1}{d^6} \{ \frac{1}{2\tau^2} \exp(-4\tau\frac{\pi}{d}) + \frac{2}{2\tau\beta} \exp(-\frac{\pi\beta}{d}) \} + ... \tag{6}$$

The β-independent terms coincides with the results of [2] and [3]. Measurements over long times are sensitive to the low energy spectrum of the fluctuations. For "thick" plates this spectrum starts at $p_0 = 0$, whereas for "thin" plates the lowest state lays at $p_0 = \pi/d$. Consequently, we see a power-like behaviour for "thick" plates and exponential suppression for "thin" plates with respect to the parameter τ.

Finally we consider the limit of very high temperature. Of course, this problem involves unphysical issues. Nevertheless, this limit is theoretically interesting. Let us perform the high temperature limit $\beta \to 0$ of the scalar Green's function for the free space

$$D_\beta^-(z) = \frac{-i}{4\pi^2} \sum_{n=-\infty}^{\infty} \frac{1}{(z_0 - i\epsilon\delta_{n0} - in\beta)^2 - \vec{z}^2} \tag{7}$$

$$= \frac{i\pi}{4\pi^2 2|\vec{z}|\beta}[\coth\frac{\pi}{\beta}(|\vec{z}| - (z_0 - i\epsilon)) + \coth\frac{\pi}{\beta}(|\vec{z}| + (z_0 - i\epsilon))]. \tag{8}$$

and obtain

$$D_\beta^-(x,x')|_{\beta\to 0} = \frac{i}{4\pi^2}\frac{\pi}{|\vec{z}|\beta}\Theta(-z^2), \quad z = x - x'. \tag{9}$$

It seems to be that at very high temperature, there are neither causal propagations nor time-dependent correlations. This result changes if we have to consider derivatives of this function. For example, we obtain for the black body radiation at very high temperature:

$$< F_{m0}(x)F_{m'0'}(x') >_{\beta\to 0} = iO^{\rho\rho'}_{m0;m'0'}g_{\rho\rho'}\frac{i}{4\pi^2}\frac{\pi}{|\vec{z}|\beta}\Theta(-z^2) \tag{10}$$

$$= \frac{1}{\beta 8\pi}\{8(\delta_{nn'}|\vec{z}| - \frac{z_n z_{n'}}{|\vec{z}|})\delta'(z^2)$$

$$+2\frac{3z_0 z_0' - \delta_{nn'}\vec{z}^2}{|\vec{z}|^5}\Theta(-z^2) + 4\frac{\delta_{nn'}\vec{z}^2 - 2z_n z_{n'}}{|\vec{z}|^3}\delta(z^2)\},$$

where $O^{\rho\rho'}_{\mu\nu;\mu'\nu'}$ is a differential operator. The differentiations of the Θ-functions create δ-functions and its derivatives. These functions are responsible for light-like correlations. This is interesting because for finite $\beta \neq 0$ the temperature-dependent part of the correlation function does not contain singularities. The singular behaviour is a consequence of the limit $\beta \to 0$. The special structure of the field strength correlations as generalized functions does not allow their multiplication because the products $\delta(z^2)\delta(z^2)$ etc. are ill-defined. This means, that within the theory at infinite temperature, we cannot investigate energy and pressure fluctuations without arbitrariness.

For the two-plate-system we determine the classical Green's function. For space-like distances we write

$$D_{2,\beta}^-(x,x')|_{\beta\to 0} = \Theta(-z^2)\frac{i}{4\pi^2}\frac{\pi}{\beta}\sum_{l=-\infty}^{+\infty}[\frac{1}{[z_\perp^2 + (x_3 - x'_3 + 2ld)^2]^{1/2}}$$

$$-\frac{1}{[z_\perp^2 + (x_3 + x'_3 + 2ld)^2]^{1/2}}].$$

Relying on this formula, we derive the Casimir effect in the Rayleigh- Jeans limit as

$$p(d,\beta)|_{\frac{d}{\beta}\to\infty} = -\frac{\zeta(3)}{4\pi\beta d^3}. \tag{11}$$

This coincides with the high temperature limit of the standard result for the Casimir effect [7], [8], [9].

Because of the bad multiplication properties of the Green's functions in the high-temperature limit an investigation of the fluctuations has to start at finite temperature. After lengthy calculations we obtain [4]

$$(\Delta T_\tau)^2|_{\substack{\beta\to 0,\,\tau\gg d\\ \text{"thin"}}} = \frac{480}{\pi^4}\zeta(6)\frac{1}{\beta^7(2\tau)} - \frac{80}{\pi^3}\zeta(5)\frac{1}{\beta^6(2\tau)d} + \frac{13}{2\pi^3}\zeta(4)\frac{1}{\beta^5(2\tau)d^2} + \cdots$$

$$(\Delta T_\tau)^2|_{\substack{\beta\to 0,\,\tau\gg d\\ \text{"thick"}}} = \frac{480}{\pi^4}\zeta(6)\frac{1}{\beta^7(2\tau)} + \frac{2}{\pi^3}\zeta(4)\frac{1}{\beta^5(2\tau)d^2} + \cdots \tag{12}$$

For both cases we find the same leading behaviour. Moreover, the distance independent leading term is twice the result for free space or the result corresponding to one plate (at the position of the plate). It seems to be reasonable then, that in the limit of very high temperature, the nonleading terms are distance dependent and reflect special assumptions.

These fluctuations diverge as $\sim (\beta)^{-7}$ for $\beta \to 0$. This concerns the pressure fluctuations in free space as well as the Casimir pressure. If we compare the power behaviour with respect to β^{-1} of the Casimir pressure $\sim (\beta)^{-1}$ with that of its fluctuations $\sim (\beta)^{-7}$ for $\beta \to 0$, then the fluctuations seem to be remarkably large.

References

[1] V.M. Mostepanenko, N.N. Trunov, "The Casimir Effect and its Application", Atomizdat, Moscow, 1990 (in Russian); G. Plunien, B. Müller and W. Greiner, *Phys. Rep.* **134** (1986) 87.

[2] G. Barton, *J. Phys. A: Math. Gen.* **24** (1991) 991; **24** (1991) 5533; New Aspects of the Casimir Effect: Fluctuations and Radiative Reactions *in* "Cavity Quantum Electrodynamics" (P.R.Berman, Ed.), Academic Press.

[3] D. Robaschik and E. Wieczorek, *Ann. Phys. (N.Y.)* **236** (1994) 43.

[4] D.Robaschik and E.Wieczorek, *Phys. Rev.* **D52** (1995) 2341.

[5] M. Bordag, D. Robaschik and E. Wieczorek, *Ann. Phys. (N.Y.)* **165** (1985) 192.

[6] A.J. Niemi and G. W. Semenoff, *Ann. Phys. (N.Y.)* **152** (1984) 105; H. Umezawa, H. Matsumoto, and M. Tachiki, "Thermo Field Dynamics and Condensed States, North-Holland Pub. Comp. Amsterdam 1982.

[7] M. Fierz, *Helv. Phys. Acta* **33** (1960) 855; J. Mehra, *Physica* **37** (1967) 145.

[8] L.S. Brown and G.J. MacLay, *Phys. Rev.* **184** (1969) 1272.

[9] K. Scharnhorst, D. Robaschik and E. Wieczorek, *Annalen der Physik* **44** (1987) 351.

Casimir Theory for the Piecewise Uniform String

Iver Brevik

The Casimir energy for the transverse oscillations of a piecewise uniform closed string is calculated. In its simplest version the string consists of two parts I and II, each having in general different tension and mass density but adjusted in such a way that the velocity of sound always equals the velocity of light. In this sense the string model is a relativistic model.

The model was introduced in 1990 [1], the energy being regularised by means of an exponential cutoff. Later, it was shown how regularization can be achieved by use of the Hurwitz ζ function [2]. A third kind of regularization is to use the technique introduced in [3], consisting in writing the Casimir energy as a contour integral in the complex frequency plane [4]. One of the virtues of the last-mentioned method is that it is immediately applicable to the case of finite temperatures.

The Casimir energy E is found to be nonpositive. The transformation of an initially uniform string into a state of nonuniformity usually diminishes its zero point energy. The negativity of E becomes more pronounced when the state of inhomogeneity becomes more pronounced, i.e., when the two string pieces are of very different length, or when the two string tensions are very different.

The general formalism for a string divided into 2N equal pieces, of alternating type I and II of material, was worked out in [5] although solved in full only in the case of $N = 2$. Also now the same basic property was found: a uniform string can lower its energy by transforming itself into a nonuniform state.

The situation is in some sense related to the case of Casimir theory in material media. If the medium is relativistic, satisfying the condition $\varepsilon\mu = 1$ where ε is the permittivity and μ the permeability, then the Casimir energy is easy to calculate [6]. Perhaps the present simple model can even help us to understand the issue of the energy of the vacuum state in two-dimensional quantum field theory.

References

[1] I. Brevik and H.B. Nielsen, Phys. Rev. D**41**, 1185 (1990)

[2] X. Li, X. Shi and J. Zhang, Phys. Rev. D**44**, 560 (1991)

[3] N.G. van Kampen, B.R.A. Nijboer and K. Schram, Phys. Lett. **26A**, 307 (1968)

[4] I. Brevik and E. Elizalde, Phys. Rev D**49**, 5319 (1994)

[5] I. Brevik and H.B. Nielsen, Phys. Rev. D**51**, 1869 (1995)

[6] I. Brevik and H. Kolbenstvedt, Ann. Phys. (NY) **149**, 237 (1983)

Casimir Force Between Real Boundaries: Contributions of Mechanical and Electrical Imperfections

Vladimir M. Mostepanenko

In this report the corrections to the Casimir force due to non-ideality of the geometrical and electrical properties of the boundary bodies are discussed. Such corrections must be taken into account in experiments on measuring the Casimir force and in some applications of the Casimir effect (e.g., in obtaining of constraints on the parameters of hypothetical long-range interactions [1, 2]). At first the known results were reminded on the Casimir force between the dielectrical plates [3] and the plates made of real metal [4]. The corrections due to the penetration of zero oscillations into the metal were calculated up to the second order in the relative penetration depth. Both the cases of normal and anomalous skin effects were considered. Also the corrections due to the dielectrical layers on the surfaces of real metal were demonstrated. The most important subject of the report is the presentation of the new general perturbative approach to the calculation of the corrections to the Casimir force for configurations with small deviations from the plane parallel geometry [5, 6]. Here all the corrections are obtained up to the fourth order in the relative deviations amplitudes. The cases of nonparallel plates, of plates with a large scale unevenness and of plates with short scale distortions which may be both periodical and nonperiodical are considered. The continuous transition is traced numerically between the cases of large scale and short scale deviations. The developed approach was applied in [7] to configurations of a spherical lens above a plane and of two crossed cylinders which are used in experiments on Casimir force measurements. The calculated corrections may achieve some tens percents of the main contribution to the Casimir force. In the last section of the report the analogical approach was suggested for the case of small stochastic deviations of the boundary surfaces from the plane parallel configuration. The perturbation theory was developed up to the fourth order in the relative dispersions of the deviations [8]. The results were applied both in the cases of stationary and nonstationary stochastic functions describing the deviations.

References

[1] V.M. Mostepanenko, I.Yu. Sokolov. Phys. Rev.D47 (1993) 2882.

[2] M. Bordag, V.M. Mostepanenko, I.Yu. Sokolov. Mod. Phys. Lett. **A9** (1994) 2671.

[3] E.M. Lifshitz, L.P. Pitaevsky. *Statistical Physics* (Pergamon, Oxford, 1980).

[4] V.M. Mostepanenko, N.N. Trunov. Sov. J. Nucl.Phys.(USA) **42** (1985) 818.

[5] M. Bordag, G.L. Klimchitskaya, V.M. Mostepanenko. Mod. Phys. Lett. **A9** (1994) 2515.

[6] M. Bordag, G.L. Klimchitskaya, V.M. Mostepanenko. Int. J. Mod. Phys. **A10** (1995) 2661.

[7] G.L. Klimchitskaya, Yu.V. Pavlov. Int. J. Mod. Phys. **A**, in print.

[8] M. Bordag, G.L. Klimchitskaya, V.M. Mostepanenko. Phys. Lett. **A200** (1995) 95.

Part II

QED and QCD in External Fields

Renormalization Group Flow of the Chern-Simons Parameter

Martin Reuter

Abstract

The effective average actions for gauge theories and their associated exact renormalization group equations are briefly reviewed and an application to Chern-Simons theory is described.

1 Introduction

The purpose of these notes is twofold. First we shall give a brief introduction to the method of the effective average actions and their associated exact renormalization group equations [1, 2, 16] and then we shall apply these ideas to a system which is also quite interesting in its own right, namely pure Chern-Simons theory in three dimensions [4].

The effective average action Γ_k can be thought of as a continuum version of the block spin action for spin systems [1]. The functional Γ_k is the action relevant to the physics at (mass) scale k. It has the quantum fluctuations with momenta larger than k integrated out already, but those with momenta smaller than k are not yet included. Γ_k interpolates between the classical action S for large values of k, and the conventional effective action for k approaching zero: $\Gamma_{k\to\infty} = S, \Gamma_{k\to 0} = \Gamma$. In many important cases where perturbation theory is inapplicable due to infrared divergences the limit $k \to 0$ exists and can be computed by various methods. This includes for instance massless theories in low dimensions or the high temperature limit of 4 dimensional theories. The functional Γ_k can be obtained by solving an exact renormalization group equation which describes its evolution while k is lowered from infinity to zero. In the approach of ref. [2], and for models with a scalar field ϕ only, this evolution equation reads

$$\frac{\partial}{\partial t}\Gamma_k[\phi] = \frac{1}{2}\mathrm{Tr}\left[\frac{\partial}{\partial t}R_k\left(\Gamma_k^{(2)}[\phi] + R_k\right)^{-1}\right]. \tag{1}$$

Here $t \equiv \ln k$ is the "renormalization group time" and $\Gamma_k^{(2)}$ denotes the matrix of the second functional derivatives of Γ_k. The operator $R_k \equiv R_k(-\partial^2)$ or, in momentum space, $R_k \equiv R_k(q^2)$ describes the details of how the small momentum modes are cut off and it is to some extent arbitrary. It has to vanish for $q^2 \gg k^2$ and to become a mass-like term proportional to k^2 for small momenta $q^2 \ll k^2$. The derivation of (1) proceeds as follows. In the euclidean functional integral for the generating functional of the connected Green functions one adds a momentum-dependent mass term (playing the role of a smooth IR cutoff) $\frac{1}{2}\int \phi R_k(-\partial^2)\phi$ to the classical action S. Then, up to an explicitly known correction term [2], the resulting k-dependent functional $W_k[J]$ is related to $\Gamma_k[\phi]$ by a conventional Legendre transformation at fixed k.

In Section 2 we shall generalize the evolution equation to gauge theories and, in Section 3, apply it to the calculation of the beta-function of the nonabelian gauge coupling. In the remaining sections we shall then apply the same strategies to the study of Chern-Simons field theory in 3 dimensions. This is an interesting theory from many points of view. It can be used to give a path-integral representation of knot and link invariants [5] and to understand many properties of 2-dimensional conformal field theories [5, 6]. Being a topological field theory the model has no propagating degrees of freedom. Canonical quantization yields a Hilbert space with only finitely many physical states which can be related to the conformal blocks of (rational) conformal field theories. Perturbative covariant quantization [7, 8, 9, 10, 11] shows that the theory is not only renormalizable but even ultraviolet finite. It is remarkable that despite this high degree of "triviality" the theory produces nontrivial radiative corrections. One-loop effects were found [12, 5] to lead to a renormalization of the parameter κ which multiplies the Chern-Simons 3-form in the action,

$$S_{\mathrm{CS}}[A] = i\kappa \, \frac{g^2}{8\pi} \int d^3x \; \varepsilon_{\alpha\beta\gamma} \, [A^a_\alpha \, \partial_\beta A^a_\gamma + \frac{1}{3} g f^{abc} A^a_\alpha A^b_\beta A^c_\gamma] \tag{2}$$

A variety of gauge invariant regularization methods, including spectral flow arguments based upon the η-invariant, predict a finite difference between the bare and the renormalized value of κ:

$$\kappa_{\mathrm{ren}} = \kappa_{\mathrm{bare}} + \mathrm{sign}(\kappa) \, T(G) \tag{3}$$

Here $T(G)$ denotes the value of the quadratic Casimir operator of the gauge group G in the adjoint representation. It is normalized such that $T(SU(N)) = N$. The shift of κ has a natural relation to similar shifts in the Sugawara construction of 2-dimensional conformal field theories. On the other hand, in standard renormalization theory a relation of the type (3) is rather unusual, and there has been some controversy in the literature about the correct interpretation of eq. (3). Following ref.   [4] we shall investigate this problem in the context of the effective average action.

2   The Renormalization Group Equation

In the case of gauge theories the derivation of an exact evolution equation faces additional complications because the inhomogeneous gauge transformation law of the Yang-Mills fields forbids a mass-type cutoff. In refs. [16, 13] this problem was overcome recently by using the background gauge technique [14] which allows us to work with a *gauge invariant* effective average action. The prize which one has to pay for this advantage is that Γ_k depends on two gauge fields: the usual classical average field A^a_μ and the background field $\bar{A}^a_\mu$. For pure Yang-Mills theory one finds the following renormalization group equation [16]

$$k\frac{d}{dk}\Gamma_k[A,\bar{A}] = \frac{1}{2}\text{Tr}\left[\left(\Gamma_k^{(2)}[A,\bar{A}] + R_k(\Delta[\bar{A}])\right)^{-1} k\frac{d}{dk}R_k(\Delta[\bar{A}])\right]$$
$$-\text{Tr}\left[\left(-D_\mu[A]\,D_\mu[\bar{A}] + R_k(-D^2(\bar{A}))\right)^{-1} k\frac{d}{dk}R_k(-D^2[\bar{A}])\right] \tag{4}$$

Eq.(4) has to be solved subject to the initial condition

$$\Gamma_\infty[A,\bar{A}] = S[A] + \frac{1}{2\alpha}\int d^dx \,\left(D_\mu^{ab}[\bar{A}]\,(A_\mu^a - \bar{A}_\mu^a)\right)^2 \tag{5}$$

where the classical action is augmented by the background gauge fixing term. Furthermore, $\Gamma_k^{(2)}[A,\bar{A}]$ denotes the matrix of the second functional derivatives of Γ_k with respect to A at fixed $\bar{A}$. Again, the function R_k specifies the precise form of the infrared cutoff, and it has the same properties as mentioned in the introduction. A convenient choice is

$$R_k(u) = Z_k\,u\,\left[\exp\left(u/k^2\right) - 1\right]^{-1} \tag{6}$$

but in some cases even a simple constant $R_k = Z_k k^2$ is sufficient. The factor Z_k has to be fixed in such a way that a massless inverse propagator $Z_k q^2$ combines with the cutoff to $Z_k(q^2+k^2)$ for the low momentum modes. Z_k may be chosen differently for different fields. In particular, different Z_k-factors are used for the gauge field fluctuations and for the Faddeev-Popov ghosts. (They give rise to the first and the second trace on the RHS of eq.(4), respectively.) Observable quantities will not depend on the form of R_k. A similar remark applies to the precise form of the operator $\Delta[\bar{A}] \equiv -D^2[\bar{A}] + ...$ which is essentially the covariant laplacian, possibly with additional nonminimal terms [16]. The rôle of Δ is to distinguish "high momentum modes" from "low momentum modes". If one expands all quantum fluctuations in terms of the eigenmodes of Δ, then it is the modes with eigenvalues larger than k^2 which are integrated out in Γ_k.

In order to understand the structure of the exact renormalization group equation (4) it is useful to realize that it can be rewritten in a form which is reminiscent of a one-loop formula:

$$\frac{\partial}{\partial t}\Gamma_k[A,\bar{A}] = \frac{1}{2}\frac{D}{Dt}\text{Tr}\ln\left[\Gamma_k^{(2)}[A,\bar{A}] + R_k\left(\Delta[\bar{A}]\right)\right]$$
$$-\frac{D}{Dt}\text{Tr}\ln\left[-D^\mu[A]D_\mu[\bar{A}] + R_k(-D^2[\bar{A}])\right] \tag{7}$$

By definition, the derivative $\frac{D}{Dt}$ acts only on the explicit k-dependence of R_k, but not on $\Gamma_k^{(2)}[A,\bar{A}]$. It is easy now to describe the relation between the effective average action Γ_k and the conventional effective action. Let us first make the approximation $\frac{D}{Dt} \to \frac{\partial}{\partial t}$ in eq. (7). This amounts to neglecting the running of Γ_k on the RHS of the evolution equation. Therefore it can be solved by simply integrating both sides of the equation from the infrared cutoff k to the ultraviolet cutoff Λ:

$$
\begin{aligned}
\Gamma_k[A, \bar{A}] \;=\;\; & \Gamma_\Lambda[A, \bar{A}] + \frac{1}{2}\mathrm{Tr}\left\{\ln\left[\Gamma_k^{(2)}[A, \bar{A}] + R_k(\Delta[\bar{A}])\right]\right. \\
& \left. - \ln\left[\Gamma_\Lambda^{(2)}[A, \bar{A}] + R_\Lambda(\Delta[\bar{A}])\right]\right\} \\
& - \mathrm{Tr}\left\{\ln\left[-D^\mu[A]D_\mu[\bar{A}] + R_k(-D^2[\bar{A}])\right]\right. \\
& \left. - \ln\left[-D^\mu[A]D_\mu[\bar{A}] + R_\Lambda(-D^2[\bar{A}])\right]\right\} + O\left(\frac{\partial}{\partial t}\Gamma_k^{(2)}\right)
\end{aligned}
\tag{8}
$$

Ultimately we shall send the ultraviolet cutoff to infinity and identify Γ_Λ with the classical action S plus the gauge fixing term. Eq.(8) has a similar structure as a regularized version of the conventional one-loop effective action in the background gauge. There are two important differences, however: (i) The second variation of the classical action, $S^{(2)}$, is replaced by $\Gamma_k^{(2)}$. This implements a kind of "renormalization group improvement". (ii) The effective average action contains an explicit infrared cutoff R_k. Because $\lim_{u\to\infty} R_k(u) = 0, \lim_{u\to 0} R_k(u) = Z_k k^2$, a mass-term is added to the inverse propagator $\Gamma_k^{(2)}$ for low frequency modes ($u \to 0$), but not for high frequency modes ($u \to \infty$). Despite the similarity of (8) with a one-loop expression, we stress that, for $k \to 0$, the solution of the original renormalization group equation (4) or (7) equals the *exact* effective action which includes contributions from all orders of the loop expansion.

The solution $\Gamma_k[A, \bar{A}]$ of (4) with (5) is gauge invariant under simultaneous gauge transformations of A and $\bar{A}$. Following the lines of the conventional background method [14] one would try to equate the two gauge field arguments of Γ_k, and work with the functional $\bar{\Gamma}_k[A] \equiv \Gamma_k[A, A]$. However, it is important to note that the evolution equation (4) cannot be rewritten in terms of $\bar{\Gamma}_k[A]$ alone, since $\Gamma_k^{(2)}$ does not involve derivatives with respect to $\bar{A}$. In fact, let us introduce the decomposition

$$
\Gamma_k[A, \bar{A}] = \bar{\Gamma}_k[A] + \Gamma_k^{\mathrm{gauge}}[A, \bar{A}]
\tag{9}
$$

This leads to

$$
\Gamma_k^{(2)}[A, \bar{A}] = \bar{\Gamma}_k^{(2)}[A] + \Gamma_k^{\mathrm{gauge}(2)}[A, \bar{A}]_{|\bar{A}}
\tag{10}
$$

where the second functional derivative $\Gamma_k^{\mathrm{gauge}(2)}$ is performed at fixed $\bar{A}$. The interpretation of (9) and (10) is as follows. Because $\bar{\Gamma}_k[A]$ is a gauge invariant functional of its argument, $\bar{\Gamma}_k^{(2)}[A]$ is necessarily singular, i.e., it has the usual gauge zero modes. They are gauge fixed by the generalized gauge fixing term $\Gamma_k^{\mathrm{gauge}}$. This is possible because $\Gamma_k^{\mathrm{gauge}}$ is not invariant under separate gauge transformations of A alone.

3   A Simple Example

The exact evolution equation is a nonlinear differential equation for a function of infinitely many variables. There seems to be little hope for finding closed-form solutions.

The successful use of this equation therefore depends crucially on the existence of an appropriate approximation scheme. This will consist in a truncation of the infinitely many invariants characterizing Γ_k to a finite number. If one makes an ansatz for Γ_k which contains only finitely many parameters (depending on k) and inserts it into (4), the functional differential equation reduces to a set of coupled ordinary differential equations for the parameter functions. The truncation should be chosen in such a way that it encapsulates the essential physics in an ansatz as simple as possible. In a second step one has to verify that upon including more terms in the truncation the results do not change significantly any more.

In this section we demonstrate the practical use of our equation by computing approximately the running of the nonabelian gauge coupling of pure Yang-Mills theory with gauge group SU(N) in arbitrary dimension d [16]. In order to approximate the solution $\Gamma_k[A, \bar{A}]$ of (4) by a functional with at most second derivatives, we make the ansatz

$$\Gamma_k[A, \bar{A}] = \int d^d x \left\{ \frac{1}{4} Z_{Fk}\, F_{\mu\nu}^a(A) F_{\mu\nu}^a(A) + \frac{Z_{Fk}}{2\alpha_k} [D_\mu[\bar{A}](A_\mu - \bar{A}_\mu)]^2 \right\} \tag{11}$$

We want to determine the running of Z_{Fk} from the flow equation. The truncation (11) leads to the Hessian

$$\frac{\delta^2 \Gamma_k[A, \bar{A}]}{\delta A_\mu^a(x) \delta A_\nu^b(x')} = Z_{Fk} \left\{ \mathcal{D}_{\mathrm{T}}[A]_{\mu\nu} + D_\mu[A]D_\nu[A] - \frac{1}{\alpha_k} D_\mu[\bar{A}]D_\nu[\bar{A}] \right\}^{ab} \delta(x - x') \tag{12}$$

where $(\mathcal{D}_T)_{\mu\nu} \equiv -D^2 \delta_{\mu\nu} + 2i\bar{g}F_{\mu\nu}$ with the color matrix F in the adjoint representation. ($\bar{g}$ denotes the bare gauge coupling.) In the following we neglect the running of α_k and restrict our discussion to $\alpha_k = 1$. Thus

$$\frac{\delta^2}{\delta A^2} \Gamma_k[A, \bar{A}] \Big|_{\bar{A}=A} = Z_{Fk} \mathcal{D}_{\mathrm{T}}(A) \tag{13}$$

and the evolution equation reads for $\bar{A} = A$:

$$\begin{aligned}
\frac{\partial}{\partial t} \Gamma_k[A, A] &= \frac{\partial Z_{Fk}}{\partial t} \int d^d x \frac{1}{4} F_{\mu\nu}^a F_{\mu\nu}^a \\
&= \frac{1}{2} \mathrm{Tr} \left[\left(\frac{\partial}{\partial t} R_k(\mathcal{D}_{\mathrm{T}}) \right) (Z_{Fk}\mathcal{D}_{\mathrm{T}} + R_k(\mathcal{D}_{\mathrm{T}}))^{-1} \right] \\
&\quad - \mathrm{Tr} \left[\left(\frac{\partial}{\partial t} R_k(\mathcal{D}_{\mathrm{S}}) \right) (\mathcal{D}_{\mathrm{S}} + R_k(\mathcal{D}_{\mathrm{S}}))^{-1} \right]
\end{aligned} \tag{14}$$

Here $\mathcal{D}_{\mathrm{T}}$ and $\mathcal{D}_{\mathrm{S}} \equiv -D^2$ depend on A_μ now. For mathematical convenience we chose $\Delta = \mathcal{D}_T$ for the cutoff operator. The function R_k is defined with $Z_k = Z_{Fk}$ in the first trace on the RHS of (14) (gluons) and with $Z_k = 1$ in the second trace (ghosts). In order to determine $\partial Z_{Fk}/\partial t$ it is sufficient to extract the term proportional to the invariant $F_{\mu\nu}^a F_{\mu\nu}^a$ from the traces on the RHS of (14). This can be done by using standard heat-kernel techniques or by inserting a simple field configuration on both sides of the equation for which the traces can be calculated easily. In either case one finds for $d > 2$ [16]

$$\frac{\partial}{\partial t} Z_{Fk} = \quad - \quad 2N \left(1 - \frac{d}{24}\right) \frac{v_{d-2}}{\pi} \bar{g}^2 \int_0^\infty dx \; x^{\frac{d}{2}-2} \frac{d}{dx} \frac{\partial_t R_k(x)}{Z_{Fk} x + R_k(x)}$$

$$\quad - \quad \frac{1}{6} N \frac{v_{d-2}}{\pi} \bar{g}^2 \int_0^\infty dx \; x^{\frac{d}{2}-2} \frac{d}{dx} \frac{\partial_t R_k(x)}{x + R_k(x)} \; \equiv \; \bar{g}^2 \, b_d \, k^{d-4} \qquad (15)$$

with $v_d \equiv \left[2^{d+1} \pi^{d/2} \Gamma(d/2)\right]^{-1}$. The second integral is due to the trace containing $\mathcal{D}_S$ with $Z_k = 1$ in $R_k(x)$. Introducing the dimensionless, renormalized gauge coupling

$$g^2(k) = k^{d-4} \, Z_{Fk}^{-1} \, \bar{g}^2 \qquad (16)$$

the associated beta function reads

$$\beta_{g^2} \equiv \frac{\partial}{\partial t} g^2(k) = (d-4)g^2 + \eta_F \, g^2 = (d-4)g^2 - b_d \, g^4. \qquad (17)$$

where $\eta_F \equiv -\partial_t \ln Z_{Fk}$ denotes the anomalous dimension. For $d = 4$ the result for the running of $g^2(k)$ becomes universal, i.e., b_4 is independent of the precise form of the cutoff function $R_k(x)$, only its behavior for $x \to 0$ enters in (15). One obtains, with $\lim_{x \to 0} R_k = Z_{Fk} k^2$ for the first term in (15) and $\lim_{x \to 0} R_k = k^2$ for the second term,

$$b_4 = \frac{N}{24\pi^2} \left[11 - 5\eta_F\right] \qquad (18)$$

In lowest order in g^2 we can neglect η_F on the RHS of (18) and obtain the standard perturbative one-loop β-function. More generally, one finds for η_F the equation $\eta_F = -g^2 b_d(\eta_F)$, which has, for $d = 4$, the nonperturbative solution

$$\eta_F = -\frac{11N}{24\pi^2} g^2 \left[1 - \frac{5N}{24\pi^2} g^2\right]^{-1} \qquad (19)$$

The resulting β-function can be expanded for small g^2

$$\beta_{g^2} = \quad -\frac{11N}{24\pi^2} g^4 \left[1 - \frac{5N}{24\pi^2} g^2\right]^{-1}$$

$$= \quad -\frac{22N}{3} \frac{g^4}{16\pi^2} - \frac{220}{9} N^2 \frac{g^6}{(16\pi^2)^2} - \; \dots \qquad (20)$$

Comparing with the standard perturbative two-loop expression

$$\beta_{g^2}^{(2)} = -\frac{22N}{3} \frac{g^4}{16\pi^2} - \frac{204}{9} N^2 \frac{g^6}{(16\pi^2)^2} \qquad (21)$$

we find a surprisingly good agreement even for the two-loop coefficient. The missing 7 % in the coefficient of the g^6-term in β_{g^2} should be due to our truncations. For a more detailed discussion of the (non-universal) β-function for general d and the running of g we have to refer to [16].

4　Chern-Simons Theory

Let us try to find an approximate solution of the initial value problem (4) with (5) for the classical Chern-Simons action (2). We work on flat euclidean space and allow for an arbitrary semi-simple, compact gauge group G. We use a truncation of the form [4]

$$\Gamma_k[A, N, \bar{A}] \;=\; i\kappa(k)\,\frac{g^2}{4\pi}\, I[A] + \kappa(k)\,\frac{g^2}{8\pi}\int d^3x\,\Big\{ iN^a D_\mu^{ab}[\bar{A}]\,(A_\mu^b - \bar{A}_\mu^b) \tag{22}$$

$$-i(A_\mu^a - \bar{A}_\mu^a) D_\mu^{ab}[\bar{A}]\,N^b + \alpha\,\kappa(k)\frac{g^2}{4\pi}N^a N^a \Big\}$$

with

$$I[A] \equiv \frac{1}{2}\int d^3x\,\varepsilon_{\alpha\beta\gamma}\,[A_\alpha^a\,\partial_\beta A_\gamma^a + \frac{1}{3}g f^{abc} A_\alpha^a A_\beta^b A_\gamma^c] \tag{23}$$

The first term on the RHS of (22) is the Chern-Simons action, but with a scale-dependent prefactor. In the second term we introduced an auxiliary field $N^a(x)$ in order to linearize the gauge fixing term. By eliminating N^a one recovers the classical, k-independent background gauge fixing term $\frac{1}{2\alpha}(D_\mu[\bar{A}](A_\mu - \bar{A}_\mu))^2$. In principle also the gauge fixing term could change its form during the evolution, but this effect is neglected here.

For $k \to \infty$, and upon eliminating N^a, the ansatz (22) reduces to (5) with the identification $\kappa(\infty) \equiv \kappa_{\text{bare}}$. We shall insert (22) into the evolution equation and from the solution for the function $\kappa(k)$ we shall be able to determine the renormalized parameter $\kappa(0) \equiv \kappa_{\text{ren}}$. We have to project the traces on the RHS of (4) on the subspace spanned by the truncation (22). This means that we have to extract only the term proportional to $I[A]$ and to compare the coefficients of $I[A]$ on both sides of the equation. In the formalism with the auxiliary field, $\Gamma_k^{(2)}$ in (4) denotes the matrix of second functional derivatives with respect to both A_μ^a and N^a, but with $\bar{A}_\mu^a$ fixed. Setting $\bar{A} = A$ after the variation, one obtains

$$\delta^2 \Gamma_k[A, N, A] \;=\; i\kappa(k)\frac{g^2}{4\pi}\int d^3x\,\Big\{ \delta A_\mu^a\,\varepsilon_{\mu\nu\alpha} D_\alpha^{ab}\,\delta A_\nu^b + \delta N^a D_\mu^{ab}\,\delta A_\mu^b$$

$$-\delta A_\mu^a D_\mu^{ab} N^b \Big\} + \alpha\,(\kappa(k)\frac{g^2}{4\pi})^2 \int d^3x\,\delta N^a\,\delta N^a \tag{24}$$

In order to facilitate the calculations we introduce three 4×4 matrices γ_μ with matrix elements $(\gamma_\mu)_{mn}$, $m=(\mu,4)=1,...,4$, etc., in the following way [11]:

$$(\gamma_\mu)_{\alpha\beta} = \varepsilon_{\alpha\mu\beta},\; (\gamma_\mu)_{4\alpha} = -(\gamma_\mu)_{\alpha 4} = \delta_{\mu\alpha},\; (\gamma_\mu)_{44} = 0 \tag{25}$$

If we combine the gauge field fluctuation and the auxiliary field into a 4-component object $\Psi_m^a \equiv (\delta A_\mu^a, \delta N^a)$ and choose the gauge $\alpha = 0$, we find

$$\delta^2 \Gamma_k[A, N, A] = i\kappa(k)\frac{g^2}{4\pi}\int d^3x\,\Psi_m^a\,(\gamma_\mu)_{mn} D_\mu^{ab}\Psi_n^b \tag{26}$$

M. Reuter

so that in matrix notation

$$\Gamma_k^{(2)} = i\kappa(k)\,\frac{g^2}{4\pi}\,\slashed{D} \tag{27}$$

Clearly $\slashed{D} \equiv \gamma_\mu D_\mu$ is reminiscent of a Dirac operator. In fact, the algebra of the γ-matrices is similar to the one of the Pauli matrices: $\gamma_\mu\gamma_\nu = -\delta_{\mu\nu} + \varepsilon_{\mu\nu\alpha}\gamma_\alpha$. Because $\gamma_\mu^+ = -\gamma_\mu$, $\slashed{D}$ is hermitian. Its square reads $\slashed{D}^2 = -D^2 - ig\,{}^*F_\mu\gamma_\mu$ where ${}^*F_\mu \equiv \frac{1}{2}\varepsilon_{\mu\alpha\beta}F_{\alpha\beta}$ is the dual of the field strength tensor. Because $\slashed{D}^2$ is essentially the covariant laplacian, it is the natural candidate for the cutoff operator Δ. With this choice, and $c \equiv g^2/4\pi$, the evolution equation (4) reads at $\bar{A} = A$:

$$\begin{aligned}
ic\,k\frac{d}{dk}\kappa(k)\,I[A] \;=\; & \frac{1}{2}\mathrm{Tr}\left[\left(ic\kappa\,\slashed{D} + R_k(\slashed{D}^2)\right)^{-1} k\frac{d}{dk}R_k(\slashed{D}^2)\right] \\
& -\mathrm{Tr}\left[\left(-D^2 + R_k(-D^2)\right)^{-1} k\frac{d}{dk}R_k(-D^2)\right]
\end{aligned} \tag{28}$$

The second trace on the RHS of (28) is due to the ghosts. It is manifestly real, so it cannot match the purely imaginary $iI[A]$ on the LHS and can be omitted therefore. For the same reason we may replace the first trace by i times its imaginary part:

$$k\frac{d}{dk}\kappa(k)\,I[A] = -\frac{1}{2}\kappa(k)\,\mathrm{Tr}\left[\slashed{D}\,\left(c^2\kappa^2\,\slashed{D}^2 + R_k^2(\slashed{D}^2)\right)^{-1} k\frac{d}{dk}R_k(\slashed{D}^2)\right] + \cdots \tag{29}$$

The trace in (29) involves an integration over spacetime, a summation over adjoint group indices, and a "Dirac trace". We shall evaluate it explicitly in the next section. Before turning to that let us first look at the general structure of eq. (29). In terms of the (real) eigenvalues λ of $\slashed{D}$ eq. (29) reads

$$\frac{d\kappa(k)}{dk^2}\,I[A] = -\frac{1}{2}\kappa(k)\sum_\lambda \frac{\lambda}{c^2\kappa^2(k)\lambda^2 + R_k^2(\lambda^2)} \cdot \frac{dR_k(\lambda^2)}{dk^2} \tag{30}$$

where we switched from k to k^2 as the independent variable. We observe that the sum in (30) is related to a regularized form of the spectral asymmetry of $\slashed{D}$.

An approximate solution for $\kappa(k)$ can be obtained by integrating both sides of eq. (30) from a low scale k_0^2 to a higher scale Λ^2 and approximating $\kappa(k) \simeq \kappa(k_0)$ on the RHS. This amounts to "switching off" the renormalization group improvement. The result is

$$[\kappa(k_0) - \kappa(\Lambda)]\,I[A] = \frac{1}{2}\kappa(k_0)\sum_\lambda \int_{k_0^2}^{\Lambda^2} dk^2\,\frac{dR_k(\lambda^2)}{dk^2} \cdot \frac{\lambda}{c^2\kappa^2(k_0)\lambda^2 + R_k^2(\lambda^2)} \tag{31}$$

Upon using R_k as the variable of integration one arrives at

$$[\kappa(k_0) - \kappa(\Lambda)]\,I[A] = \frac{1}{2c}\,\mathrm{sign}(\kappa(k_0))\sum_\lambda \mathrm{sign}(\lambda)\,G(\lambda; k_0, \Lambda) \tag{32}$$

with

$$G(\lambda; k_0, \Lambda) \equiv \arctan\left[c\, |\kappa(k_0)\lambda| \, \frac{R_\Lambda(\lambda^2) - R_{k_0}(\lambda^2)}{c^2 \kappa(k_0)^2 \lambda^2 + R_\Lambda(\lambda^2)\, R_{k_0}(\lambda^2)} \right] \tag{33}$$

Recalling the properties of R_k we see that in the spectral sum (32) the contributions of eigenvalues $|\lambda| \ll k_0$ and $|\lambda| \gg \Lambda$ are strongly suppressed, and only the eigenvalues with $k_0 < |\lambda| < \Lambda$ contribute effectively. Ultimately we would like to perform the limits $k_0 \to 0$ and $\Lambda \to \infty$. In this case the sum over λ remains without IR and UV regularization. This means that if we want to formally perform the limits $k_0 \to 0$ and $\Lambda \to \infty$ in eq. (32), we have to introduce an alternative regulator. In order to make contact with the standard spectral flow argument [5] let us briefly describe this procedure. We avoid IR divergences by putting the system in a finite volume and imposing boundary conditions such that there are no zero modes. In the UV we regularize with a zeta-function-type convergence factor $|\lambda/\mu|^{-s}$ where μ is an arbitrary mass parameter. Thus the spectral sum becomes $\lim_{s\to 0} \sum_\lambda \operatorname{sign}(\lambda)\, |\lambda/\mu|^{-s}\, G(\lambda; k_0, \Lambda)$. Now we interchange the limits $k_0 \to 0$, $\Lambda \to \infty$ and $s \to 0$. By construction, only finite ($|\lambda| \le \mu$) and nonzero eigenvalues contribute. For such λ's we have $G(\lambda; 0, \infty) = \pi/2$ irrespective of the precise form of R_k. Therefore (32) becomes

$$[\kappa(0) - \kappa(\infty)]\, I[A] = \frac{2\pi^2}{g^2}\, \operatorname{sign}(\kappa(0))\, \eta[A] \tag{34}$$

where $\eta[A] \equiv \lim_{s\to 0} \frac{1}{2} \sum_\lambda \operatorname{sign}(\lambda)\, |\lambda/\mu|^{-s}$ is the eta-invariant. If we insert the known result [5] $\eta[A] = (g^2/2\pi^2)\, T(G)\, I[A]$ we recover eq.(3): $\kappa(0) = \kappa(\infty) + \operatorname{sign}(\kappa(0))\, T(G)$. Obviously R_k has dropped out of the calculation. The parameter κ is universal: it does not depend on the form of the IR cutoff.

5 Evolution of the Chern-Simons Parameter

Next we turn to an explicit evaluation of the trace in eq. (29) which keeps the full k-dependence of κ on the RHS, i.e., the renormalization group improvement. To start with we use the constant cutoff [1] $R_k = k^2$ for which eq. (29) assumes the form

$$\frac{d}{dk^2} \kappa(k)\, I[A] = -\frac{1}{2c^2\kappa(k)}\, \operatorname{Tr}\left[\slashed{D} \left(\slashed{D}^2 + l(k)^2 \right)^{-1} \right] \tag{35}$$

where

$$l(k) \equiv \frac{k^2}{c\, |\kappa(k)|} \tag{36}$$

If we extract from the trace the term quadratic in A and linear in the external momentum and equate the coefficients of the $A\,\partial A$-terms on both sides of (35) we obtain

$$\frac{d\kappa(k)}{dk^2} \int d^3x \; \varepsilon_{\alpha\beta\gamma}\, A_\alpha^a\, \partial_\beta A_\gamma^a = -\frac{g^2 T(G)}{c^2 \kappa(k)} \int d^3x \; \varepsilon_{\alpha\beta\gamma}\, A_\alpha^a \Pi_k(-\partial^2) \partial_\beta A_\gamma^a + O(A^3) \tag{37}$$

---

[1] As the Faddeev-Popov ghosts do not contribute to the effect under consideration we may set $Z_k = 1$ also in the cutoff for the gauge field.

The function Π_k is given by the Feynman parameter integral

$$\Pi_k(q^2) = 8 \int_0^1 dx \ x(1-x) \int \frac{d^3p}{(2\pi)^3} \frac{q^2}{[p^2 + l^2 + x(1-x)q^2]^3} \tag{38}$$

Expanding $\Pi_k(-\partial^2) = \Pi_k(0) - \Pi'_k(0)\partial^2 + ...$, we see that only for the term with $\Pi_k(0)$ the number of derivatives on both sides of eq.(24) coincides. Therefore one concludes that

$$\frac{d\kappa(k)}{dk^2} = -\frac{g^2 T(G)}{c^2 \kappa(k)} \Pi_k(0) \tag{39}$$

where $\Pi_k(0)$ depends on $\kappa(k)$ via (36). Equation (39) is the renormalization group equation for $\kappa(k)$ which we wanted to derive. Formally it is similar to the evolution equation in Section 3 or the ones of the abelian Higgs model [13]. The special features of Chern-Simons theory, reflecting its topological character, become obvious when we give a closer look to the function $\Pi_k(q^2)$. Assume we fix a non-zero value of k ($l \neq 0$) and let $q^2 \to 0$ in (38). Because the l^2-term prevents the p-integral from becoming IR divergent, we may set $q^2 = 0$ in the denominator, and we conclude that the integral vanishes $\sim q^2$. This means that the RHS of (39) is zero and that $\kappa(k)$ keeps the same value for all strictly positive values of k. However, $\Pi_k(0)$ really vanishes only for $k > 0$. If we set $l = 0$ in (38) we cannot conclude anymore that $\Pi_k \sim q^2$, because in the region $p^2 \to 0$ the term $x(1-x)q^2$ provides the only IR cutoff and may not be set to zero in a naive way. In fact, $\Pi_k(0)$ has a δ-function-like peak at $k = 0$. To see this, we first perform the integrals in (38):

$$\Pi_k(q^2) = \frac{1}{\pi} \left[\frac{1}{2|q|} \arctan\left(\frac{|q|}{2|l|}\right) - \frac{|l|}{q^2 + 4l^2} \right] \tag{40}$$

As q^2 approaches zero, this function develops an increasingly sharp maximum at $l = 0$. Integrating (40) against a smooth test function $\Phi(l)$ it is easy to verify that

$$\lim_{q^2 \to 0} \int_0^\infty dl \ \Phi(l) \ \Pi_k(q^2) = \frac{1}{4\pi}\Phi(0) \tag{41}$$

This means that on the space of even test functions $\lim_{q^2 \to 0} \Pi_k(q^2) = \delta(l)/2\pi$. Even though the value of $\kappa(k)$ does not change during almost the whole evolution from $k = \infty$ down to very small scales, it performs a finite jump in the very last moment of the evolution, just before reaching $k = 0$. This jump can be calculated in a well-defined manner by integrating (39) from $k^2 = 0$ to $k^2 = \infty$:

$$\kappa(0) - \kappa(\infty) = 4\pi \ T(G) \lim_{q^2 \to 0} \int_0^\infty dl \ \text{sign}(\kappa(l)) \cdot \left[1 - cl\frac{d}{dk^2}|\kappa(k)|\right]^{-1} \Pi_k(q^2) \tag{42}$$

The term $\sim d|\kappa|/dk^2$ is a Jacobian factor which is due to the fact that l depends on $\kappa(k)$. This factor is the only remnant of the $\kappa(k)$-dependence of the RHS of the evolution equation. As we saw in Section 3, this dependence of the RHS on the running couplings is the origin of the renormalization group improvement. If we use (41) in (42), $l\ d|\kappa|/dk^2$ is set to zero and we find

$$\kappa(0) = \kappa(\infty) + \mathrm{sign}(\kappa(0))\,T(G), \tag{43}$$

which is precisely the 1-loop result. It is straightforward to check that the shift (43) is independent of the choice for R_k.

6 Discussion and Conclusion

We investigated the renormalization group flow of the Chern-Simons parameter by using a simple truncation of the space of actions. In general this method yields non-perturbative answers which require neither an expansion in the number of loops nor in the gauge coupling. The approximation involved here is that during the evolution the mixing of the Chern-Simons term with other operators is neglected.

It is quite instructive to compare the situation in Chern-Simons theory with what we found for ordinary Yang-Mills theory in Section 3. Like κ, also the gauge coupling in QCD_4 is a universal quantity. Its running is governed by a R_k-independent β-function which leads to a logarithmic dependence on the scale k. The Chern-Simons parameter κ, on the other hand, does not run at all between $k = \infty$ and any infinitesimally small value of k. Only at the very end of the evolution, when k is very close to zero, κ jumps by a universal, unambiguously calculable amount $\pm T(G)$. Though surprising in comparison with non-topological theories, this feature is precisely what one would expect if one recalls the topological origin of a non-vanishing η-invariant [5]. If $\eta[A] \neq 0$ for a fixed gauge field A, some of the low lying eigenvalues of $\slashed{D}[A]$ must have crossed zero during the interpolation from $A = 0$ to A. However, this spectral flow involves only that part of the spectrum which, in the infinite volume limit, is infinitesimally close to zero.

Another unusual feature of Chern-Simons theory is the absence of any renormalization group improvement beyond the 1-loop result. This should be contrasted with the running of g in QCD_4 where the truncation of Section 3 leads to a nonperturbative β-function involving arbitrarily high powers of g. We emphasize that our evolution equation with the truncation (22) potentially goes far beyond a 1-loop calculation. It is quite remarkable therefore that in Chern-Simons theory all higher contributions vanish. From the discussion following eq. (42) it is clear that this is again due to the unusual discontinuous behavior of κ which reflects the topological field theory nature of the model. While it is not possible to translate a "nonrenormalization theorem" for a given truncation into a statement about the nonrenormalization at a given number of loops, our results point in the same direction as ref. [8] where the absence of 2-loop corrections was proven.

References

[1] K.G. Wilson, I.G. Kogut, Phys.Rep. 12(1974)75; F. Wegner, A. Houghton, Phys.Rev. A8(1973) 401; J. Polchinski, Nucl.Phys.B231(1984)269; B. Warr, Annals of Physics 183(1988) 1 and 59; G. Mack et al., Proceedings Schladming 1992, H.Gausterer, C. B.Lang (eds.), Springer, Berlin, 1992; U. Ellwanger, L. Vergara, Nucl. Phys. B398(1993)52; M. Bonini, M. D'Attanasio, G. Marchesini, Nucl.Phys. B409(1993)441, B418(1994)81, B421(1994)429; U. Ellwanger, M. Hirsch, A. Weber, LPTHE Orsay 95-39

[2] C. Wetterich, Phys.Lett. B301 (1993) 90

[3] M. Reuter, C. Wetterich, Nucl.Phys. B417(1994)181

[4] M. Reuter, Preprint DESY 95-111

[5] E. Witten, Commun.Math.Phys. 121(1989)351

[6] G. Moore, N. Seiberg, Phys.Lett. B220(1989)422;
M. Bos, V.P. Nair, Phys.Lett. B223(1989)61;
J.M.F. Labastida, A.V. Ramallo, Phys.Lett. B227(1989)92

[7] L. Alvarez-Gaumé, J.M.F. Labastida, A.V. Ramallo, Nucl.Phys. B334(1990)103

[8] G. Giavarini, C.P. Martin, F.Ruiz Ruiz, Phys.Lett. B314(1993)328;
Nucl.Phys. B381(1992)222; Preprint hep-th/9406034

[9] M. Asorey, F. Falceto, Phys.Lett. B241(1990)31; M. Asorey, F. Falceto, J.L. Lopez, G. Luzon, Phys.Rev. D49(1994)5377; Nucl.Phys. B429(1994)344

[10] E. Guadagnini, M. Martellini, M. Mintchev, Phys.Lett. B227(1989)111;
Nucl.Phys. B330(1990)557

[11] M.A. Shifman, Nucl.Phys. B352(1991)87;
M.A. Shifman, A.I. Vainshtein, Nucl.Phys. B365(1991)312

[12] R.D. Pisarski, S. Rao, Phys.Rev. D32(1985)2081

[13] M. Reuter, C. Wetterich, Nucl.Phys. B391(1993)147;
Nucl.Phys. B408(1993)91; Nucl.Phys. B427(1994)291

[14] L.F. Abbott, Nucl.Phys. B185(1981)189;
W. Dittrich, M. Reuter, Selected Topics in Gauge Theories, Springer, Berlin, 1986

The Effective Lagrangian of Arbitrary Inhomogeneous Electromagnetic Field

Vladimir V. Skalozub and Andrei Yu. Tishchenko

Abstract

A new approach for the construction of an effective Lagrangian of the electromagnetic field with arbitrary spatial configuration in a dense medium is proposed. The constructed effective Lagrangian has been used for the investigation of the interaction between charged fermions in a dense environment. The possibility for formation of metastable electron bound states in a presence of external magnetic field is shown.

1 The effective Lagrangian of the electromagnetic field

The method of the effective Lagrangians has been well developed for the cases of homogeneous [1] and/or smoothly varying [2] external fields. However, in some problems, such for example as a photon splitting in an electron-positron plasma [3], it is necessary to consider rapidly varying in space or time gauge fields. For such a type of conditions the calculation of the effective Lagrangian (EL) is a non-trivial mathematical task.

In general, the effective action of the electromagnetic field induced by a vacuum/medium polarization can be calculated as the infinite series of the multiphoton vertex functions [4]

$$
\begin{aligned}
\mathcal{S}^{(n)}(A) &= \frac{(-1)^n e^n}{n} \int \hat{A}(x_1) G(x_2 - x_1)...\hat{A}(x_n) G(x_1 - x_n) d^3 x_1 d^3 x_2 ... d^3 x_n = \\
&\quad \frac{(-1)^n e^n}{n(2\pi)^{6n}} \int \hat{A}(k_1) G(p_1)...\hat{A}(k_n) G(p_n) e^{ik_1 x_1} e^{ip_1(x_2 - x_1)} ... \times \\
&\quad e^{ik_n x_n} e^{ip_n(x_1 - x_n)} d^3 x_1 ... d^3 x_n d^3 k_1 ... d^3 k_n d^3 p_1 ... d^3 p_n = \\
&\quad \frac{(-1)^n}{n(2\pi)^{3(n-1)}} \int A_{\mu_1}(k_1)...A_{\mu_n}(k_n) \Pi^{\mu_1 \cdots \mu_n}(k_1...k_n) \delta(\sum_{i=1}^{n} k_i) d^3 k_1 ... d^3 k_n = \\
&\quad \frac{(-1)^n}{n(2\pi)^{3n}} \int A_{\mu_1}(k_1)...A_{\mu_n}(k_n) \Pi^{\mu_1 \cdots \mu_n}(k_1...k_n) e^{ix \sum_{i=1}^{n} k_i} d^3 k_1 ... d^3 k_n d^3 x, \quad (1)
\end{aligned}
$$

where A_μ is the potential and G is the Green function of the electromagnetic field, $\Pi^{\mu_1 \cdots \mu_n}(k_1...k_n)$ are polarization tensors with n external photon lines carrying momenta k_i. For $\Pi^{\mu_1 \cdots \mu_n}$ being arbitrary functions of momenta the integration over k_i in eq.(1) is impossible.

However, if the polarization tensors occur to be some constant in a wide interval of momenta, the integration in eq.(1) can be performed to derive EL in a coordinate space:

$$\mathcal{L}'(A) = \sum_{n=1}^{\infty} \frac{(-1)^n}{n} A_{\mu_1}(x)...A_{\mu_n}(x)\Pi^{\mu_1\cdots\mu_n}, \tag{2}$$

where A_{μ_i} should be considered as arbitrary functions of space-time coordinate x_μ.

The situation of the described type means that it is possible to neglect the space and the time dispersions in a medium. In principle, it is a unique case because in general $\Pi^{\mu_1\cdots\mu_n}(k_1...k_n)$ are complicated functions of momenta k_i. However in a dense fermionic medium with chemical potential μ such a behaviour is realized.

In a dense medium due to non-zero chemical potential $\mu \neq 0$ the Farry theorem is violated and vertices with odd number of photon lines are non-zero [17]. In [6] the first three terms $\Pi_\mu(k_1), \Pi_{\mu\nu}(k_1, k_2), \Pi_{\mu\nu\lambda}(k_1, k_2, k_3)$ have been calculated for static case $k_0 = 0, \mathbf{k} \neq 0$ and important properties were observed. In the limit of $\mu \gg m, |\mathbf{k}|$ the tensors tend to constants which are proportional to certain degrees of μ. Moreover, these degrees of μ occurred to be a decreasing function of the number of external photon lines. So, only few terms with positive degrees contribute in eq. (2).

For example, in the QED_{2+1} case considered below we have for leading non-zero asymptotic terms, corresponding to the interval of momenta $|\mathbf{k}| \subset [0; 2\sqrt{\mu^2 - m^2}]$ [6]:

$$\Pi_0 = \frac{e}{2\pi}(\mu^2 - m^2)\theta(\mu^2 - m^2), \tag{3}$$

$$\triangle\Pi_{00} = -\frac{e^2}{2\pi}\theta(\mu^2 - m^2)(\mu - m), \quad \triangle\Pi_{ij} = \left(\delta_{ij} - \frac{k_i k_j}{\mathbf{k}^2}\right)\triangle\Pi_{00}, \tag{4}$$

$$\Pi_{000} = \frac{e^3}{\pi}\theta(\mu^2 - m^2), \quad \Pi_{i0j} = -\left(\delta_{ij} - \frac{k_i k_j \mathbf{k}'^2 + k_i' k_j' \mathbf{k}^2 - k_i k_j'(\mathbf{kk}')}{\mathbf{k}^2\mathbf{k}'^2}\right)\frac{\Pi_{000}}{4}, \tag{5}$$

where $\triangle\Pi_{ij}$ is the statistical part of the polarization tensor which completely determines its properties in the limit considered, $\theta(x)$ is the step function. In eq.(5) the momentum conservation, $\mathbf{k}_1 + \mathbf{k}_2 + \mathbf{k}_3 = 0$, has been taken into account.

It is important to note the following features of the calculation procedure. As it has occurred, the standard Feynman parametrization being applied in the calculations of $\Pi_{\mu\nu}(k_1, k_2)$, $\Pi_{\mu\nu\lambda}(k_1, k_2, k_3)$, ... with $\mu \neq 0$ destroys the transversality of the tensors. This unexpected fact was not noticed in ref. [7] where this parametrization has been applied and intensively used not only in actual calculations but also even in proving the transversality of $\Pi_{\mu\nu\lambda}$. So, a correct function has not been derived. In our approach to the calculation of the transversal tensors no parametrization has been used and straightforward calculations resulted in the manifestly transversal expressions $\Pi_{\mu\nu}(k_1, k_2)$, $\Pi_{\mu\nu\lambda}(k_1, k_2, k_3)$ given below. Hence, the gauge invariance of EL is guaranteed.

Substituting expressions (3)-(5) into eq.(2) one finds EL

$$\mathcal{L}'(x) = B_1 A_0(x) - C A_0^2(x) - C A_i(x)(\delta^{ij}\Delta - \partial^i\partial^j)\tilde{A}_j(x)$$
$$+ D_1 A_0^3(x) + D_2 A_0(x)\left((\delta^{ij}\Delta - \partial^i\partial^j)\tilde{A}_j(x)\right)^2, \tag{6}$$

where the notation $B = \Pi_0$, $C = \frac{e^2}{4\pi}\theta(\mu^2 - m^2)(\mu - m)$, $D_1 = \frac{e^3}{3\pi}\theta(\mu^2 - m^2)$, $D_2 = -\frac{e^3}{12\pi}\theta(\mu^2 - m^2)$ and $\tilde{A}_j(x) = \frac{1}{(2\pi)^2}\int \frac{A_j(\mathbf{k})}{\mathbf{k}^2}e^{i\mathbf{kx}}d\mathbf{k}$ is introduced. This EL leads to non-linear field equations and can be used for $A_\mu(x)$ arbitrary dependent on coordinates.

2 Static charge interaction in a dense medium

As an application of expression eq.(6), let us first consider the case when $A_1 = A_2 = 0$. From the total Lagrangian $\mathcal{L}_0 + \mathcal{L}'$, where $\mathcal{L}_0 = -\frac{1}{4\gamma}F_{\mu\nu}^2$ (γ is dimensional constant caused by 2-dimensional nature of theory), one obtains the equation for electric potential, $A_0 \equiv \phi$

$$\phi'' + \frac{1}{\rho}\phi' + \gamma\frac{e^2}{\pi}(e\phi^2 - (\mu - m)\phi) = 0. \tag{7}$$

This equation adequately describes the Coulomb law modification due to the multi-photon interaction in the interval of circle radia $\rho \subset [\frac{1}{2\sqrt{\mu^2 - m^2}}; \infty]$, which is the wider the bigger value of μ is reached in a medium. Introducing the dimensionless variable $\phi \to \frac{\phi}{\gamma e}$ and the parameters $\varepsilon = 2e^2\gamma/(\mu - m)$ and $\lambda = e\sqrt{(\mu - m)\gamma/2\pi}$, one can rewrite eq.(7) as follows

$$\phi'' + \frac{1}{\rho}\phi' + \lambda^2(\varepsilon\phi^2 - \phi) = 0. \tag{8}$$

Now, considering ε as a small value it is easy to obtain the perturbative solution of eq.(8):

$$\phi \approx K_0(\lambda\rho)(1 - \varepsilon\pi), \tag{9}$$

where $K_0(\lambda\rho)$ is the modified Hankel function.

As it is seen, the three-photon vertex, being taken into account, leads to the additional multiplicative weakening of the usual Debye screening in a medium. This decreasing of the preexponent factor has been obtained in the case of small ε. However, in two dimensional models of quantum field theory where coupling constant may not be small the three-photon interaction, probably, plays a more important role. So, it is interesting to take it into account more accurately.

3 Inter-electron electrostatic potential in a magnetic field

Since three-photon interaction owing to the violation of the Farry theorem in a dense medium is not zero, it can affect an interaction between charges due to various external conditions. Especially this should be perceptible in the case of external magnetic field applied.

Let us, for the sake of simplicity, consider an external homogeneous magnetic field described by the potential $A_\varphi = \rho H/2$, $A_\rho = 0$. To investigate this problem it is necessary to study the field equations following from the whole Lagrangian, including $\mathcal{L}_0$, $\mathcal{L}_{CS} = \frac{m_{CS}}{4}\epsilon^{\mu\nu\alpha}F_{\mu\nu}A_\alpha$ (m_{CS}-Chern-Simons mass) and $\mathcal{L}'$ from (6). It is possible to show that in the case of dense medium the magnetic field generated in the system can be neglected for distances $\rho > 1/\lambda \sim \frac{2}{e}\sqrt{\frac{\pi}{\gamma\mu}}$. So, it is sufficient to consider only the equation for the electrostatic potential:

$$\Delta\phi + m_{CS}H - 2\gamma C\phi + \gamma D_2\left(\frac{1}{\rho}\int_0^\rho A_\varphi(\rho')\rho'd\rho'\right)^2 = 0 \qquad (10)$$

For this equation one can derive an approximate solution, which describes the electric potential produced by the point charge:

$$\phi(z) \approx \left(1 - \frac{\pi m_{CS}}{\gamma e^2}\left(\frac{H}{\gamma e\mu}\right)z^2 + \frac{\pi}{72}\left(\frac{H}{\gamma e\mu}\right)^2 z^4\right)\gamma eK_0(z), \qquad (11)$$

where $z = \rho/\lambda \approx \frac{1}{e}\sqrt{\frac{2\pi}{\gamma\mu}}\rho$. If the Chern-Simons mass is induced and $\frac{H}{\gamma e\mu} > 1$ the three-photon interaction term dominates in (11). As it is seen (Fig.1), the potential (11) has a local minimum, which provides an attraction between the electrons at distances depending on H and μ.

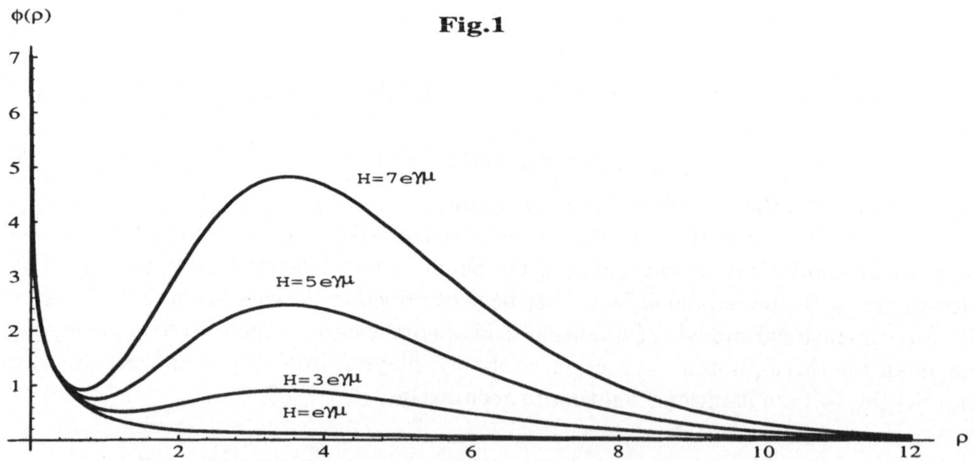

This interesting picture shows a potential ability of the formation bound electronic states which would be interesting for high temperature superconductivity.

4   Discussions

The main result of the present investigation is the construction of the gauge invariant EL for static electromagnetic field with an arbitrary dependence on coordinates. This remarkable possibility can be realized due to an important property of the polarization tensors [6]: in a dense medium the degree of μ giving asymptotics of the tensors decreases with an increasing of the number of external photon lines. It provides a rapid convergence of the series of one-loop diagrams determining EL owing to the presence of the small parameter $\sim 1/\mu$. Moreover, just this property leads to the fact that a few first diagrams in eq.(1) adequately describe an effective non-linear interaction of the electromagnetic field. Obviously, such a dependence has a general character and is not conditioned by the number of the spatial dimensions but by the structure of the fermionic propagator in the medium. A similar procedure can be realized in other gauge theories: QED_{3+1}, QCD. The only condition for this is a presence of a dense environment.

The most interesting application of the proposed EL considered here is a modification of the electrostatic potential in the presence of an external magnetic field. For some range of H and μ it is possible to form metastable electronic bound states in planar structures.

References

[1] W. Heisenberg, H. Euler, *Z. Phys.* **98**, 714 (1936)

[2] J. Schwinger, *Phys.Rev.* **82**, 664 (1951)

[3] D.B. Melrose, *Plasma Phys.* **16**, 845 (1974)

[4] C. Itzykson and J.-B. Zuber, *Quantum Field Theory*, (McGrow-Hill, 1980)

[5] E.S. Fradkin, *Proceedings of Lebedev Physical Institute* **29**, 7 (1965)

[6] V.V. Skalozub and A.Yu. Tishchenko, *JETP* **77**, 889 (1993)

[7] V. de la Incera, E. Ferrer, A.E. Shabad, *Proceedings of Lebedev Physical Institute* **169**, 183 (1986)

Gauge Theories in a Bag

Andreas Wipf

Abstract

I summarize some recent results, obtained with S. Duerr in [1], on multi-flavour gauge theories confined in the chiral limit enclosed in $d = 2n$-dimensional Euclidean bags. The dependence of the fermionic correlators and determinants on the boundary conditions is discussed. The chiral condensate in multi-flavour QED_2 is found. We prove that the condensate in QCD_2 decreases with increasing bag radius R at least as $\sim R^{-1/N_c N_f}$.

1 Introduction

From strong interaction phenomenology or supersymmetric Ward identities [2] we know that the chiral symmetry in QCD or its minimal supersymmetric extension is broken at low energy/temperature,

$$\langle \bar{u}u \rangle \neq 0.$$

A understanding of this breaking from first principles is still missing. One runs into the following paradox: In the chiral limit the generating functional for the fermionic Green's functions on a compact spacetime without boundary,

$$
\begin{aligned}
Z[\eta, \bar{\eta}] &= \int \mathcal{D}(A, \psi)\, e^{-S_{YM} + \int \bar{\psi} i \not{D} \psi + \int \bar{\eta}\psi + \bar{\psi}\eta} \\
&= \sum_N \int DA^N\, e^{-S_{YM}} \prod_{k=1}^{N} (\bar{\eta}, \psi_k)(\bar{\psi}_k, \eta)\det' i\not{D}\, e^{\int \bar{\eta} S' \eta},
\end{aligned}
\tag{1}
$$

where the gauge fields A^N support N zero modes $\psi_1, \ldots, \psi_N$ of $i\not{D}$, gets contributions from sectors with non-zero instanton numbers [3]

$$q = \frac{1}{32\pi^2} \int d^4x\, F_{\mu\nu}^a \, {}^*F_{\mu\nu}^a. \tag{2}$$

If we only allow for smooth configurations on S^4 or $S^3 \times R$ then q is an integer and the number of zero modes

$$
N = \begin{cases} N_f q & \text{for } N_f\text{-flavour } QCD \\ N_c q & \text{for supersymmetric } QCD \end{cases}
$$

is an integer multiple of N_f or N_c. Thus neither the topologically trivial sector contributes to the chiral condensate

$$\langle \bar{\psi}\psi \rangle = \frac{1}{Z} \frac{\delta^2}{\delta\eta\delta\bar{\eta}} Z\big|_{\eta=\bar{\eta}=0},$$

since S' in (1) is chirality conserving, nor the nontrivial sectors since there are too many zero modes. Hence the condensate vanishes. When switching on small quark masses one arrives at the same conclusion on a compact spacetime without boundary, since $\det(i\not{D}+m) \sim m^{N_f}$.

Possible ways out have been suggested by t'Hooft [4], who introduced twisted instantons on the 4-dimensional torus, and by Zhitnitsky [5], who considered singular gauge fields on S^4. Both constructions produce configurations with fractional instanton numbers and may resolve the above mentioned paradox. However, for $O(N>4)$ susy-YM-theories, which give rise to a nonvanishing chiral condensate [6], the centre is too small and these constructions do not work. Recently Shifman and Smilga have introduced yet another type of configuration, they called them fractons, which may generate a chiral condensate [7]. It remains to be seen whether these fractons solve the puzzle posed by the chiral condensate in $O(N)$-susy theories.

Instead of triggering the breaking by small quark masses one may put the fields in a finite box, impose symmetry breaking boundary conditions and then perform the thermodynamic limit $V \to \infty$. This is wellknown from spin models. For the Ising model both a constant magnetic field or Z_2-breaking, say spin-up, boundary conditions trigger a mean magnetization at low temperature and big volume.

When imposing $SU_A(N_f)$-breaking bag boundary conditions [8, 1]:

1. The configuration space of gauge-inequivalent gauge potential is topologically trivial and the instanton number is not quantized.

2. The Dirac operator $i\not{D}$ never possesses zeromodes.

3. The boundary conditions break the γ_5-symmetry and hence the spectrum of $i\not{D}$ is not symmetric about the origin.

4. The boundary conditions are a substitute for small quark-masses for the investigated models if one identifies the inverse bag-radius with the mass of the lightest particle in the theory.

5. The fracton results for multi-flavour QED_2 [7] is easily reproduced.

6. The measure $d\mu(A)$ for QCD_2-type theories factorizes and one can derive upper bounds for chiral condensates.

7. The boundary conditions depend on a real parameter θ. We obtain the explicit dependence of the fermionic determinants and Green's functions on θ and find a dynamical generation of the 'topological' θ-term.

2   The Bag Boundary Conditions

In the chiral limit of massless 'quarks' in the fundamental representation of $SU(N_c)$ the Euclidean action in a bag $\mathcal{M}$ is

$$S[A,\psi] = S_{YM}[A] \;+\; S_D[A,\psi], \quad \text{where} \quad S_D = \sum_{p=1}^{N_f} \int_{\mathcal{M}} \psi_p^\dagger i\!\!\not{D}\psi_p,$$

is invariant under global $SU_V(N_f) \times SU_A(N_f)$ rotations of the fermions. We shall impose the following local boundary conditions, which relate the different spin components on the bag boundary,

$$(B(\theta) \times I_f \times I_c)\psi = \psi \quad \text{on} \quad \partial\mathcal{M}. \tag{3}$$

They should break the $SU_A(N_f)$-symmetry but should respect the colour and vector-flavour symmetries such that the $\det i\!\!\not{D}$ is gauge invariant and the same for all flavours. We choose the boundary conditions such that

1. The partition function is real, which is guaranteed if $i\!\!\not{D}$ is symmetric.

2. The boundary conditions are 'lorentzinvariant' for an invariant bag.

A necessary condition for selfadjointness is that

$$n_\mu j^\mu = \psi^\dagger n^\mu \gamma_\mu \psi = 0$$

on the bag boundary, where $n^\mu(x)$ is the outward oriented normal vectorfield on $\partial\mathcal{M}$. Imposing further euclidean lorentzinvariance for a spherical bag one is led to a 2-parametric family of local boundary conditions. We take the following 1-parametric subfamily

$$\psi = B_\theta\psi \quad \text{on} \quad \partial\mathcal{M} \qquad \text{with} \quad B_\theta = i\bar{\gamma}e^{\theta\bar{\gamma}}\gamma_n \times I_f \times I_c. \tag{4}$$

Here $\bar{\gamma} = \text{diag}(1_n, -1_n)$ is the generalization of γ_5 in $d = 2n$ dimensions. These boundary conditions break the axial-flavour symmetry. The spectrum of the Diracoperator,

$$i\!\!\not{D}\psi_m(\theta) = \lambda_m(\theta)\psi_m(\theta), \qquad B_\theta\psi_m(\theta) = \psi_m(\theta)|_{\partial\mathcal{M}}, \tag{5}$$

is discrete but exhibits no pairing, i.e. $\{\lambda_n, -\lambda_n\}$ are not both eigenvalues.

3   On the spectrum of $i\!\!\not{D}$ in a bag

One can show [1] that

1. $i\not{D}$ has no zero-modes and hence expectation values of gauge invariant observables simplify to (besides gauge fixing and ghosts)

$$\langle O \rangle = \int d\mu_\theta(A)\,\langle O \rangle_A \quad , \quad d\mu_\theta(A) = \frac{1}{Z}e^{-S_{YM}[A]}\det{}_\theta(i\not{D})\,\mathcal{D}A,$$

   where the measure of functional integration over the gauge field configurations after the fermions have been integrated out is

$$d\mu_\theta(A) = \frac{1}{Z}e^{-S_{YM}[A]}\det{}_\theta i\not{D}\,\mathcal{D}A$$

   and $\langle O \rangle_A$ denotes the expectation value of O in a fixed background gauge field A.

2. Under a parity transformation $A(x) \to \tilde{A}(\tilde{x})$, $\psi_m(x) \to \bar{\gamma}\gamma_0\psi_m(\tilde{x})$, where $\tilde{x} = (x^0, -x^i)$, the eigenvalues change signs and $\theta \to -\theta$. In other words, $\lambda_m(\tilde{A}, \theta) = -\lambda_m(A, -\theta)$, and this property constrains the fermionic determinants and Green's functions as

$$\begin{aligned} \det i\not{D}(A, \theta) &= \det i\not{D}(\tilde{A}, -\theta). \\ S^\theta(x, y; A) &= -\gamma_0\bar{\gamma}S^{-\theta}(\tilde{x}, \tilde{y}; \tilde{A})\bar{\gamma}\gamma_0. \end{aligned} \tag{6}$$

3. A boundary Hellmann-Feynman formula relating the variation of the eigenvalues λ_m with a variation of θ can be derived:

$$\frac{d}{d\theta}\lambda_m = \frac{i}{2}\oint \psi_m^\dagger \gamma_n \bar{\gamma}\psi_m = -\lambda_m(\psi_m, \bar{\gamma}\psi_m). \tag{7}$$

4. When calculating correlators of 'quark' fields in a bag one needs the fermionic Green's functions. Their θ-dependence read

$$S^\theta = \begin{pmatrix} e^\theta S^0_{++} & S^0_{+-} \\ S^0_{-+} & e^{-\theta}S^0_{--} \end{pmatrix},$$

   where the subscripts indicate the chiral projections. The θ-dependent diagonal entries $S_{\pm\pm}$ lead to chirality violating amplitudes.

   The free Green's functions in a spherical bag of dimension $d = 2n$ is found to be

$$S^\theta(x, y; 0) = S_0(x, y) + \frac{\Gamma(n)}{2R\pi^n}\bar{\gamma}\,e^{\theta\bar{\gamma}}\frac{R^2 - (x, \gamma)(y, \gamma)}{(R^2 - 2xy + \frac{x^2y^2}{R^2})^n}, \tag{8}$$

   where $S_0(x, y)$ is the free Green's function in d-dimensional Euclidean spacetime.

4 θ-dependence of the fermionic Determinants

For the explicit calculations we have employed the gauge invariant ζ-function definition of the determinants [9] and found

$$\log \frac{\det_\theta i \not{D}}{\det_0 i \not{D}} = \frac{-\theta}{n!(4\pi)^n} \int_{\mathcal{M}} \epsilon_{\mu_1 \ldots \mu_d} F_{\mu_1 \mu_2} \cdots F_{\mu_{d-1}\mu_d} - \int_0^\theta d\theta' \oint_{\partial \mathcal{M}} \operatorname{tr} b_{d/2}(\bar{\gamma}),$$

where $b_{d/2}$ is the surface Seeley-deWitt coefficient. We see that the θ variation is proportional to the parity-odd instanton number q which is not quantized in a bag. In 2 and 4 dimensions $\operatorname{tr} b_{d/2}(\bar{\gamma}) = 0$ and thus

$$d\mu_\theta(A) = e^{\theta q} \, d\mu_{\theta=0}(A). \tag{9}$$

Since the Dirac operator in a bag is hermitean its determinant is real and positive and thus, to make contact with the θ-worlds in QCD, we would have to continue θ in (9) to $i\theta$.

5 2-dimensional Gauge Theories

In a 2d-bag a gauge potential can be written as [10, 11]

$$A_z \equiv A_0 - iA_1 = ig^{-1}(\partial_0 - i\partial_1)g \equiv ig^{-1}\partial_z g \tag{10}$$

with g from the complexified gauge group G^c, e.g. $g \in GL(n, C)$ for $U(n)$-gauge theories. Now it is easy to see that

$$\not{D} = G^\dagger \not{\partial} G, \quad \text{where} \quad G = \begin{pmatrix} g^{-1\dagger} & 0 \\ 0 & g \end{pmatrix}, \quad \not{\partial} = \begin{pmatrix} 0 & \partial_z \\ \partial_{\bar{z}} & 0 \end{pmatrix}$$

and we made the matrix-forms in spinor space explicit. The Green's functions are then related as

$$S^\theta(x, y; A) = G^{-1}(x) S^\theta(x, y; 0) G^{-1\dagger}(y).$$

The Yang-Mills action is easily expressed in terms of the gauge invariant field $J = gg^\dagger$.

To determine the fermionic effective action one introduces a τ-dependent family $g(x, \tau)$ as

$$g(x, 0) = I \quad, \quad g(x, 1) = g(x) \quad \text{and} \quad \frac{d}{d\tau}g(\tau) \equiv \dot{g}(\tau) = -g(\tau)a(\tau).$$

The τ-variation of the fermionic determinant is given by the $U(1)$-anomaly and can be computed. The result reads

$$\log \frac{\det_\theta i\slashed{D}}{\det_\theta i\slashed{\partial}} = -\frac{1}{8\pi} \int_{\mathcal{M}} \mathrm{tr}\left(J^{-1}\partial J J^{-1}\bar{\partial}J\right) + \frac{i}{12\pi} \int_{\mathcal{Z}} \mathrm{tr}\,(J^{-1}d_3 J)^3$$
$$+ \frac{\theta}{4\pi}\int_{\mathcal{M}} \mathrm{tr}\,\bar{\partial}(J^{-1}\partial J). \tag{11}$$

In the Wess-Zumino term in the middle on the right hand side $J = J(x,\tau)$ and thus $\mathcal{Z} = \mathcal{M} \times [0,1]$ is the finite cylinder over the bag. Let us now suppose that $G = U(1) \times SU(N_c)$. We represent the gauge potential $A = \tilde{A} + \hat{A}$ as in (10) and factorize the $U(1)$ field, that is we set $g = \tilde{g}\hat{g}$. We parametrize the $U(1)$-part as $\tilde{g} = e^{-e\varphi - ie\lambda}0$ and then

$$A_\mu = \tilde{A}_\mu + \hat{A}_\mu = -e\epsilon_{\mu\nu}\partial_\nu\varphi + e\partial_\mu\lambda + \hat{A}_\mu \quad \text{and} \quad F_{01} = e\Delta\varphi + \hat{F}_{01}.$$

For this particular case we find

$$d\mu_\theta(A) = d\mu_\theta(\tilde{A})\,d\mu(\hat{A}) = \frac{e^{-\Gamma_\theta[\varphi]}}{\tilde{Z}_\theta}\mathcal{D}\tilde{A}\,\frac{e^{-\Gamma[\hat{A}]}}{\hat{Z}}\mathcal{D}\hat{A}. \tag{12}$$

For the N_f-flavour model the θ-dependent effective action for the $U(1)$-sector and the θ-independent one for the $SU(N)$ sector read

$$\Gamma_\theta[\varphi] = \frac{N_c}{2}\left\{ \int (\Delta\varphi)^2 - m_\eta^2 \int \varphi\Delta\varphi + \frac{e\theta N_f}{\pi}\oint \partial_n\varphi \right\}$$
$$\Gamma[\hat{A}] = S_{YM}[\hat{A}] + \frac{N_f}{8\pi}\int_{\mathcal{M}} \mathrm{tr}\,(\hat{J}^{-1}\partial\hat{J}\hat{J}^{-1}\bar{\partial}\hat{J}) - \frac{iN_f}{12\pi}\int_{\mathcal{Z}} \mathrm{tr}\,(\hat{J}^{-1}d_3\hat{J})^3. \tag{13}$$

Note that due to the wellknown Schwinger mechanism the mass $m_\eta^2 = N_f e^2/\pi$, which is the analog of the η'-mass in QCD, has been induced in the abelian subsector of the theory.

6 Chiral Symmetry Breaking in 2d-Gauge Theories

Since the fermionic Green's functions in a bag has chirality violating entries we expect a nonvanishing chiral condensate although there are no fermionic zeromodes and no instantons. In 2 dimensions we know the fermionic Green's function and determinant explicitly in terms of the field $g(x)$ and inserting them we arrive at $(u = \psi_1)$

$$\langle \bar{u}P_+ u\rangle(x) = -\frac{e^\theta}{2\pi R}\frac{1}{1 - r^2/R^2}\int d\mu_\theta(A)\,\mathrm{tr}\,J(x) \tag{14}$$

and it remains to calculate the colour trace average of the gauge invariant field J.

7   Multi-flavour QED_2

For abelian gauge groups the effective fermionic action and Yang-Mills action are both quadratic in the gauge potential and the functional integral can be performed, although one must be careful in treating the boundary terms correctly. The result for N_f flavours reads [1]

$$\langle \bar{u} P_+ u \rangle(x) = -\frac{m_\eta e^\gamma}{4\pi} \left(\frac{m_\eta R e^\gamma}{2} [1 - \frac{r^2}{R^2}] \right)^{-1+1/N_f} e^{\theta I_e + F_e/N_f}, \tag{15}$$

where

$$F_e(r, R) = \sum \epsilon_n \frac{K_n(m_\eta R)}{I_n(m_\eta R)} I_n^2(m_\eta r) \;, \quad I_e(r, R) = \frac{I_0(m_\eta r)}{I_0(m_\eta R)}. \tag{16}$$

The function F_e has the asymptotic expansions

$$F_e(r, R) \sim \begin{cases} e^{-m_\eta R} & \text{for } 1 \ll m_\eta R \gg m_\eta r \\ -\log \frac{1}{2} m_\eta R e^\gamma [1 - \frac{r^2}{R^2}] & \text{for } m_\eta R \ll 1. \end{cases} \tag{17}$$

Thus for large and small bags or equivalently for strong and weak coupling constant e the condensate simplifies to

$$\langle \bar{u} P_+ u \rangle \sim \begin{cases} -\frac{m_\eta e^\gamma}{4\pi} \left(\frac{1}{2} m_\eta R e^\gamma \right)^{-1+1/N_f} & \text{for } 1 \ll m_\eta R \gg m_\eta r \\ -\frac{e^\theta}{2\pi R} (1 - r^2/R^2)^{-1} & \text{for } m_\eta R \ll 1. \end{cases} \tag{18}$$

As expected, for weak couplings and/or small bags the condensate tends to the chirality violating entry $-S_{++}^\theta(x, x; 0)$ of the free Green's function (8).

For *one flavour* and large bags we recover the wellknown value for the condensate in the Schwinger model [21]

$$\langle \bar{u} P_+ u \rangle = -\frac{m_\eta}{4\pi} e^\gamma. \tag{19}$$ We

stress that this result has been obtained without doing any instanton physics. The calculations in a bag are actually much simpler as compared with those on a torus [13, 14, 15] or sphere [16], where a careful treatment of the different instanton sectors is required to find the result (19).

For *several flavours* the condensate inside the bag, e.g. at the center of a large bag,

$$\langle \bar{u} P_+ u \rangle(0) = -\frac{1}{2\pi R} \left(\frac{m_\eta R e^\gamma}{2} \right)^{1/N_f} \tag{20}$$

decreases with increasing R and vanishes for $R \to \infty$.

Let us compare our result with that of Smilga [17] who calculated the condensate

in multiflavour QED_2 for small 'quark' masses. Using bosonization techniques he found that the condensate depends
on the mass μ of lightest particle in the theory as

$$\langle \bar{\psi}\psi \rangle \sim \mu \left(\frac{m_\eta}{\mu} \right)^{1/N_f}. \tag{21}$$

Comparing with (20) we see that the bag- and small quark mass results coincide if we identify μ with $1/R$.

8 Multi-flavour nonabelian gauge theories.

Due to the factorization of the measure for the gauge bosons, (12), the chiral condensate (14) in $U(N_c)$ gauge theories factorizes as

$$\langle \bar{u}P_+ u \rangle_{U(N_c)} = e^{\theta(I_e - 1)} \left(\frac{m_\eta R \, e^{\gamma + F_e}}{2} [1 - \frac{r^2}{R^2}] \right)^{1/N_c N_f} \langle \bar{u}P_+ u \rangle_{SU(N_c)}.$$

Using the asymptotic expansion of F_e for small arguments, (17), we see that for $e \to 0$ the $U(N_c)$ result reduces to the $SU(N_c)$ one, as expected.

Assuming that the $U(N_c)$ condensate has a smooth thermodynamic limit we conclude at once that

$$\langle \bar{u}P_+ u \rangle_{SU(N_c)} \leq \text{const} \cdot R^{-1/N_c N_f}. \tag{22}$$

Only when we take the limit in which the number of colours tends to infinity *before* we perform the thermodynamic limit $R \to \infty$ can a quark condensate survive.

Finally we note, that the expectation values of arbitrary powers of the topological charge are gotten by differentiating the partition function sufficiently often with respect to θ. The correlators are reproduced by the following Gaussian distribution for the topological charge:

$$d\mu(q) = \sqrt{\frac{N_c N_f}{\pi \sigma}} \, e^{-N_c N_f \sigma [q + \theta/2\sigma]^2} \, dq \quad , \quad \sigma = \frac{I_0(m_\eta R)}{m_\eta R I_1(m_\eta R)}. \tag{23}$$ The

expectation value of the instanton number vanishes for vanishing θ, but its fluctuation does not. In the semiclassical regime of small volumes and/or weak coupling is the instanton number distribution sharply peaked about $q=0$ as can be seen by inspection from (23) or from

$$\langle |q| \rangle = \begin{cases} 0 & \text{for } m_\eta R \to 0 \\ \sqrt{\frac{eR}{\pi N_c}} (\pi N_f)^{-1/4} & \text{for } m_\eta R \to \infty. \end{cases} \tag{24}$$

For big volumes and/or strong coupling those fields with $q^2 \sim 1/\sqrt{N_f}$ dominate the functional integral.

9 Finite temperature bags

To be a model for a hadron at finite temperature, $\mathcal{M}$ must be a bag in space and hence $[0, \beta] \times \mathcal{M}$ a subspace of the Euclidean spacetime. The gluon (quark) fields must then be periodic (antiperiodic) in the Euclidean time with period $\beta = 1/T$. In [18] we have studied multi-flavour QED_2 at finite temperature enclosed in a spatial bag $[0, L]$. Besides the finite temperature boundary conditions we imposed the bag boundary conditions $B_\theta \psi = \psi$ at $x^1 = 0$ and $x^1 = L$. By applying the methods developed in this paper we found in the low temperature limit $T \ll 1/L \ll m_\eta$ [18]

$$\langle \bar{u} P_+ u \rangle = -\frac{1}{4L} \, e^{\gamma/N_f} \left(\frac{m_\eta L}{\pi} \right)^{1/N_f}. \tag{25}$$

In particular, for 2 flavours

$$\langle \bar{u} P_+ u \rangle = -\left(\frac{e^\gamma m_\eta}{16 \pi L} \right)^{1/2} \tag{26}$$

and this result is identical to that of Shifman and Smilga [7] when they allowed for fracton configurations.

References

[1] S. Duerr and A. Wipf, Nucl. Phys. **B443** (1995) 201

[2] V. Novikov, M. Shifman, A. Vainshtain and V. Zakharov, Nucl. Phys. **B229** (1975) 85.

[3] G. t'Hooft, Phys. Rev. Lett. **37** (1976) 8; Phys. Rev. **D14** (1976) 3432;77) 309.

[4] G. t'Hooft, Commun. Math. Phys. **81** (1981) 267; E. Witten, Nucl. Phys. **B202** (1982) 253.

[5] A.R. Zhitnitsky, Nucl. Phys. **B340** (1990) 56; **B374** (1992) 183.

[6] M.A. Shifman and A.I. Vainstein, Nucl. Phys. **B296** (1988) 445.

[7] M.A. Shifman and A.V. Smilga, Phys. Rev. **D50** (1994) 7659.

[8] J. Balog and P. Hrasko, Nucl. Phys. **245** (1984) 118.

[9] J.S. Dowker and R. Critchley, Phys. Rev. **D13** (1976) 3224; S.W. Hawking, Commun. Math. Phys. **55** (1977) 133.

[10] S. Blau, M. Visser and A. Wipf, Int. J. Mod. Phys., **A4** (1989) 1467.

[11] C.N. Yang, Phys. Rev. Lett. **38** (1977) 1377;

[12] J.H. Loewenstein and J.A. Swieca, Ann. Phys. **68** (1961) 172; N.K. Nielsen and B. Schroer, Nucl. Phys. **B120** (1977) 62.

[13] I. Sachs and A. Wipf, Helv. Phys. Acta **65** (1992) 652.

[14] H. Joos, Helv. Phys. Acta **63** (1990) 670; Nucl. Phys. (Proc. Suppl.) **B17** (1990) 704; H. Dilger and H. Joos, Nucl. Phys. (Proc. Suppl.) **B34** (1994) 195; H. Joos and S.I. Azakov, preprint DESY-94-142.

[15] A.V. Smilga, Phys. Lett. **B278** (1992) 371.

[16] C. Jayewardena, Helv. Phys. Acta **61** (1988) 636.

[17] A.V. Smilga, Phys. Lett. **B278** (1992) 1992.

[18] S. Duerr and A. Wipf, in preparation.

Instantons in the Bag Model:
Matrix Elements of Scalar Propagator

Alexei A. Abrikosov Jr.

Abstract

I consider the interaction of instantons with scalar matter in the bag model. First the exact Green function in the bag with instanton is found for massless scalar particle in fundamental representation. Matrix elements of this function describe interaction of matter with the instanton. The Green function is expanded in wave functions that far from the instanton coincide with perturbative eigenfunctions. This gives the matrix elements in question.

1   Introduction.

It passed already 20 years since the discovery of instantons [1, 2] but still we do not understand clearly what is their role in strong interactions. Instantons come into play near confinement scales and their effects depend on the unknown physics in that region. However that means that they may serve as a bridge to QCD at large distances.

In Minkowski space instantons are interpreted as tunneling between topologically different gauge vacua. In Euclidean formulation the instanton is a gauge field configuration that satisfies equations of motion being a local minimum of the action. Topology guarantees that it is stable with respect to deformations of the field.

An argument in favour of instantons is that they break chiral invariance and offer a beautiful solution of the $U(1)$ problem [3]. Interaction with instantons generates masses and explains nonzero vacuum expectation values of $\langle \bar{q}q \rangle$ for massless quarks.

There was an attempt to construct a theory of strong interactions [4] using instantons. The idea was that there are two stable phases of QCD vacuum. The gas of pseudoparticles is dilute in the vicinity of quarks but far enough another phase is favourable. Quarks are locked in the bubble as it happens in the MIT bag model [5]. The difference of energy densities of the phases provides the bag constant B. The physical picture looks fair and has inspired the present work.

Among the reasons that attract attention to instantons now are the growth of instanton contributions to weak and strong processes at high energies [6]. Instantons are important for electroweak baryon-number violating processes [7] which define the baryon asymmetry of the Universe. Let us stress that even not being a reason of confinement instantons certainly are significant for hadron physics [8].

2 Instantons and bag model.

When one tries to include instantons into models of confinement, *e. g.* the bag model [5] the question about properties of instantons in closed domains arises. Up to now the influence of boundaries was analyzed half qualitatively.

We can single out the following aspects. First of all deformations of pseudoparticles and of the bag itself take place. They depend on boundary conditions for gauge fields that in turn are defined by confinement physics. Fortunately if the radius of the pseudoparticle ρ is smaller than the bag radius R_B these effects are of secondary influence.

Secondly the spectrum of quantum fluctuations differs from the unbounded case (Casimir effect). Obviously this affects the behaviour of the coupling constant and leads to freezing of $\alpha(\rho)$ at $\rho \sim R_B$. An interesting way to address this problem would be to use the projection of instanton onto $S^3 \otimes R$ obtained recently by Smilga [9].

Finally, instantons interact with matter, *i. e.* quarks. Fermionic states are quantized and via this quantization pseudoparticles feel the boundary even though their own deformations are negligible. Here we shall address the problem of interaction of an instanton with scalar matter in fundamental representation. This is the first step in approaching the fermionic case.

Let us show what could be the effect of instantons on properties of hadrons. In the bag model framework the latter are defined by the bag constant B. Taking into account instantons inside the bag we notice that B becomes smaller. In the dilute gas approximation the correction is (the factor 2 stands for antiinstantons):

$$B - B^0 = -2 \int_{B_3} d\rho \, \frac{\langle N | e^{-S^I_{QCD}} | N \rangle}{\langle N | e^{-S^0_{QCD}} | N \rangle} \tag{1}$$

The ratio of matrix elements in the RHS takes into account quantum fluctuations and the interaction of pseudoparticles with their hadronic environment. It modifies the density of pseudoparticles comparing to the vacuum value $d_0(\rho)$ and generates an addition δS_N to the instanton action

$$d_N(\rho) = d_0(\rho) e^{-\delta S_N}; \qquad \delta S_N = \delta S_d + \delta S_C + \delta S_m. \tag{2}$$

The three components in δS_N correspond to the listed above factors, *i. e.* deformations, Casimir and matter effects. We are interested in the last component, that may be shown to be:

$$\delta S_m = -\lim_{\beta \to \infty} \log \frac{C^*_{\alpha\beta\gamma}[S^\alpha_{\alpha'}]^I[S^\beta_{\beta'}]^I[S^\gamma_{\gamma'}]^I C^{\alpha'\beta'\gamma'}}{C^*_{\alpha\beta\gamma}[S^\alpha_{\alpha'}]^0[S^\beta_{\beta'}]^0[S^\gamma_{\gamma'}]^0 C^{\alpha'\beta'\gamma'}}. \tag{3}$$

In this equality $[S^\alpha_{\alpha'}]^{I,0}$ are matrix elements of quark propagator with and without instanton. We assume that the wave function of the hadron is $C^*_{\alpha\beta\gamma}\psi^\alpha\psi^\beta\psi^\gamma$, where the

ψ's are fields of quarks. The bag sweeps in the Euclidean space a cylinder with base B_3 that is $\beta \to \infty$ high in time. We see that the problem of interaction of instantons with matter was reduced to the calculation of matrix elements:

$$[S_{\alpha'}^{\alpha}]^{0,I} = \frac{\langle 0|\psi^{\alpha}(\beta)\psi^{*}_{\alpha'}(0)e^{-S_{QCD}^{0,I}(\beta)}|0\rangle_B}{\langle 0|e^{-S_{QCD}^{0,I}(\beta)}|0\rangle_B}. \tag{4}$$

The main difficulty is that the instanton field is strong and the problem is essentially nonperturbative. However, for massless particles the exact Green function can be found. The method will be the generalization of that constructed previously for quark matter [10]. We shall consider solutions which at time infinities tend to the *in-* and *out-* perturbative states. This makes physical interpretation more transparent and greatly simplifies calculation of matrix elements.

3   Instanton

I shall work with $SU(2)$ gauge fields. Vector potential is A_{μ}^{a} and the field strength is $F_{\mu\nu} = F_{\mu\nu}^{a}\tau^{a}$ where τ^{a} stands for Pauli matrix. The Lagrangian is:

$$L = \int d^4x \frac{\operatorname{tr} F_{\mu\nu}^2}{8g^2} = \frac{1}{4g^2}\int d^4x(\partial_{\mu}A_{\nu}^{a} - \partial_{\nu}A_{\mu}^{a} + \epsilon^{abc}A_{\mu}^{b}A_{\nu}^{c})^2. \tag{5}$$

Equations of motion of the gauge field have an instanton solution. In conventional singular gauge the field of the instanton of radius ρ located at the origin is:

$$A_{\mu}^{a} = \frac{2\rho^2\bar{\eta}_{\mu\nu}^{a}x_{\nu}}{x^2(x^2+\rho^2)} = -\bar{\eta}_{\mu\nu}^{a}\partial_{\nu}\log\Pi(x). \tag{6}$$

Here $\Pi(x) = 1 + \rho^2/x^2$ and $\eta_{\mu\nu}^{a}$, $\bar{\eta}_{\mu\nu}^{a}$ denote dual and antiselfdual 'tHooft's symbols [11]. This choice makes possible an easy generalization onto the $5N$-parametric 'tHooft's multiinstanton [12] and thermal "caloron" solutions [13]. It is convenient to make use of the following notation (latin indices stand for "spatial" dimensions).

$$\tau_{\mu} = (\tau_i, i); \qquad \tau_{\mu}^{\dagger} = (\tau_i, -i); \tag{7}$$

$$\tau_{\mu}^{\dagger}\tau_{\nu} = \delta_{\mu\nu} + i\eta_{\mu\nu}^{a}\tau_a; \qquad \tau_{\mu}\tau_{\nu}^{\dagger} = \delta_{\mu\nu} + i\bar{\eta}_{\mu\nu}^{a}\tau_a; \tag{8}$$

From here one can easily deduce properties of η-symbols. Convolutions of τ-matrices with 4-vectors are denoted by hats:

$$\hat{x} = x_{\mu}\tau_{\mu}; \qquad \hat{x}^{\dagger} = x_{\mu}\tau_{\mu}^{\dagger}. \tag{9}$$

Deformation of the pseudoparticle caused by boundaries can be estimated as the part of the action that would come from the region outside the bag. The latter in our case is of the order $(\rho/R)^4 \ll 1$. Thus the direct effect of the bag onto pseudoparticles is small and the interaction is mediated by quarks.

4   Scalar Green function

The classical instanton field is not small and cannot be taken into account perturbatively. It was shown that exact Green functions of fermions, gauge bosons and ghosts in the instanton field can be expressed in terms of the propagator of scalar particles in fundamental representation Δ^I, [14]. The Lagrangian of matter and the equation for the Green function are:

$$L_m = -\int \phi^* \nabla_\mu^2 \phi \, d^3x, \qquad -\nabla_\mu^2 \Delta^I(x,y) = \delta(x-y). \tag{10}$$

The covariant derivative in fundamental representation is $\nabla_\mu = \partial_\mu - \frac{i\tau^a A_\mu^a}{2}$.

We shall assume that the matter satisfies zero boundary conditions, $i.e.$ $\phi(x \in \partial B_3) = 0$. For the propagator this implies that when either of x, y lie on the border

$$\Delta_B(x,y) = 0 \quad \text{for} \quad x,y \in \partial B_3. \tag{11}$$

In the infinite Euclidean space the propagator is given by:

$$\Delta_\infty^I(x,y) = \Pi^{-\frac{1}{2}}(x) \frac{1 + \rho^2 \frac{\hat{x}\hat{y}^\dagger}{x^2 y^2}}{4\pi^2(x-y)^2} \Pi^{-\frac{1}{2}}(y). \tag{12}$$

The covariant d'Alembert operator in the field of the instanton Eq. (6) can be converted to the form

$$\nabla_\mu^2 = \Pi^{\frac{1}{2}}(x) \hat{\partial} \frac{1}{\Pi(x)} \hat{\partial}^\dagger \Pi^{\frac{1}{2}}(x). \tag{13}$$

It is a straightforward observation that if one denotes by $1/\partial^2$ the Green function of the ordinary d'Alembertian,

$$\partial_\mu^2 \frac{1}{\partial^2}(x,y) = \delta(x-y), \qquad \Delta_\infty^0 = -\frac{1}{\partial^2}(x,y) = \frac{1}{4\pi^2(x-y)^2}, \tag{14}$$

then the function

$$\Delta^I(x,y) = -\Pi^{-\frac{1}{2}}(x) \left[\int d^4z \, \hat{\partial} \frac{1}{\partial^2}(x,z) \, \Pi(z) \, \hat{\partial}^\dagger \frac{1}{\partial^2}(z,y)\right] \Pi^{-\frac{1}{2}}(y) \tag{15}$$

satisfies the Eq. (10).

It is easy to check that substitution of the free propagator (14) into the Eq. (15) gives the expression (12). However one can put into this formula other scalar Green functions including Δ_B obtained for closed domains as well. Note that due to the relation (15) propagator in the instanton field Δ_B^I satisfies conditions (11) if Δ_B^0 does. This condition eliminates surface terms which could appear after integrating by parts and all operators we shall encounter (*e.g.* $i\partial$) will be Hermitean.

5   Wave functions in the instanton field

In the absence of a classical field the Green function can be Fourier-expanded in 3-dimensional harmonics. We shall show that for massless scalar particle in the instanton field an analogous system of functions can be constructed. Far from the instanton their asymptotes are the same as in the main vacuum.

Let me remind that the Fourier expansion of the free scalar propagator has the form (we shall always take $E > 0$), which I call the Schrödinger representation:

$$\frac{1}{-\partial^2 + m^2} = \sum_n \left[\theta(x_4 - y_4) \frac{\phi_n^0(\vec{x})\phi_n^0(\vec{y})^\dagger}{2E_n} e^{-E_n(x_4 - y_4)} + (x_4 \leftrightarrow y_4) \right]. \qquad (16)$$

Here $\vec{x}$, $\vec{y}$ are spatial 3-vectors. The eigenfunctions $\phi_n^0(\vec{x})$, $\phi_n^0(\vec{y})^\dagger$ satisfy the $3d$-equations:

$$(-\overrightarrow{\partial}_i^2 + m^2)\phi_n^0(\vec{x}) = E_n^2 \phi_n^0(\vec{x}); \qquad \phi_n^0(\vec{y})^\dagger(-\overleftarrow{\partial}_i^2 + m^2) = E_n^2 \phi_n^0(\vec{y})^\dagger. \qquad (17)$$

We shall call them the "Schrödinger basis". Obviously the functions $\phi_{n\pm}^0(x) = e^{\mp E_n x_4} \phi_n^0(\vec{x})$ and $\phi_{n\pm}^0(y)^\dagger = e^{\pm E_n y_4}\phi_n^0(\vec{y})^\dagger$ are solutions of the Klein–Gordon equations,

$$(-\overrightarrow{\partial}^2 + m^2)\phi_n^0(x) = \phi_n^0(y)^\dagger(-\overleftarrow{\partial}^2 + m^2) = 0. \qquad (18)$$

The similar system of functions for massless scalar particles in the field of the instanton in singular gauge Eq. (6) is:

$$\phi_{n\pm}^I(x) = \Pi^{-\frac{1}{2}}(x)\left[1 + \frac{\rho^2}{2x^2} \frac{\hat{x}\overrightarrow{\hat{\partial}}_*^{\,\dagger}}{x\overrightarrow{\partial}_*} \right] e^{\mp E_n x_4} \phi_n^0(\vec{x}); \qquad (19)$$

$$\phi_{n\pm}^I(y)^\dagger = \phi_n^0(\vec{y})^\dagger e^{\pm E_n y_4}\left[1 + \frac{\overleftarrow{\hat{\partial}}_* \hat{y}^\dagger}{\overleftarrow{\partial}_* y} \frac{\rho^2}{2y^2} \right] \Pi^{-\frac{1}{2}}(y); \qquad (20)$$

Here the operators $\overrightarrow{\partial}_*$, $(x\overrightarrow{\partial}_*)^{-1}$ and $\overleftarrow{\partial}_*$, $(\overleftarrow{\partial}_* y)^{-1}$ act only on the functions outside the brackets. Hence one can handle them like constant light-like 4-vectors,

$$\partial_*^2 = \hat{\partial}_* \hat{\partial}_*^{\,\dagger} = \hat{\partial}_*^{\,\dagger} \hat{\partial}_* = 0. \qquad (21)$$

This is the only property we need here and we may be not puzzled by the particular appearance of $(x\overrightarrow{\partial}_*)^{-1}$ and $(\overleftarrow{\partial}_* y)^{-1}$. For the simplest case of the plane waves $\phi \propto e^{i\vec{k}\vec{x}}$ one readily finds $\partial_* = (i\vec{k}, \mp k)$.

With the help of the representation Eq. (13) and properties of τ-matrices Eq. (8) it is easy to show that these functions indeed satisfy the covariant d'Alembert equations,

$$\overrightarrow{\nabla}^2 \phi_{n\pm}^I(x) = \phi_{n\pm}^I(y)^\dagger \overleftarrow{\nabla}^2 = 0. \qquad (22)$$

At small times $|x_4| \sim \rho$ the solutions Eqs. (19), (20) take at the border values $\phi^I(x \in B) \sim \rho^2/R^2$. Far from the instanton at $|x_4| \to \infty$ they approach their free counterparts ϕ^0 from Eq. (18).

6 Schrödinger representation of Δ^I

Now we shall derive the expansion of the Green function Eq. (15) in terms of the functions Eqs. (19), (20). The latter play the role of the Schrödinger basis in the instanton field. The field of the pseudoparticle leads to transitions between different states and the expansion contains nondiagonal terms proportional to ρ^2.

We shall split the propagator in the following manner:

$$\Delta^I(x, y) = \Delta_1^I(x, y) + \Delta_2^I(x, y). \tag{23}$$

The first part comprises only leading diagonal elements and looks exactly like the free Green function, Eq. (16) with ϕ^0 substituted by ϕ^I. This choice simplifies the comparison with the no instanton case since all the additional terms are put together in Δ_2^I. The direct subtraction of Δ_1^I with wave functions Eqs. (19), (20) from Eq. (15) gives a rather cumbersome expression for Δ_2^I. Fortunately with the help of Eq. (8) and the identity (21) it can be simplified:

$$\Delta_2^I(x, y) = \Pi^{-\frac{1}{2}}(x) \left[\int d^4z \, \frac{1}{\partial^2}(x, z) \, \hat{\partial} \frac{\rho^2}{z^2} \hat{\partial}^\dagger \, \frac{1}{\partial^2}(z, y) \right] \Pi^{-\frac{1}{2}}(y). \tag{24}$$

We see that ρ^2 pieces in the wave functions ϕ^I led to the change of sign of ρ^2/z^2 relative to that in $\Pi(z)$ in Eq. (15).

The last thing to do is to express Δ_2^I in terms of the functions ϕ^I. With the help of some algebra analogous to that used above one can prove that the addition to the propagator induced by the pseudoparticle field has the form

$$\Delta_2^I(x, y) = \sum_{\substack{n\pm \\ m\pm}} \phi_{n\pm}^I(x) M_{\substack{n\pm \\ m\pm}} \phi_{m\pm}^I(y)^\dagger. \tag{25}$$

The matrix elements M are given by the integrals

$$M_{\substack{n\pm\diamond \\ m\pm\spadesuit}}(x_4, y_4) = \int d^4z \, [\theta(\pm x_4 \mp z_4) \frac{\phi_{n\pm}^0(z)^\dagger}{2E_n}]_\diamond \hat{\partial}_z \frac{\rho^2}{z^2} \hat{\partial}_z^\dagger [\theta(\pm z_4 \mp y_4) \frac{\phi_{m\pm}^0(z)}{2E_m}]_\spadesuit. \tag{26}$$

All the integrals are convergent since θ functions ensure the exponential fall of the integrand at $|z_4| \to \infty$. Wave functions ϕ^0 are square integrable and the singularity z^{-2} does not spoil the fact.

Now it is easy to do the final step and to calculate matrix elements. Only the $(++)$ component of M contributes and we obtain *e. g.* for the diagonal component (remember that we are working with scalar matter):

$$\Delta_{nn}^{I} = \frac{e^{-E_n\beta}}{2E_n}\left(1 + \frac{\pi^2\rho^2}{E_n}\phi_n^0(\vec{z}^0)^2 - i\bar{\eta}_{4j}^{a}\tau^a\frac{\pi\rho^2}{2}\frac{\partial}{\partial z_j^0}\int_{B_3}d^3z\frac{\phi_n^0(\vec{z})^2}{|\vec{z} - \vec{z}^0|}\right). \tag{27}$$

where $\vec{z}^0$ stands for the space coordinate of the pseudoparticle center. Note that the correction is proportional to ρ^2, that justifies omitting the higher terms in ρ^2/R^2.

This completes the mathematical stage of the analysis of the interaction of an instanton with scalar quarks in the closed volume.

7   Conclusion.

We constructed the Green function of scalar particles in the fundamental representation of $SU(2)$ gauge group in a finite volume taking into account the instanton field. This is a first step in studying nonperturbative QCD effects in the bag model..

We expanded the propagator in terms of exact solutions of the covariant Klein-Gordon equation. These correspond to *in-* and *out*-states in the main perturbative vacuum. The expansion contains nondiagonal terms that describe scattering in the field of a pseudoparticle.

We calculated the matrix elements of the scalar Green function in the instanton field. It contains positive corrections proportional to ρ^2. The sign of the correction tells that the instanton action diminishes, *i. e.* that matter favours pseudoparticles. The value of the effect evidences selfconsistency of our approximation.

More involved spin $\frac{1}{2}$ and spin 1 Green functions can be expressed in terms of $\Delta^I(x, y)$. Our results can be easily generalized to higher gauge groups, multiinstantons and thermal instantons.

The next step will be to examine spinor quarks. First of all one could expect suppression of instantons analogous to that in quark plasma. The second question concerns fermionic zero modes. They are of prior importance for the chiral invariance breaking in QCD but their presence in the bag is directly related to boundary conditions [15]. On the other hand the bag itself explicitly breaks chiral invariance. One could expect interesting interference of these effects.

It should be noted that there is a physical question of boundary conditions for matter and gauge fields. Even leaving instantons away this is a serious problem.

Independently of these problems the approach announced here is useful for investigation of QCD or electroweak pseudoparticles interacting with matter in closed domains or in some nonhomogeneous environment. These problems appear in the theory of strong interactions, thermal field theories and in physics of the early universe.

Acknowledgements

I would like to thank the Organizers of the conference and personally Dr. M. Bordag for their kind attention and financial support. I'm greatly indebted to E. V. Shuryak for indicating me the existence of this and connected problems.

References

[1] A.A. Belavin, A.M. Polyakov, A.S. Schwartz, Yu. S. Tyupkin, Phys. Lett. B **59** (1975) 85.

[2] *Instantons in gauge theories*, ed. by M.A. Shifman, World Scientific, Singapore, 1994.

[3] G. 'tHooft, Phys. Repts. **142** (1986).

[4] C.G. Callan, R.F. Dashen, D.G. Gross, Phys. Rev. D **19** (1979) 1826.

[5] A. Chodos, R.L. Jaffe, C.B. Thorn, and V.F. Weisskopf, Phys. Rev. D **9** (1974) 3471.

[6] A. Ringwald, Nucl. Phys. B **330** (1990) 1.

[7] L. McLerran, A.I. Vainshtein, M.B. Voloshin, Phys. Rev. D **42** (1990) 171.

[8] E.V. Shuryak, Nucl. Phys. B **203** (1982) 93, 116, 140.

[9] A.V. Smilga, hep-th 9504117.

[10] A.A. Abrikosov (Jr.), Nucl. Phys. B **218** (1983) 459.

[11] G. 'tHooft, Phys. Rev. D **14** (1976) 3432.

[12] R. Jackiw, C. Nohl, C. Rebbi, Phys. Rev. D **15** (1977) 1642.

[13] B.J. Harrington, H.K. Sheppard, Phys. Rev. D **17** (1978) 1583.

[14] L.S. Brown, R.D. Carlitz, D.B. Creamer and Ch. Lee, Phys. Rev. D **17** (1978) 1583.

[15] The proof of this relation was communicated to me by A. Wipf.

QED in the External Aharonov–Bohm Field

Jürgen Audretsch, Uwe Jasper and <u>Vladimir D. Skarzhinsky</u>

Summary:

In the framework of QED we investigate quantum processes in the presence of a magnetic string, carrying the magnetic flux $\phi = N + \delta$, $(0 < \delta < 1)$ (in units of the magnetic flux quantum). The exact electron and positron solutions of the Dirac equation are found and discussed. They are eigenfunctions of a complete set of commuting operators marked by quantum numbers of the energy, the z - components of linear, p_3, and of total angular momentum, $l + N + 1/2$, and the parameter s which describes the spin polarization in the magnetic field. The singular mode with $l = 0$ is taken in accordance with the sign of spin-field interaction by a limiting procedure from a model with the distributed flux. Based on these solutions we evaluate differential and total cross sections of the bremsstrahlung from an electron passing by the magnetic string, and of the electron-positron pair production by a single photon.

The partial analysis of the processes shows a rather unexpected feature. The processes turn out to be forbidden unless charged particles have total angular momentum projections of opposite signs. In the framework of a semiclassical picture it means that charged particles need to pass by the magnetic string in opposite directions, encircling the magnetic flux.

The dependence of the resulting cross sections on the energy, direction and polarization of the involved particles is rather complicated but it gets simplified at low and high energies.

Bremsstrahlung:

The cross section at low electron energy reads

$$\frac{d\sigma}{d\omega_k} = \frac{r_0 \sin^2 \pi\delta}{\pi} \frac{v}{\omega_k} \left[\left(1 - \frac{\omega_k}{E_p - M}\right)^{1+s\delta} + \left(1 - \frac{\omega_k}{E_p - M}\right)^{-s\delta} \right].$$

where $r_0 = e^2/4\pi M$ is the classical electron radius.

Pair production:

The total cross section above the pair production threshold is following

$$\sigma \approx \frac{r_0 \sin^2 \pi\delta}{\sqrt{2\pi}} \left(\frac{\omega_k - 2M}{2M}\right)^{\frac{3}{2}-2\delta}.$$

This work was supported by the Deutsche Forschungsgemeinschaft.

Quantum Fields in a Conical Background[1]

Edisom S. Moreira Jnr.

In this work the *proper time* method to study second quantization on the cone is generalized to arbitrary dimensions (N), arbitrary twist angles $(2\pi\sigma)$, higher spins (s) and massive fields (M).

The proper time representation of the spin-0 Feynman propagator is obtained using the eigenfunctions of the operator $\Delta + M^2$. These eigenfunctions are chosen to be finite at the tip of the cone $(\rho = 0)$. It is remarked that other boundary conditions, mildly divergent at $\rho = 0$, are also available.

The spin-1/2 Feynman propagator is shown to be obtained from the scalar propagator twisted by angles $2\pi\sigma \pm \mathcal{D}/2$, where $\mathcal{D}$ is the cone deficit angle. The N-bein corresponds to a spin connection which vanishes everywhere but on an arbitrary ray. These and other Aharonov-Bohm like features are discussed. Considering known results in the context of quantum mechanics on the cone, the spin-s Feynman propagator is conjectured to be obtained from the scalar propagator twisted by angles $2\pi\sigma \pm \mathcal{D}s$.

Using the point-splitting method in the propagators, some vacuum averages are determined. They are expressed in terms of the generalized hypergeometric functions $_1F_2\left[a; b, c; z = (M\rho)^2\right]$ which converge for all values of z. For a scalar field, in the limit $M\rho \to 0$, the vacuum fluctuations $\langle\phi^2(x)\rangle$ and the energy momentum tensor $\langle T^\mu{}_\nu(x)\rangle$ are shown to behave as $1/\rho^{N-2}$ and $1/\rho^N$ respectively. $\langle T^\mu{}_\nu(x)\rangle$ is covariantly conserved and traceless when conformally coupled with the curvature scalar.

The dependences on σ and $\mathcal{D}$ are shown for $N = 4$. Finite and logarithmic divergent mass corrections to $\langle\phi^2(x)\rangle$ are obtained. The energy density of a spin-1/2 field is also evaluated when $M\rho \to 0$.

It is remarked that this work can be extended to consider a spinning cone. It is also remarked that the results can be used in the context of related backgrounds (Rindler space, black holes) and that the boundary conditions mentioned above are worth investigating.

---

[1] Talk based on the paper: E.S. Moreira Jnr., Nuclear Physics **B 451** (1995) 365 (Archive: hep-th/9502016) and references therein.

Magnetic Response of the Many-Particle Chern-Simons Gauge Theory in (2+1) Dimensions

Efrain J. Ferrer and Vivian de la Incera

An investigation of the magnetic response of the many-particle ($\mu \neq 0$) anyon fluid in the low density limit ($n_e < Nm^2$) is presented. We start from the criterion that in a many-particle system the self-consistent zero temperature formulation is obtained as the $T \to 0$ limit of the thermal theory. Contrary to what occurs in anyon QFT ($T = 0$); in the $T \to 0$ limit of the many-particle theory, superconductivity arises only in the case that an electric field, transverse to the supercurrent in the plane of the two dimensional fluid, is considered. We interpret this transverse electric field as a simulation of the vortex effects in charged anyon fluids. This boundary Hall potential in the effective linear theory can be seen as a simple way to model vortex effects in anyon superconductivity. In this way our simplified model serves to reveal that, even at zero temperature ($T \to 0$), superconductivity in a charged anyon fluid can take place only in the presence of vortices.

It is also found that for $B_{ext} > \alpha E_H$; where α depends on the polarization operators of the many-particle anyon system in the static, $k_4 = 0$, and slowly varying, $\mathbf{k} \to 0$, limits, and E_H is the Hall electric field; the strength of the applied magnetic field B_{ext} exceeds the capacity of the anyon system to reaccommodate its vortex structure. In this case a magnetic field sufficiently strong to cancel the applied one cannot be produced. As a consequence, a partial Meissner effect takes place.

Finally, it should be pointed out that in the zero temperature limit ($T \to 0$), the penetration depth in the superconducting phase is dependent on the filling factor ν, contrary to what takes place at $T = 0$, where the penetration depth coincides with the London penetration depth in BCS superconductors.

Self-Interaction Force for Charged Particle in the Space-Time of Supermassive Cosmic String

Nail R. Khusnutdinov

Cosmic strings may have resulted from phase transitions in the early Universe [1, 2]. It is well-known that cosmic strings have no Newtonian potential and the space-time is flat.

At the same time many interesting effects exist which are connected with topological charge of the strings. Among of them there is a repulsive self interaction force between the string and a charged particle [3]. In this article we calculate in different ways and analyse in more detail the force of self interaction of a charged particle near a straight cosmic string [4, 5]. We obtain this force for an arbitrary trajectory of the particle and for an arbitrary deficite of angle of the string up to supermassive strings. The force can build up to high values for supermassive strings.

First of all we calculate the electromagnetic force acting on the test charged particle (b) from origin particle (a) in the space time of infinitely thin cosmic string. Next we situate the test particle (b) on the world line of charge (a) and then tend test particle (b) to the charge (a).

For a particle at rest the self interaction force has the only component

$$\mathcal{F}^{\rho} = \frac{1}{2} L \frac{e^2}{\rho^2} \, , \quad L = \sum_{n=1}^{[\nu/2]} \int_0^1 \frac{dq}{\sqrt{q}} \delta(q - \sin^2(\pi n/\nu))$$

$$- \frac{\nu \sin(\pi\nu)}{2\pi} \int_0^{+\infty} \frac{dy}{\cosh(y/2)} \frac{1}{\cosh(\nu y) - \cos(\pi\nu)} \, .$$

The first term occurs owing to the property that in the cone space time any two events may be connected by several geodesic lines.

If ν is close to unity then

$$L \approx \frac{\pi(\nu - 1)}{4} \, , \quad \nu \to 1 \qquad \left[L \approx \frac{2.5(\nu - 1)}{2\pi} \, [3] \right]$$

When $\nu \to \infty$ (supermassive string) the function L is approximated by the following expression

$$L(\nu) \approx \frac{\nu}{\pi} \ln \frac{2\nu}{\pi} \, , \quad \nu \to \infty \, .$$

References

[1] T.W.B. Kibble 1976 J. Phys. A. **9** 1387

[2] A. Vilenkin 1981 Phys. Rev. D. **23** 852

[3] B. Linet 1986 Phys.Rev. D. **33** 1833

[4] N.R. Khusnutdinov 1994 Class. Quant. Grav. **11** 1807

[5] N.R. Khusnutdinov 1995 Theor. Math. Phys. **103** 339

Exact Euler-Heisenberg Effective Action for Chiral Fermions in Some Special External Fields

Gautam Bhattacharya

The quantity of central importance in quantum field theory is just not the action of its classical counterpart but the partition function that contains it and also the path measures. As a consequence even if the action part is trivial, i.e., it is equivalent to a free theory by redefining the phases of the complex fields, the quantum theory may not still be trivial. The path measures may acquire a nontrivial Jacobian because of this redefinition of the variables. Analogous thing was the approach of Fujikawa [1] to study the axial anomaly, except that in those cases only the infinitesimal variation (Ward identities) of the effective actions are calculable. The contention of the present talk is that for the classically trivial theories involving chiral fermions in any even space time dimensions the complete effective actions can be exactly evaluated. The techniques involved are well known [2] but the final results are interesting. Below is a short prescription.

The starting point is the action of chiral fermions in an external pure gauge potential in an even space-time dimension D

$$S = \int d^D x \bar{\psi} \gamma^\mu (\partial_\mu + V_\mu + \gamma^5 A_\mu) \psi \tag{1}$$

where

$$V_\mu = \frac{1}{2}(R_\mu + L_\mu) \qquad \text{and} \qquad A_\mu = \frac{1}{2}(R_\mu - L_\mu) \tag{2}$$

$$\begin{aligned} \text{with} \quad & \partial_\mu L_\nu - \partial_\nu L_\mu + [L_\mu, L_\nu] = 0 \\ \text{and} \quad & \partial_\mu R_\nu - \partial_\nu R_\mu + [R_\mu, R_\nu] = 0. \end{aligned} \tag{3}$$

Clearly one can write $L_\mu = U\partial_\mu U^{-1}$ and $R_\mu = V^{-1}\partial_\mu V$ where U and V are two arbitrary elements of the gauge group G.

The action of eq.(1) is trivial since the potentials can be gauged away by redefining $\psi'_L = U^{-1}\psi_L$ and $\psi'_R = V\psi_R$ where $\psi_{L,R} = \frac{1}{2}(1 \mp \gamma^5)\psi$ are the chiral components of the fermions. However, because of the asymmetric variations of the left and right handed fermion measures, the Jacobian will be no-trivial. To evaluate one has to proceed through infinitesimal steps of change of fermionic variables in the path integral to maintain consistency with the underlying Schwinger-Dyson equation.

Such continuous infinitesimal steps are achieved through the homotopy parameters ξ and η:

$$U(x,\xi) \equiv e^{i\xi\theta(x)} \tag{4}$$

$$V(x,\eta) \equiv e^{i\eta\phi(x)} \tag{5}$$

$$\text{so that} \quad U(x,0) = V(x,0) = 1$$

$$\text{and} \quad U(x,1) = U(x); \quad V(x,1) = V(x).$$

The partition function is now written as-

$$e^{\Gamma(L_\mu(x,\xi),R_\mu(x,\eta))} = \int d\psi d\bar\psi e^{iS(\psi,\bar\psi L_\mu(x,\xi),R_\mu(x,\eta))}, \tag{6}$$

changing of variables

$$\psi'_L = e^{i\delta\xi\theta(x)}\psi_L$$
$$\psi'_R = e^{-i\delta\eta\phi(x)}\psi_R$$

will make the right hand side of eq.(6) -

$$\int d\psi' d\bar\psi' e^{iS(\psi',\bar\psi',L'_\mu(x,\xi),R'_\mu(x,\eta))} \mathcal{J},$$

where $\mathcal{J}$ is the Jacobian due to the change of measure and is formally of the form

$$\mathcal{J} = e^{-i\text{Tr}\left[\gamma^5(\theta\delta\xi+\phi\delta\eta)\right]}. \tag{7}$$

In eq.(7) Tr is the trace operation over all possible configurations, Dirac spinor labels and the generators of the Lie algebra in the representation in which the chiral fermions belong to. L'_μ and R'_μ are expressed in terms of the unprimed configurations as

$$L'_\mu = e^{i\delta\xi\theta(x)} L_\mu e^{-i\delta\xi\theta(x)} + e^{i\delta\xi\theta(x)}\partial_\mu e^{-i\delta\xi\theta(x)} \tag{8}$$

$$\text{and} \quad R'_\mu = e^{-i\delta\eta\phi(x)} R_\mu e^{i\delta\eta\phi(x)} + e^{-i\delta\eta\phi(x)}\partial_\mu e^{i\delta\eta\phi(x)}. \tag{9}$$

The expression for the Jacobian in eq.(7) is formally infinite and hence requires a regularization that preserves the maximal set of invariances of the theory. Here we will use the ζ-function regularization invoking the asymptotic heat kernel expansion [2]. Thus the regularised expression for $\mathcal{J}$ is

$$\mathcal{J} = e^{-i\int d^D x\,\text{tr}\left[\gamma^5(\theta\delta\xi+\phi\delta\eta)\zeta(x,x,0)\right]}, \tag{10}$$

where tr is the ordinary finite dimensional trace over the Dirac spinor labels and the generators of the Lie algebra. The generalized ζ-function in in eq.(10) is related to the heat kernel coefficient

$$\zeta(x,x,0) = \frac{1}{(4\pi)^{D/2}} a_{D/2}(x,x). \tag{11}$$

Each of the coefficients $a_n(x,x)$ can be uniquely obtained in terms of the potentials $L_\mu(x,\xi)$ and $R_\mu(x,\eta)$ and their derivatives [2].

Eq.(6) thus leads to

$$\frac{\partial\Gamma}{\partial\xi} = \frac{i}{(4\pi)^{D/2}} \int d^D x\,\mathrm{tr}(\gamma^5\theta(x)a_{D/2}(x,x,\xi,\eta)), \tag{12}$$

$$\frac{\partial\Gamma}{\partial\eta} = \frac{i}{(4\pi)^{D/2}} \int d^D x\,\mathrm{tr}(\gamma^5\phi(x)a_{D/2}(x,x,\xi,\eta)). \tag{13}$$

a priori it is not obvious if these two relations are compatible or not. The integrability clearly depends upon the specific form of $a_{D/2}(x,x)$. In a sense, it is equivalent to the Wess-Zumino consistency condition [3] that determines the structure of anomaly in a gauge like theory.

Once the integrability in the homotopy space is established one can conveniently choose a path from $(0,0)$ to $(1,1)$ to decouple the fermions completely and express the effective action as a $D+1$-dimensional integral. Since the gauge potentials are pure, the terms involving the derivatives of them can be easily integrated by parts to give rise to unique local Lagrangian contributions. Others will give rise to the Wess-Zumino terms by interpreting the path parameter variable as an additional space dimension with the resultant manifold having a boundary.

In two space-time dimensions the Abelian models of the types discussed above are the massless Thirring models and a_1 would give rise to a quadratic derivative terms of bosons. Thus fermion decoupling generates the bosonic kinetic energy. This is precisely the bosonisation phenomenon. Another interesting example is the QED where effectively, because of the kinematics of two dimensions, even a nontrivial vector gauge theory can be written in terms of pure axial vector gauge theory. The resultant quadratic terms of derivatives of spin zero bosons are now interpreted as the mass term of the gauge fields – the well known Schwinger mechanism [4]. The non-Abelian generalization of the models prescribed above are also easily executable. The case equivalent to Thirring model will be the Wess-Zumino action in the conformal limit [5, 6]. That also shows that such models are indeed, in some sense, equivalent to free fermion theories, just as the Thirring and Schwinger models are. For the non-Abelian vector gauge theories (QCD_2), one will recover the Polyakov-Weigmann results [7].

The purpose of going through this rehash in two dimensions was just to show that the local terms arising out of Jacobian evaluation are just as important as the anomaly terms. There is a traditional approach of throwing away the local terms in the garb of renormalization. But then one may loose a lot of informations. It will be interesting to see what new information one can have in four dimensions out of a_2. In Skyrme model these are known to give rise to the quartic local terms [8, 9], part of which makes the solitons stable. It would be interesting to study these models in different contexts, say the parity violating theories involving chiral quarks, i.e., the weak interactions of quarks.

References

[1] K. Fujikawa, Phys. Rev. **D24**, 134 (1981)

[2] A.P. Balachandran *et al.*, Phys. Rev. **D26**, 2134 (1981)

[3] J.Wess and B.Zumino, Phys. Lett **B105**, 256 (1979)

[4] G. Bhattacharya *et al.* Phys. Rev **D50**, 4183 (1994)

[5] E. Witten, Comm. Math. Phys. **92** 455 (1984)

[6] G. Bhattacharya and S.G. Rajeev, Nuclear Phys. **B246** 157 (1984)

[7] A.M. Polyakov and P.B. Weigmann, Phys. Lett **131B**, 223 (1983)

[8] G. Bhattacharya and S.G. Rajeev, Solitons in Nuclear and Elementary Particle Physics, p.229, ed. A. Chodos *et al.* *World Scientific* (1985)

[9] G. Bhattacharya, Proceedings of VII HEP Symposium of DAE, India, ed. A.K. Raina and M.V.S. Rao, p.114 (1985)

Energy-Momentum Tensor of Particles Created by an External Field

Anatoli I. Nikishov

In Minkowski space-time the energy-momentum tensor (EMT) of particles, which arise after turning off the external field, is defined by normal ordering of out-operators. In this way the finite expression for the expectation value $\langle 0^{in}| : T_{\mu\nu} : |0^{in}\rangle$ is obtained. No regularization is needed and EMT is treated on equal footing with other observables such as current or the number of created particles. This means that an expectation value of an observable is well defined only after the process of its formation is finished. In application to particles produced by a mirror, which moves with acceleration during a finite time interval in 1+1 space-time, the value $\langle 0^{in}| : T_{\mu\nu} : |0^{in}\rangle$ does not coincide with the regularized value, $\langle 0^{in}|T_{\mu\nu}|0^{in}\rangle_{reg}$, although the integrals over all space of their 00-components give the same total energy of produced particles.

Nonlocality, Self-Adjointness of the Hamiltonian and Vacuum Polarization in Spaces with Nontrivial Topology

Yurii A. Sitenko and S.A. Yushchenko

Polarization effects in the fermionic vacuum under the influence of external conditions are considered. It is known that external fields can polarize vacuum locally, which results in the appearance of the nontrivial vacuum quantum numbers depending on the global chracteristics of the background in the whole (e.g. fermion number fractionization). The new effect discovered by one of the authors (Yu.A.S.) in 1990 is that vacuum can be polarized by the nonlocal action of external fields. The latter is due to the nonvanishing asymptotics of the axial-vector currents on the topologically nontrivial base manifolds. So, if one excludes some spatial regions in order to get a space with nontrivial topology, then external fields bounded to excluded regions will nevertheless polarize the vacuum in the remaining regions. The induced vacuum quantum numbers depend periodically on the global characteristics of the background in the excluded regions. Our approach allows one to consider singular configurations of the background fields, as well as regular ones, deleting the points (or lines) of singularity from the physical space. We also consider the most general conditions on the boundary between the excluded and the remaining regions, which are compatible with the self-adjointness of the Dirac Hamiltonian, and find the dependence of the induced vacuum quantum numbers on the parameters of self-adjoint extensions.

Hydrogen Atom in the Spacetime of a Cosmic String

Sven Falkenberg[1]

In the following article we analyze the energy level scheme of a hydrogen atom in the spacetime of a cosmic string. We will obtain a specific spectrum, which differs from that of other cylindrically symmetric problems and which is suitable for the spectroscopic detection of strings.

The exterior metric of an infinitely long straight and static string reads in spherical coordinates

$$ds^2 = dr^2 + r^2 \, d\vartheta^2 + b^2 \, r^2 \sin^2 \vartheta \, d\varphi^2 - dt^2 \tag{1}$$

with $0 < r < \infty$, $0 < \vartheta < \pi$, $0 \leq \varphi \leq 2\pi$, $-\infty < t < \infty$ and the string parameter b, which runs in the interval $(0, 1]$. The string is situated on the z-axes. In the special case $b = 1$ we obtain the Minkowski space in spherical coordinates.

To solve the Klein-Gordon equation or the Dirac equation for the hydrogen atom, first of all we need the electromagnetic potential of the atomic nucleus. We consider the atomic nucleus as a point charge and as a test particle. Our charge distribution rests at a fixed point in the space.

With the metric (1) we find from the Maxwell system the potential equation

$$\triangle \, U(x^m) \;=\; \frac{\varrho(x^m)}{\varepsilon_0}, \tag{2}$$

$$\triangle \;=\; \frac{1}{r^2} \frac{\partial}{\partial r}\left(r^2 \frac{\partial}{\partial r}\right) + \frac{1}{r^2 \sin \vartheta} \frac{\partial}{\partial \vartheta}\left(\sin \vartheta \frac{\partial}{\partial \vartheta}\right) + \frac{1}{b^2 \, r^2 \sin^2 \vartheta} \frac{\partial^2}{\partial \varphi^2} \; .$$

The charge distribution of a point charge reads $\varrho = Z \, e \, \delta(\vec{r} - \vec{r_0})$, where Z is the atomic number. Solving this equation we obtain for the case $r_0 = 0$ the potential of a point charge $U(\vec{r}) = -\frac{Ze}{4\pi\varepsilon_0 \, r}$.

We analyze the Klein-Gordon equation first. Although we neglect the spin of the electron, we get all essential effects. We restrict ourselves to the case that the atomic nucleus is located on the string. This permits us to use spherical coordinates where the zero point is the nucleus. Therefore we consider a smoothed model without singularity where the string has a finite but small radius compared with the atomic radius. Then we get negligible corrections of the potential and the probability density only.

The time-independent Klein-Gordon equation reads for our problem

$$\triangle \, \varphi + \left[(E_k - e \, U)^2 - m_0^2\right] \varphi = 0 \qquad . \tag{3}$$

---

[1] I thank Dr. Lotze for the idea to this theme and for stimulating discussions.

For bound states we find the energy levels (α_s is the fine structure constant)

$$E_k = \frac{m_0}{\sqrt{1 + \dfrac{Z^2\,\alpha_s^2}{\left[\tilde{n}+\frac{1}{2}+\sqrt{\left(l+\frac{1}{2}\right)^2 - Z^2\,\alpha_s^2}\,\right]^2}}} \quad , \tag{4}$$

where $\tilde{n} \in N_0$, $l = |m| + \tilde{l}$, $m\,b \in Z$ and $\tilde{l} \in N_0$. If we introduce a main quantum number n, we get $n = \tilde{n} + l + 1$, $n \in N$ in the flat space. In (1) n is no longer an integer but depends on m.

Because all quantum numbers depend on m, the number m enters the energy formula indirectly. Only levels with the same $|m|$ can have the same energy. Levels without a z-component of the angular momentum are not shifted relative to the Minkowski case. Except these levels all the other ones are twofold degenerated. The absolute value of the splitting is about one order above the hyperfine structure splitting of the hydrogen atom, if $b \simeq 1 -- 10^{-5}$ and amounts to about $10^{-4}eV$.

We represent now the results for the Dirac equation. Now as before the atomic nucleus is located on the string. The time-independent Dirac equation and thus the energy eigenvalue problem for the hydrogen atom in the metric (1) reads

$$(E_k - H)\,\psi = 0 \quad \text{with} \quad H = i\,\gamma^4\left[\gamma^m\,\frac{\partial}{\partial x^m} + \frac{i}{2}\,(1-b)\,\gamma\right] + e\,U + i\,m_0\,\gamma^4 \tag{5}$$

and the abbreviation $\gamma = \gamma^3 S_3$. The γ_μ are the Dirac matrices, $S_3 = \mathrm{diag}(1,-1,1,-1)$ the operator of the z-component of the spin.

For the energy eigenvalues, we find two different cases. The first case is

$$E_k = \frac{m_0}{\sqrt{1 + \dfrac{Z^2\,\alpha_s^2}{\left[\tilde{n}+\sqrt{\left(l+1\right)^2 - Z^2\,\alpha_s^2}\,\right]^2}}} \quad , \tag{6}$$

where $\tilde{n} \in N_0$, $l = |m| + \tilde{l}$, $\tilde{l} \in N_0$ and $m = \frac{1}{2}\left(\pm\frac{1}{b} - 1\right)$, $\frac{1}{2}\left(\pm\frac{3}{b} - 1\right), \ldots$. In the second case we have to exchange $l \to (l-1)$, and $\tilde{l} \in N$. All levels are shifted ($m = 0$ is not allowed). The reason is that the spin of the electron couples with the string. Also the degeneration is different. The highest level in each group $(n, \mathrm{int}(l))$ is twofold degenerated, the lower ones fourfold.

References

[1] A. Vilenkin, *Phys. Rep.* **121**, 263–315 (1985).

[2] B. Linet, *Gen. Rel. Grav.* **17**, 1109–15 (1985).

[3] R. Gott, *Astrophys. Journ.* **288**, 422-427 (1985).

[4] A.N. Aliev, D.V. Galtsov, *Ann. Phys.* **193**, 142–165 (1989).

[5] J. Audretsch, A. Economou, *Phys. Rev. D* **44**, 980–90 (1991).

[6] J. Audretsch, R. Müller, *Phys. Rev. D* **49**, 6566–75 (1994).

[7] V.D. Skarzhinsky, D.D. Harari, U. Jasper, *Phys. Rev. D* **49**, 755–762 (1994).

[8] L. Iliadakis, U. Jasper, J. Audretsch, preprint, to appear in *Phys. Rev. D* (1994).

Part III

Ground State in External Fields

Functional Determinants on Möbius Corners

J. Stuart Dowker

1 Historical remarks

My title and subject matter are entirely appropriate for a talk delivered in Leipzig where Möbius spent most of his distinguished academic career. Furthermore, the fundamental quantum field theory paper by Heisenberg and Euler originated here and the classic book by Pockels on the Helmholtz equation was published in 1891 by Teubner's in their famous Handbuch series.

2 Brief motivation

A functional determinant is the determinant of an operator, thus extending the standard notion from the finite to the infinite dimensional situation. The reason for its physical importance is that the one-loop effective action, or the effective action in a background field, is determined, in the simplest case, by the determinant of the propagating (differential) operator.

In such, or equivalent, language this result dates to Feynman, Neumann and Schwinger in the 40's and 50's although the idea, and evaluation, of an effective action is much older and can be traced back to the beginnings of quantum field theory (QED) being associated with Uehling, Serber, Heisenberg & Euler and Weisskopf. Nowadays, the most popular way of doing things in field, instanton and string theory is via path-integration.

If the background field is a gravitational one, the determinant is a function of the metric and therein lies its mathematical importance. Usually the operator is a geometric one such as the Laplacian, and the determinant is a *geometric invariant*. In fact the alternating sum (with coefficients) of form determinants on a manifold is a *topological* invariant – the *analytic torsion*.

Functional determinants have been employed in connection with the uniformisation theorem of Riemann surfaces and with the isospectral problem. The functional determinant depends on, and is determined by, the eigenvalues of the operator. It is thus a *spectral invariant.*

3 The zeta function

A convenient but by no means the only way of organising the eigenvalues is via the Minakshisundaram ζ–function,

$$\zeta(s) = \sum \frac{1}{\lambda^s}$$

so that formally (*à la* Euler),

$$\operatorname{Det} D = \prod \lambda = \exp\left(\sum \ln \lambda\right) = e^{-\zeta'(0)}$$

and I will hereafter consider the problem of finding the functional determinant as synonymous with finding $\zeta'(0)$. My emphasis is already turning to the mathematical side.

The central quantity in this approach is thus the ζ–function. The sum definition converges only for s is a certain region, typically $s > d/2$ where d is the dimension of the manifold $\mathcal{M}$, if D is the Laplacian for example. This makes a continuation to $s = 0$ (and may be to other values) necessary.

One more standard thing. The heat-kernel, K (the propertime kernel of Fock, Nambu, Feynman and Schwinger), and the ζ–function are related by a Mellin transform and contain the same information, modulo zero modes. Some formulae will follow.

The coefficients in the short-time expansion of K are important quantities determining the divergences of the field theory and its scaling behaviour. If D is a geometric operator, they are geometric invariants. Some explicit general forms exist for the early coefficients and specific algorithms for determining all the coefficients can be found in the contribution by Schimming to this conference.

The analytic structure, *i.e.* the poles, and the values of the ζ–function at specific s, is related to these coefficients whose values in any particular case can then be found from the ζ–function. They can be checked against the general forms, or, more usefully, can be used in the determination of any unknown coefficients.

The calculations I shall later report on are to be taken in this spirit of 'Special Case Evaluation'.

We will hear more things about the ζ–function in the following talk by Elizalde.

4   Summary of methods.

A number of possible approaches are available for finding the ζ–function.

First, if the eigenvalues are known explicitly, *e.g.* $\lambda = n$, one might look at $\sum(1/\lambda^s)$ and seek a continuation, often by reducing the sum, in a more or less direct way, to known (*i.e.* named) ζ–functions, (in the above case to the Riemann ζ–function, $\zeta_R(s)$) for which the continuation has already been done for us. This would be the ideal situation, if all we want is the answer. Alternatively, there might be special function properties, contour representations and summation formulae that can be used, requiring a certain amount of ingenuity to effect the continuation.

Second, if the eigenvalues are known only *implicitly*, the above approach is not possible and one might not be able to find $\zeta(s)$ for all s. However, sufficient information might still be available for our purposes. This is the situation I want to consider in a specific case later.

I will introduce such a method via the modified ζ–function

$$\zeta(s, m^2) = \sum \frac{1}{(\lambda + m^2)^s}$$

which bears the same relation to $\zeta(s)$ that the Hurwitz ζ–function does to the Riemann one and

$$\zeta(s, m^2) = \frac{1}{\Gamma(s)} \int_0^\infty \tau^{s-1} e^{-m^2 \tau} K(\tau)\, d\tau \tag{1}$$

$$= \int_{c-i\infty}^{c+i\infty} ds' (m^2)^{s-s'} \frac{\Gamma(s')\Gamma(s-s')}{\Gamma(s)} \zeta(s', 0), \tag{2}$$

$d/2 < c < (d+1)/2$. (Dikii.)

$\zeta(s, m^2)$ can be evaluated at an s, say q, where the sum converges, the variable m^2 then providing the access to the required information. (In this sense the construction of $\zeta(q, m^2)$ is an alternative regularisation process to that of continuing $\zeta(s)$. Dikii, Gelfand).

In physical terms, m^2 is a mass and, as such, was separated by Feynman in his proper-time approach. In the heat-kernel, the mass produces convergence at the upper, infinite limit of the Mellin transform allowing one to substitute the short-time expansion of K. This yields an alternative method of deducing the analytic properties of the ζ–function and results in an asymptotic expansion of $\zeta(s, m^2)$, valid for large m^2, whose coefficients are obviously the heat-kernel expansion coefficients, and provides a method of finding these coefficients (Dikii, Moss). I am not concerned with this aspect here, being more interested in the functional determinant and will present a general method of obtaining this that also involves the asymptotic behaviour as $m \to \infty$.

5   Weierstrass regularisation

The first step is to regularise the sum definition of $\zeta(s, m^2)$ in the following way

$$\zeta^*(s, m^2) = \sum\nolimits^* \frac{1}{(\lambda + m^2)^s} \equiv \sum \left(\frac{1}{(\lambda + m^2)^s} - \frac{1}{\lambda^s} - \sum_{k=1}^M \binom{-s}{k} \frac{m^{2k}}{\lambda^{k+s}} \right)$$

where sufficient terms in the Taylor series, for any particular value of s, have been removed to ensure convergence. I refer to this as *Weierstrass regularisation*, the reason being that if the expression is differentiated with respect to s, and s set to zero I find

$$\zeta^{*\prime}(0, m^2) = -\sum \left(\ln\left(1 + m^2/\lambda\right) + P(m^2/\lambda) \right) = -\sum\nolimits^* \ln\left(1 + m^2/\lambda\right)$$

where the polynomial P is

$$P(x) = x + \frac{x^2}{2} + \ldots + \frac{x^{2[d/2]}}{[d/2]}$$

and we recognise

$$e^{-\zeta^{*\prime}(0, m^2)} = \prod \left(1 + \frac{m^2}{\lambda}\right) e^{P(m^2/\lambda)}$$

as a *Weierstrass canonical product.* In quantum field theory, a modified determinant occurs already in the work of Schwinger.

Let me now perform the summation over λ in $\zeta^*(s, m^2)$ to give the continuation

$$\zeta^*(s, m^2) = \zeta(s, m^2) - \zeta(s, 0) - \sum_{k=1}^{M} \binom{-s}{k} m^{2k} \zeta(s+k, 0) \bigg).$$

In particular

$$\zeta^{*'}(0, m^2) = \zeta'(0, m^2) - \zeta'(0, 0) - \frac{\partial}{\partial s} \sum_{k=1}^{[d/2]} \binom{-s}{k} m^{2k} \zeta(s+k, 0) \bigg|_{s=0}.$$

Combining these two expressions for $\zeta^{*'}(0, m^2)$, and turning the equation around gives the quantity I am seeking,

$$\zeta'(0, 0) = \zeta'(0, m^2) + \sum{}^* \ln\left(1 + m^2/\lambda\right) - \frac{\partial}{\partial s} \sum_{k=1}^{[d/2]} \binom{-s}{k} m^{2k} \zeta(s+k, 0) \bigg|_{s=0}.$$

(*cf* Voros, Quine, Heydari & Song, Jorgenson & Lang.) The last term can be carried further, but I don't need to here.

The essential, rather trivial, point now is that the right-hand side of this equation must be mass-independent and so can be calculated at any convenient value of m, in particular at $m = \infty$. It is not necessary to check that all m-dependence does disappear. One just needs to keep the constant, mass-independent parts of the various asymptotic limits. This is the method.

The asymptotic expansion of $\zeta'(0, m^2)$ referred to earlier shows that

$$\zeta'(0, m^2) \sim 0 + O(\ln m^2)$$

then I arrive at the final, working formula

$$\zeta'(0) \approx \lim_{m \to \infty} \sum{}^* \ln\left(1 + m^2/\lambda\right)$$

or

$$\lim_{m \to \infty} \sum{}^* \ln\left(1 + m^2/\lambda\right) = \zeta'(0) + O(\ln m).$$

This is a perfectly general equation. I want to illustrate the method by applying it to the situation where the eigenvalues are known only implicitly. Instead of giving a general treatment, I proceed now to the specific calculation I wish to report on at this meeting.

I should say that related but different, methods have been developed by Barvinsky, Kamenshchik, Karmazin and Mishakov and by Bordag, Geyer, Kirsten and Elizalde, a group based mostly here at Leipzig.

6 The geometry and the calculation

Apart from any physical motivation, the plan is to look for geometrical domains on which there is sufficient knowledge of the eigenvalue problem to enable the functional determinant to be evaluated in 'closed form'. The operator is the Laplacian on scalars. Other fields can be treated of course but I just want to illustrate the technique. Pockels' book contains valuable information on this eigenvalue problem. (Incidentally, this book contains probably the first discussion of the eigenvalue counting function.)

The basic geometry I have in mind is that of the *Euclidean* $(d+1)$-*ball*. Although I have been at pains to say that I have no immediate physical motivation in mind, I should point out that a considerable amount of work has been performed on such, and related, geometries in the context of quantum cosmology by *e.g.* D'Eath, Esposito, Schleich, Moss, Pollifrone, Vassilevich and the Russian group, so my calculation may not be entirely sterile.

The functional determinants (and heat-kernel coefficients) on the full ball have been determined by the Leipzig group in the scalar case, and by Kirsten and Cognola for other fields using methods that parallel, but are in detail different to, my own. Of course, I like my technique better.

Actually I will not be dealing with the full ball here, but rather with a *bounded, generalised cone* of a special sort. This is, loosely speaking, the solid angle subtended at the origin of the ball by a finite domain on its surface. It is not possible to treat arbitrary shapes, I think, and I shall be restricted to domains with special symmetry associations which I will describe in a moment.

The reason for the calculation is purely one of aesthetics and not governed by any physical motivation. At the formal level, I wish to advertise the usefulness of the Barnes ζ- and multiple Γ-functions.

The bounded, generalized cone is defined as the space $R^+ \times \mathcal{N}$ with the hyperspherical metric

$$ds^2 = dr^2 + r^2 d\Sigma^2$$

where $d\Sigma^2$ is the metric on the manifold $\mathcal{N}$, and r runs from 0 to 1. For us, $\mathcal{N} \sim S^d$ locally and $d\Sigma^2$ is the metric on the d-sphere. (*cf* Cheeger.)

To set up the formalism, I outline some very standard textbook material. The Laplacian can be written

$$\Delta = \frac{\partial^2}{\partial r^2} + \frac{d}{r}\frac{\partial}{\partial r} - \frac{1}{r^2}\Delta_S$$

where Δ_S here is the Laplacian on the sphere.

The eigenmodes of Δ, with eigenvalue $-\alpha^2$, are of the form

$$\frac{J_\nu(\alpha r)}{r^{(d-1)/2}} Y(\Omega)$$

where the harmonics on $\mathcal{N}$ satisfy

$$\Delta_S Y(\Omega) = -\bar{\lambda}^2 Y(\Omega)$$

and

$$\nu^2 = \bar{\lambda}^2 + (d-1)^2/4.$$

At this point I restrict the region $\mathcal{N}$ to be a fundamental domain of the complete symmetry group, Γ, of a $d+1$-dimensional polytope acting on its circumscribing sphere S^d. I have used such regions several times before and, following Terras, have referred to them as *Möbius corners*, $\mathcal{C}(\mathbf{d})$. In the present case, when the range of r is limited, the corner will be a bounded one. An interesting aspect of these spaces is that their boundaries are only piecewise smooth.

In three dimensions, the infinite corner can be realised as three planar mirrors intersecting at one point to form a *trihedral kaleidoscope* (Möbius' term). To qualify as a genuine Möbius corner, the dihedral angles must be such as to allow an integral number of similar corners to fill out the entire 4π solid angle when arranged around a common point of intersection.

Incidentally, if the domains are identified to form an orbifold, their edges (running through the origin) form three angular defects (cosmic strings). The Casimir effect has been calculated in the infinite corner. (Dowker and Chang.)

The eigenfunctions and eigenvalues on $\mathcal{N}$ are determined by group and invariant theory. The usual spherical harmonics are filtered (symmetry adapted) by the condition of being invariant under Γ. The important fact for my calculation is that the eigenvalues $\bar{\lambda}$ are given by

$$\bar{\lambda}_{\mathbf{n}}^2 = (\mathbf{n}.\mathbf{d} + a)^2 - (d-1)^2/4$$

where $\mathbf{d}$ is a d-dimensional vector of the integer degrees associated with the tiling group, Γ, and where the integers n_i range from zero to infinity. The degeneracy of any particular eigenvalue depends on the number of coincidences as $\mathbf{n}$ varies. The parameter a determines the type of boundary conditions on the *sides* of the cone and, in what is effectively an image-based approach, I can, unfortunately, treat only Dirichlet and Neumann conditions with a equalling $\sum d_i - (d-1)/2$ and $(d-1)/2$ respectively.

You now see the nice circumstance that the order of the Bessel function is still an integer or a half odd-integer, depending on the oddness or evenness of the dimension d,

$$\nu_{\mathbf{n}} = \mathbf{n}.\mathbf{d} + a.$$

The boundary condition on the spherical end of the cone at $r = 1$ has still to be fixed. Just as for the full ball, one can impose Dirichlet, Neumann or Robin conditions. For Dirichlet the implicit eigenvalue condition is

$$J_{\mathbf{n}.\mathbf{d}+a}(\alpha) = 0 \,,$$

while Robin boundary conditions yield,

$$\beta J_{\mathbf{n}.\mathbf{d}+a}(\alpha) + \alpha J'_{\mathbf{n}.\mathbf{d}+a}(\alpha) = 0,$$

where β is a constant. (Actually β could be a function on $\mathcal{N}$ but I won't discuss this here.)

In order to extract the eigenvalue properties from these equations, I follow previous workers and use Euler's representation of the Bessel function in terms of its zeros, to write, for example,

$$\sum_p \ln\left(2^p p! m^{-p} I_p(m)\right) = \sum_{p,\alpha_p} \ln\left(1 + \frac{m^2}{\alpha_p^2}\right) = \sum_\lambda \ln\left(1 + \frac{m^2}{\lambda}\right).$$

Then

$$\zeta'(0) \approx \lim_{m\to\infty} {\sum_p}^* \ln\left(2^p p! m^{-p} I_p(m)\right),$$

where $p = \mathbf{n}.\mathbf{d} + a$ and the sum over p means a multiple sum over the integers $\mathbf{n}$.

The further analysis of this expression involves some algebra and will be of interest only for those currently working in this field. However let me give you a flavour of the calculation.

The asymptotic behaviour of the Bessel function is known (Olver) and the $\ln p!$ term can be replaced by an integral representation. Then

$$\zeta'(0) \sim {\sum_{\mathbf{n}}}^* \left[p \ln \frac{2p}{p+\epsilon} + (\epsilon - p) - \frac{1}{2}\ln\frac{\epsilon}{p} + \sum_{l=1}^{\infty} \frac{T_l(t)}{\epsilon^l} \right.$$
$$\left. + \int_0^\infty \left(\frac{1}{2} - \frac{1}{\tau} + \frac{1}{e^\tau - 1}\right)\frac{e^{-\tau p}}{\tau}\, d\tau \right], \qquad (3)$$

where the $T_l(t)$ are Olver's polynomials in $t = p/\epsilon$ and $\epsilon^2 \equiv p^2 + m^2$.

I will indicate how to deal with the first three terms. One simply applies the Weierstrass regularisation again. Thus

$$ {\sum_{\mathbf{n}}}^* \frac{p^N}{\epsilon^{2s}} = \sum_{\mathbf{n}} p^N \left[\frac{1}{\epsilon^{2s}} - \frac{1}{p^{2s}} - \sum_{h=1}^{M} \binom{-s}{h} \frac{m^{2h}}{p^{2h+2s}} \right] \qquad (4)$$

from which, if s is given certain values, possibly after differentiating, the required limits can be found.

I remind you that $p = \mathbf{n}.\mathbf{d} + a$. If the sums are done you can see that an important role is played by the Barnes ζ–function defined by

$$\zeta_d(s, a \mid \mathbf{d}) = \frac{i\Gamma(1-s)}{2\pi} \int_L d\tau \frac{\exp(-a\tau)(-\tau)^{s-1}}{\prod_{i=1}^{d}(1 - \exp(-d_i\tau))}$$
$$= \sum_{\mathbf{n}=0}^{\infty} \frac{1}{(a + \mathbf{n}.\mathbf{d})^s}, \qquad \operatorname{Re} s > d.$$

By general theory, the asymptotic limit of the first term in (2) is

$$\sum_{\mathbf{n}} \frac{1}{((\mathbf{n}.\mathbf{d} + a)^2 + m^2)^s} \sim \sum_{k=0,1/2,1,\dots} (m^2)^{d/2-k-s}\frac{\Gamma(s - d/2 + k)}{\Gamma(s)} C_k$$

where the C_k are the coefficients in the short-time expansion of the heat-kernel associated with the Barnes ζ–function, $\zeta_d(2s, a \mid \mathbf{d})$ *e.g.*

$$(-1)^k k! C_{d/2+k} = \zeta_d(-2k, a \mid \mathbf{d}),$$

with

$$\zeta_d(-q, a \mid \mathbf{d}) = \frac{(-1)^d q!}{(d+q)! \prod d_i} B_{d+q}^{(d)}(a \mid \mathbf{d})$$

in terms of generalized Bernoulli functions.

Actually I do not need these explicit expressions. Applying the Weierstrass regularisation only the second term in (2) contributes and one easily obtains the useful limits

$$\sum_n p^N (\epsilon - p) \sim -\zeta_d(-N-1, a \mid \mathbf{d}) + O(\ln m),$$

$$\sum_n p^N \ln \left(\frac{2p}{p+\epsilon}\right) \sim -\zeta_d'(-N, a \mid \mathbf{d}) + \ln 2\, \zeta_d(-N, a \mid \mathbf{d}) + O(\ln m),$$

$$\sum_n p^N \ln \left(\frac{\epsilon}{p}\right) \sim \zeta_d'(-N, a \mid \mathbf{d}) + O(\ln m),$$

enabling the first three terms in equation (1) to be found. I won't inflict the remainder of the calculation on you, although a number of interesting technical points arise. I will just write down the final answer, which is reasonably compact,

$$\zeta_{\mathcal{C}(\mathbf{d})}'(0) \;=\; \zeta_{d+1}'(0, a+1 \mid \mathbf{d}, 1) + \ln 2 \left(\zeta_d(-1, a \mid \mathbf{d}) + \sum_{l=1}^{d} T_l(1)\, N_l(d) \right)$$

$$+ \sum_{l=1}^{d} N_l(d) \int_0^1 t^{l-1} T_l''(t)\, dt + \frac{1}{2} \sum_{l=1}^{d} T_l(1)\, N_l(d) \sum_{k=1}^{(l-1)/2} \frac{1}{k}.$$

The $N_l(d)$ are the residues of the Barnes ζ–function and, like the $T_l(1)$, are given by Bernoulli functions.

Everything is quite explicit, apart from the derivative of the Barnes ζ–function which, in any particular case, could be reduced to a number of Hurwitz ζ–functions. I have thus achieved my aim of finding a closed form. More explicit expressions exist for the full sphere case and agree with those of the Leipzig group when evaluated at specific dimensions.

7   The Robin case

A little more involved, and therefore more interesting, is the case of Robin conditions on the spherical end of the cone. I am certainly not going to go through the details systematically but I do want to point out the following little bit of formalism. This is mostly self indulgence but some of you might be interested.

A new term in this case is

$$\sum_{\mathbf{n}}^{*} \ln\left(1 + \frac{\beta}{\mathbf{n}.\mathbf{d} + a}\right)$$

which is the Weierstrass product associated with the eigenvalues $\mathbf{n}.\mathbf{d} + a$ and can be dealt with according to the general result given earlier which, in this case, relates it to the derivative of the Barnes ζ–function, $\zeta_d'(0, a \mid \mathbf{d})$. Applying this relation, I find

$$\sum_{\mathbf{n}}^{*} \ln\left(1 + \frac{\beta}{\mathbf{n}.\mathbf{d} + a}\right) = \ln\left(\frac{\Gamma_d(a)}{\Gamma_d(a + \beta)}\right) + \sum_{l=1}^{d} \frac{\beta^l}{l!} \psi_d^{(l)}(a),$$

in terms of the multiple Γ- and ψ-functions. In fact this generalises Barnes' product for the multiple Γ-function.

The power series

$$\sum_{\mathbf{n}}^{*} \ln\left(1 + \frac{\beta}{\mathbf{n}.\mathbf{d} + a}\right) = \sum_{l=d}^{\infty} \frac{(-1)^l \beta^{l+1}}{l+1} \zeta_d(l+1, a \mid \mathbf{d})$$

is also easily derived.

I give the final determinant expression for completeness,

$$
\begin{aligned}
\zeta_{\mathcal{C}(\mathbf{d})}'(0, \beta) \;=\; & \zeta_{d+1}'(0, a \mid \mathbf{d}, 1) - \ln\left(\frac{\Gamma_d(a)}{\Gamma_d(a + \beta)}\right) - \sum_{l=2,\ldots}^{d} \frac{\beta^l}{l} N_l(d) \sum_{k=1}^{l/2} \frac{1}{2k-1} \\
& + \; \frac{1}{2} \sum_{l=1,3\ldots}^{d} R_l(\beta, 1) \, N_l(d) \sum_{k=1}^{(l-1)/2} \frac{1}{k} + \sum_{l=1}^{d} N_l(d) \int_0^1 t^{l-1} R_l''(\beta, t) \, dt \\
& + \; \ln 2\left(\zeta_d(-1, a \mid \mathbf{d}) + \sum_{l=1,3,\ldots}^{d} R_l(\beta, 1) \, N_l(d)\right).
\end{aligned}
$$

It is not much more complicated than the Dirichlet one. The $R(\beta, t)$ are the relevant Olver asymptotic polynomials (Moss).

Using the definition of the multiple Γ-function and Γ-modular form, ρ, I can rewrite part of this expression,

$$
\begin{aligned}
\zeta_{d+1}'(0, a \mid \mathbf{d}, 1) - \ln\left(\frac{\Gamma_d(a)}{\Gamma_d(a + \beta)}\right) \;=\; & \ln\left(\frac{\Gamma_{d+1}(a)}{\rho_{d+1}(\mathbf{d}, 1)} \frac{\Gamma_d(a + \beta)}{\Gamma_d(a)}\right) \\
\;=\; & \ln\left(\frac{\rho_d(\mathbf{d})}{\rho_{d+1}(\mathbf{d}, 1)} \frac{\Gamma_d(a + \beta)}{\Gamma_{d+1}(a + 1)}\right).
\end{aligned}
$$

Numerically, the difficulty lies in the evaluation of the multiple Γ-functions and the Γ-modular forms. The definitions are,

$$\lim_{\epsilon \to 0} \zeta_r'(0, \epsilon \mid \mathbf{d}) = -\ln \epsilon - \ln \rho_r(\mathbf{d}), \qquad \zeta_r'(0, a \mid \mathbf{d}) = \ln\left(\frac{\Gamma_r(a)}{\rho_r(\mathbf{d})}\right).$$

120   J.S. Dowker

8   Note on the Barnes ζ–function

The original papers of Barnes contain a lot of information about this function, which can be looked upon as a multidimensional generalisation of the Hurwitz ζ–function. It forms a happy playground for those of us amused by special functions, recursion formulae, curious identities *etc.*

I will not attempt to summarize this material here. All I want to do is to point out that the Barnes ζ–function arises 'naturally' when considering the d-dimensional harmonic oscillator with operator

$$-\Delta_{\text{HO}} = -\nabla^2 + \sum_{i=1}^{d} \omega_i^2 x_i^2$$

whose eigenvalues are $\lambda = 2\mathbf{n}.\omega + \sum_i \omega_i$. The corresponding ζ–function is

$$\zeta_{\text{HO}}(s) = \sum_{\mathbf{n}=0}^{\infty} \frac{1}{(2\mathbf{n}.\omega + \sum \omega_i)^s} = \zeta_d(s, \textstyle\sum \omega_i \mid 2\omega).$$

The diagonal heat-kernel can also be written down immediately

$$K_{\text{HO}}(\mathbf{x}, \tau; \mathbf{x}, 0) = \prod_i \frac{\exp\left(-\omega_i x_i^2 \tanh \omega_i \tau\right)}{2\pi \sinh 2\omega_i \tau}.$$

Since a constant magnetic field is more or less mathematically equivalent to a multidimensional harmonic oscillator, the Barnes ζ–function will occur here also.

9   Conclusion

There is no conclusion because I am still pursuing the calculation, trying to make sense of some special cases, and I haven't even mentioned Neumann conditions or other fields like spin-1/2 and Maxwell.

The domains I have been engaged with form a discrete series and it would be nice to have some continuously varying parameter so that a graph or two could be drawn. It is possible to discuss a spherical wedge with an arbitrary opening angle (in which case one can have Robin conditions on the sides) and it may be possible to treat a standard spherical ice-cream cone, *i.e.* one whose surface domain, $\mathcal{N}$, is a *spherical* d-ball or cap. The Möbius cone can also be truncated at an *inner* radius giving a portion of a *shell*.

10   Added note

I have learnt at this conference that Weierstrass regularisation has been used by Wipf in an interesting discussion of tunnel determinants.

References

[1] E.W. Barnes *Trans. Camb. Phil. Soc.* **19** (1903) 374, 426.

[2] A.O. Barvinsky, Yu.A. Kamenshchik and I.P. Karmazin *Ann. Phys.* **219** (1992) 201 219

[3] M. Bordag, E. Elizalde and K. Kirsten *Heat kernel coefficients of the Laplace operator on the D-dimensional ball* UB-ECM-PF 95/3; hep-th/9503023.

[4] M. Bordag, B. Geyer, K. Kirsten and E. Elizalde *Zeta function determinant of the Laplace operator on the D-dimensional ball* UB-ECM-PF 95/10; hep-th /9505157.

[5] J. Cheeger *J. Diff. Geom.* **18** (1983) 575.

[6] J. Cheeger and M. Taylor *Comm. pure and Appl. Math.* **35** (1982) 275,487.

[7] P.D. D'Eath and G.V.M. Esposito *Phys. Rev.* **D43** (1991) 3234.

[8] P.D. D'Eath and G.V.M. Esposito *Phys. Rev.* **D44** (1991) 1713.

[9] L.A. Dikii *Usp. Mat. Nauk.* **13** (1958) 111.

[10] J.S. Dowker *Robin conditions on the Euclidean ball* MUTP/95/7; hep-th/9506042.

[11] J.S. Dowker and Peter Chang *Phys. Rev.* **D46** (1992) 3458.

[12] J.S. Dowker *Spin on the 4-ball* MUTP/95/13; hep-th/9508082.

[13] W. Heisenberg and H. Euler *Z. f. Phys.* **98** (1936) 714.

[14] J. Jorgenson and S. Lang Lect. Notes in Math. 1564 Springer-Verlag, Berlin 1993.

[15] Yu.A. Kamenshchik and I.V. Mishakov *Phys. Rev.* **D47** (1993) 1380.

[16] Yu.A. Kamenshchik and I.V. Mishakov *Int. J. Mod. Phys.* **A7** (1992) 3265.

[17] K. Kirsten and G. Cognola *Heat-kernel coefficients and functional determinants for higher spin fields on the ball* UTF354. hep-th/9508088.

[18] I.G. Moss *Class. Quant. Grav.* **6** (1989) 659.

[19] F. Pockels *Über die partielle Differentialgleichung* $\Delta u + k^2 u = 0$, B.G.Teubner, Leipzig 1891.

[20] J.R. Quine, S.H. Heydari and R.Y. Song *Trans. Am. Math. Soc.* **338** (1993) 213.

[21] J. Schwinger *Phys. Rev.* **93** (1953) 615.

[22] A. Voros *Comm. Math. Phys.* **110** (1987) 110.

[23] A. Wipf *Nucl. Phys.* **B269** (1986) 24; and earlier references here.

Applications of Zeta Function Regularization in QFT

Emilio Elizalde

Abstract

Zeta function regularization [1] is proving to be a quite powerful method [2, 3]. Uses in QFT of quite non-trivial formulas, as some extensions of the Chowla-Selberg equation [4] to the non-inhomogeneous case and to the case of a truncated spectrum have recently been obtained. They are here given, together with some applications, and accompanied by a number of references that will lead the reader through more detailed descriptions of this subject.

1 Extensions of the Chowla-Selberg Formula

We just quote here, without further comment for lack of space, the following generalizations of the Chowla-Selberg formula [4] that have been obtained in [5, 6], by using Jacobi's fundamental identity for the theta function θ_3,

$$\theta_3\left(z\,|\tau\right) = 1 + 2\sum_{n=1}^{\infty} q^{n^2}\cos(2nz), \qquad q = e^{\pi i\tau},\ \ |q| < 1,\ \ \tau \in \mathbf{C}, \tag{1}$$

that is

$$\theta_3\left(z\,|\tau\right) = \frac{1}{\sqrt{-i\tau}}e^{z^2/(\pi i\tau)}\theta_3\left(\frac{z}{\tau}\,\bigg|\,\frac{-1}{\tau}\right), \tag{2}$$

or,

$$\sum_{n=-\infty}^{+\infty} e^{n^2\pi i\tau + 2niz} = \frac{1}{\sqrt{-i\tau}}\sum_{n=-\infty}^{+\infty} e^{(z-n\pi)^2/(\pi i\tau)} \tag{3}$$

(for a popular reference see, for instance, Wittaker & Watson [7], p. 476). Here z and τ are arbitrary complex numbers, $z, \tau \in \mathbf{C}$, with the only restriction that Im $\tau > 0$ (in order that $|q| < 1$). For subsequent application, it turns out to be better to recast the Jacobi identity as follows (with $\pi i\tau \to -t$ and $z \to \pi z$):

$$\sum_{n=-\infty}^{+\infty} e^{-n^2 t + 2\pi i n z} = \sqrt{\frac{\pi}{t}}\sum_{n=-\infty}^{+\infty} e^{-\pi^2(n-z)^2/t}, \tag{4}$$

equivalently

$$\sum_{n=-\infty}^{+\infty} e^{-(n+z)^2 t} = \sqrt{\frac{\pi}{t}}\left[1 + \sum_{n=1}^{\infty} e^{-\pi^2 n^2/t}\cos(2\pi n z)\right], \tag{5}$$

where $z, t \in \mathbf{C}$, Re $t > 0$. Using it, we obtain the following expression (a consequence of the Jacobi and gamma function identity)

$$\sum_{n=-\infty}^{+\infty} \left[a(n+c)^2 + q \right]^{-s} = \sqrt{\frac{\pi}{a}} \frac{\Gamma(s-1/2)}{\Gamma(s)} q^{1/2-s} \tag{6}$$

$$+ \frac{4\pi^s}{\Gamma(s)} a^{-1/4-s/2} q^{1/4-s/2} \sum_{n=1}^{\infty} n^{s-1/2} \cos(2\pi nc) K_{s-1/2}(2\pi n \sqrt{q/a}),$$

where K_ν is the modified Bessel function of the second kind. This equation is very useful, since it provides the analytic continuation of what can be called inhomogeneous, generalized Epstein series in one index, in terms of a series that turns out to be very quickly convergent for extense ranges of values of the parameters, in particular, when Im $c < \sqrt{q/a}$, a case that frequently appears in the applications. An example of a result that can only be obtained by explicit, non-trivial interchange of the summation indices and subsequent contour integration —and that it is *not* a consequence of the Jacobi identity— is the following:

$$F(s; a, c; q) \equiv \sum_{n=0}^{\infty} \left[a(n+c)^2 + q \right]^{-s}$$

$$\sim \frac{q^{-s}}{\Gamma(s)} \sum_{m=0}^{\infty} \frac{(-1)^m \Gamma(m+s)}{m!} \left(\frac{q}{a} \right)^{-m} \zeta_H(-2m, c) \tag{7}$$

$$+ \sqrt{\frac{\pi}{a}} \frac{\Gamma(s-1/2)}{2\Gamma(s)} q^{1/2-s} + \frac{2\pi^s}{\Gamma(s)} a^{-1/4-s/2} q^{1/4-s/2}$$

$$\cdot \sum_{n=1}^{\infty} n^{s-1/2} \cos(2\pi nc) K_{s-1/2}(2\pi n \sqrt{q/a}).$$

This function is studied in detail in Ref. [6] (with numerical tables, plots, and a couple of explicit applications). Notice, however, that this is not a convergent series but an asymptotic one [8]. By observing that $\zeta_H(-2m, c) + \zeta_H(-2m, 1-c) = 0$, for $m \in \mathbf{N}$, and that $G(s; a, c; q) = F(s; a, c; q) + F(s; a, 1-c; q)$, it is very easy to obtain (6) as a particular case of (7).

Those formulas give explicit answers to many situations that remained unsolved in zeta-function regularization. Only the case $c = 0$ had been dealt with satisfactorily (through the corresponding expression obtained just putting $c = 0$ in (6)). As a particular case of Eq. (6), one has

$$\sum_{n=0}^{\infty} \left[a(n+1/2)^2 + q \right]^{-s} = \frac{1}{2} \sqrt{\frac{\pi}{a}} \frac{\Gamma(s-1/2)}{\Gamma(s)} q^{1/2-s} \tag{8}$$

$$+ \frac{2\pi^s}{\Gamma(s)} a^{-1/4-s/2} q^{1/4-s/2} \sum_{n=1}^{\infty} (-1)^n n^{s-1/2} K_{s-1/2}(2\pi n \sqrt{q/a}).$$

More involved multiple series of this kind, such as the diagonal, inhomogeneous, generalized Epstein (or Epstein-Hurwitz) multiple series

$$
E_k(s; a_1, \ldots, a_k; c_1, \ldots, c_k; c)
$$
$$
\equiv \sum_{n_1, \ldots, n_k \in \mathbf{Z}} \left[a_1(n_1 + c_1)^2 + \cdots + a_k(n_k + c_k)^2 + c \right]^{-s}, \tag{9}
$$

can be treated in a recurrent way, starting from (6). The general recurrence is

$$
E_k(s; a_1, \ldots, a_k; c_1, \ldots, c_k; c)
$$
$$
= \sqrt{\frac{\pi}{a_k}} \frac{\Gamma(s - 1/2)}{\Gamma(s)} E_{k-1}(s; a_1, \ldots, a_{k-1}; c_1, \ldots, c_{k-1}; c)
$$
$$
+ \frac{4\pi^s}{\Gamma(s)} a_k^{-s/2-1/4} \sum_{n_1, \ldots, n_{k-1} \in \mathbf{Z}} \left[\sum_{j=1}^{k-1} a_j(n_j + c_j)^2 + c \right]^{-s/2+1/4} \tag{10}
$$
$$
\cdot \sum_{n_k=1}^{\infty} n_k^{s-1/2} \cos(2\pi n_k c_k) K_{s-1/2} \left(\frac{2\pi n_k}{\sqrt{a_k}} \sqrt{\sum_{j=1}^{k-1} a_j(n_j + c_j)^2 + c} \right).
$$

This exact recurrence is very appropriate for numerical computation, since from the second term on the rhs, owing to the rapid exponential convergence of the Bessel function, only a few first terms need to be taken into account to achieve a good approximation. The recurrence is then implementable in any of the algebraic computational packages commonly available.

Now let us consider

$$
E(s; a, b, c; q) \equiv \sum_{m,n \in \mathbf{Z}}{}' (am^2 + bmn + cn^2 + q)^{-s}. \tag{11}
$$

With $q \neq 0$ (in general), the parenthesis in (11) is the inhomogeneous quadratic form

$$
Q(x, y) + q, \quad Q(x, y) \equiv ax^2 + bxy + cy^2, \tag{12}
$$

restricted to the integers. In the general theory that deals with the homogeneous case, one assumes that $a, c > 0$ and that the discriminant

$$
\Delta = 4ac - b^2 > 0 \tag{13}
$$

(see [4]). Here we will impose the additional condition that q be such that $Q(m, n) + q \neq 0$, $\forall m, n \in \mathbf{Z}$. In the usual applications of the theory, those conditions are indeed satisfied. After some rather non-trivial manipulations, the following expression has been obtained [5]

$$
E(s; a, b, c; q) = 2\zeta_{EH}(s, q/a)\, a^{-s} + \frac{2^{2s}\sqrt{\pi}\, a^{s-1}}{\Gamma(s)\Delta^{s-1/2}} \Gamma(s - 1/2)\zeta_{EH}(s - 1/2, 4aq/\Delta)
$$

$$+\frac{2^{s+5/2}\pi^s}{\Gamma(s)\sqrt{a}} \sum_{n=1}^{\infty} n^{s-1/2} \cos(n\pi b/a) \sum_{d|n} d^{1-2s} \left(\Delta + \frac{4aq}{d^2}\right)^{-s/2+1/4} \tag{14}$$

$$\cdot K_{s-1/2}\left(\frac{\pi n}{a}\sqrt{\Delta + \frac{4aq}{d^2}}\right),$$

where $d|n$ means sum over the divisors d of n, and where the function $\zeta_{EH}(s,p)$ (the one dimensional Epstein-Hurwitz or inhomogeneous Epstein series) is given by

$$\zeta_{EH}(s;p) = \sum_{n=1}^{\infty} \left(n^2 + p\right)^{-s} = \frac{1}{2}\sum_{n\in\mathbf{Z}}{}' \left(n^2 + p\right)^{-s} \tag{15}$$

$$= -\frac{p^{-s}}{2} + \frac{\sqrt{\pi}\,\Gamma(s-1/2)}{2\,\Gamma(s)} p^{-s+1/2} + \frac{2\pi^s p^{-s/2+1/4}}{\Gamma(s)} \sum_{n=1}^{\infty} n^{s-1/2} K_{s-1/2}(2\pi n\sqrt{p}),$$

as a particular case of Eq. (6) (or (7)).

Eq. (14) provides the analytical continuation of the inhomogeneous Epstein series, in the variable s, as a meromorphic function in the complex plane. Its pole structure is explicitly given in terms of the well-known pole structure of $\zeta_{EH}(s,p)$ (see, for instance, [6]). It contains the Chowla-Selberg formula as the particular case $q = 0$, i.e.,

$$E(s; a, b, c; 0) = 2\zeta(2s)\, a^{-s} + \frac{2^{2s}\sqrt{\pi}\, a^{s-1}}{\Gamma(s)\Delta^{s-1/2}} \Gamma(s-1/2)\zeta(2s-1) + \frac{2^{s+3/2}\pi^s}{\Gamma(s)\Delta^{s/2-1/4}\sqrt{a}}$$

$$\cdot \sum_{n=1}^{\infty} n^{s-1/2}\sigma_{1-2s}(n) \cos(n\pi b/a) \int_0^{\infty} dt\, t^{s-3/2} \exp\left[-\frac{\pi n\sqrt{\Delta}}{2a}(t + t^{-1})\right], \tag{16}$$

where

$$\sigma_s(n) \equiv \sum_{d|n} d^s, \tag{17}$$

namely the sum over the s-powers of the divisors of n. (Notice that there is an error in the transcription of formula (16) in Ref. [9]).

The good convergence properties of expression (16), that were so much prised by Chowla and Selberg, are shared by its non-trivial extension (14). This renders the use of the formula quite simple. In fact, the two first summands are still rather nice —under the form (15)— while the last one (impressive in appearance) is even more quickly convergent than in the case of Eq. (16), and thus absolutely harmless in practice. Only a few first terms of the three series of Bessel functions in (14), (15) need to be calculated, even if one demands good accuracy. We should also notice that the pole of (16) at $s = 1$ appears through $\zeta(2s - 1)$ in the second term, while for $s = 1/2$, the apparent singularities of the first and second cancel each other and no pole is formed. Analogously, the pole at $s = 1/2$ in (14) appears only from the first term. It is remarkable that Eq. (14) also has these good properties, *for any non-negative value of q*. In fact, for large q the convergence properties of the series of Bessel functions are clearly enhanced, while for q small we get back to the case of Chowla and Selberg. Notice, however, that this is not obtained through the high-q expansion (6), but using the low-q, binomial expansion

$$\sum_{n=0}^{\infty} \left[a(n+c)^2 + q\right]^{-s} = a^{-s} \sum_{m=0}^{\infty} \frac{(-1)^m \Gamma(m+s)}{\Gamma(s)\, m!} \left(\frac{q}{a}\right)^m \zeta_H(2s+2m, c), \qquad (18)$$

which is convergent for $q/a \leq 1$. For $q \to 0$ it reduces to $a^{-s}\zeta_H(2s, c)$. Actually, formula (14) is still valid in a domain of negative q's, namely for $q > -\min(a, c, a - b + c)$.

2   Applications

Let us consider the effective potential of a quantum theory that has been studied recently, the phase-transition structure of which we want to analyze, in terms of the dimension D of the space where we are working. D is one of the variables of the theory, ranging from $D = 2$ to $D = 3$ (see, for instance, [10]). For what concerns us here, we just need to know that the effective potential is given by the expression

$$V^{\beta\mu}(\sigma) = \frac{\sigma^2}{2\lambda} - \frac{\sqrt{\pi}}{\beta\,(2\pi)^{D/2}} \Gamma\left(\frac{1-D}{2}\right) \qquad (19)$$

$$\cdot \sum_{n=-\infty}^{+\infty} \left\{ \left[\left(\frac{2n+1}{\beta}\pi - i\mu\right)^2\right]^{(D-1)/2} + \left[\left(\frac{2n+1}{\beta}\pi - i\mu\right)^2 + \sigma^2\right]^{(D-1)/2} \right\}.$$

By direct application of the first of the equations considered above, Eq. (6), we obtain

$$\sum_{n=-\infty}^{+\infty} \left[\left(\frac{2n+1}{\beta}\pi - i\mu\right)^2 + \sigma^2\right]^{(D-1)/2} = \frac{\beta}{2\sqrt{\pi}} \frac{\Gamma(-D/2)}{\Gamma((1-D)/2)} \sigma^D$$

$$+ \frac{\beta^{1-D/2}(2\sigma)^{D/2}}{\sqrt{\pi}\,\Gamma((1-D)/2)} \sum_{n=1}^{\infty} (-1)^n n^{-D/2} \left(e^{\beta\mu n} + e^{-\beta\mu n}\right) K_{-D/2}(\beta\sigma n), \qquad (20)$$

and

$$\sum_{n=-\infty}^{+\infty} \left[\left(\frac{2n+1}{\beta}\pi - i\mu\right)^2\right]^{(D-1)/2} \qquad (21)$$

$$= \left(\frac{2\pi}{\beta}\right)^{D-1} \left[\zeta\left(1 - D, \frac{1}{2} - i\frac{\beta\mu}{2\pi}\right) + \zeta\left(1 - D, -\frac{1}{2} + i\frac{\beta\mu}{2\pi}\right) - \left(\frac{1}{2} - i\frac{\beta\mu}{2\pi}\right)^{D-1}\right].$$

The effective potential becomes

$$V^{\beta\mu}(\sigma) = \frac{\sigma^2}{2\lambda} - \frac{\sqrt{\pi}}{\beta\,(2\pi)^{D/2}} \left\{ \Gamma\left(\frac{1-D}{2}\right) \left(\frac{2\pi}{\beta}\right)^{D-1} \left[\zeta\left(1-D, \frac{1}{2} - i\frac{\beta\mu}{2\pi}\right) \right.\right.$$

$$\left.+ \zeta\left(1-D, -\frac{1}{2} + i\frac{\beta\mu}{2\pi}\right) - \left(\frac{1}{2} - i\frac{\beta\mu}{2\pi}\right)^{D-1} \right] + \frac{\beta}{2\sqrt{\pi}}\Gamma\left(\frac{-D}{2}\right)\sigma^D$$

$$+ \frac{\beta^{1-D/2}(2\sigma)^{D/2}}{\sqrt{\pi}} \sum_{n=1}^{\infty} (-1)^n n^{-D/2} \left(e^{\beta\mu n} + e^{-\beta\mu n}\right) K_{D/2}(\beta\sigma n) \right\}. \tag{22}$$

The equation for the extrema of the potential leads us to the critical points (phase transitions), i.e.

$$\frac{\partial V^{\beta\mu}(\sigma)}{\partial\sigma} = 0. \tag{23}$$

This yields

$$\frac{\sigma}{\lambda} - (2\pi)^{-D/2}\left[\frac{1}{2}\Gamma\left(\frac{-D}{2}\right) D\sigma^{D-1} + \beta^{-D/2}D(2\sigma)^{D/2-1}\right.$$

$$\sum_{n=1}^{\infty}(-1)^n n^{-D/2}\left(e^{\beta\mu n} + e^{-\beta\mu n}\right) K_{D/2}(\beta\sigma n) \tag{24}$$

$$\left. + \beta^{1-D/2}(2\sigma)^{D/2}\sum_{n=1}^{\infty}(-1)^n n^{1-D/2}\left(e^{\beta\mu n} + e^{-\beta\mu n}\right) K'_{D/2}(\beta\sigma n)\right] = 0.$$

The series are convergent when $|\mu| \leq \sigma$, for any D. In the first approximation, we have just

$$\frac{\sigma}{\lambda} - \frac{\Gamma\left(\frac{-D}{2}\right)D}{2\,(2\pi)^{D/2}}\sigma^{D-1} = 0, \tag{25}$$

that is

$$\sigma_0 = \left[-\frac{2\,(2\pi)^{D/2}\Gamma(D/2+1)\sin(D\pi/2)}{\pi\,D\,\lambda}\right]^{1/(D-2)}. \tag{26}$$

In the particular case when the number of dimensions is 5/2 (an intermediate situation), we obtain

$$\sigma_0|_{D=5/2} = \frac{\Gamma(1/4)^2\sqrt{\pi}}{2^{5/2}\lambda^2} = \frac{4.1187}{\lambda^2}. \tag{27}$$

The self-consistency condition is $\sigma_0 \geq |\mu|$, namely,

$$|\mu| \leq \left[-\frac{2\,(2\pi)^{D/2}\Gamma(D/2+1)\sin(D\pi/2)}{\pi\,D\,\lambda}\right]^{1/(D-2)}. \tag{28}$$

(Notice that, for $D = 5/2$, this means $|\mu| < 4.12$ for $\lambda = 1$).

The second approximation yields, with $\sigma = \sigma_0 + \sigma_1$

$$\frac{\sigma_1}{\lambda} - \frac{D(D-1)\Gamma(-D/2)}{2(2\pi)^{D/2}}\sigma_0^{D-2}\sigma_1 + \frac{\sigma_0^{(D-3)/2}}{\sqrt{2}(\pi\beta)^{(D-1)/2}}\left(\frac{D}{2\beta} - \sigma_0\right)e^{-\beta(\sigma_0-|\mu|)} \simeq 0. \quad (29)$$

The consistency check to second order is $\sigma_1 << \sigma_0$. We have

$$\frac{\sigma_1}{\sigma_0} \simeq \frac{\lambda}{(2-D)\sqrt{2}}\left(1 - \frac{D}{2\beta\sigma_0}\right)\left(\frac{\sigma_0}{\pi\beta}\right)^{(D-1)/2}\frac{1}{\sigma_0}e^{-\beta(\sigma_0-|\mu|)} << 1. \quad (30)$$

This is satisfied at high temperature T, if $\lambda \sim \beta = (kT)^{-1}$. In particular, the exponential is then $\sim \exp(-2kT)$, which is very good for the quick convergence of the series of Bessel functions. For the particular case $D = 5/2$, substituting σ_0 into the last expression, we get

$$\begin{aligned}
\left.\frac{\sigma_1}{\sigma_0}\right|_{D=5/2} &\simeq -\left(1 - \frac{5\sqrt{2/\pi}\lambda^2}{\Gamma(1/4)^2\beta}\right)\frac{\lambda^{3/2}}{(2\pi)^{7/8}\sqrt{\Gamma(1/4)}\beta^{3/4}}e^{-\beta(\sigma_0-|\mu|)} \\
&= -0.1052\left(1 - 0.3035\frac{\lambda^2}{\beta}\right)\frac{\lambda^{3/2}}{\beta^{3/4}}e^{-\beta(4.1187/\lambda^2-|\mu|)}. \quad (31)
\end{aligned}$$

Summing up, we obtain in this way the explicit dependence of the critical points on the parameters of the theory.

Let us consider, as a second example, a (1+2)-dimensional spacetime with the topology $\mathbf{R}\times \mathbf{T}^2$ [11], with the geometry of the space $\Sigma \simeq \mathbf{T}^2$ locally flat. A flat 2-geometry is endowed on Σ by giving it a metric $ds^2 = h_{ab}d\xi^a d\xi^b$, where

$$h_{ab} = \frac{1}{\tau_2}\begin{pmatrix} 1 & \tau_1 \\ \tau_1 & |\tau|^2 \end{pmatrix}, \quad (32)$$

with (τ_1, τ_2) the Teichmüller parameters, independent of the spatial coordinates (ξ_1, ξ_2), and $\tau = \tau_1 + i\tau_2$, $\tau_2 > 0$ [12, 13]. The Laplace-Beltrami operator of this metric is

$$\Delta = -\frac{1}{\tau_2}(|\tau|^2\partial_1^2 - 2\tau_1\partial_1\partial_2 + \partial_2^2), \quad (33)$$

and has the eigenvalues

$$\lambda_{n_1,n_2} = \frac{4\pi^2}{\tau_2}(|\tau|^2 n_1^2 - 2\tau_1 n_1 n_2 + n_2^2). \quad (34)$$

In the massive case, $m \neq 0$ the spectrum runs over $n_1, n_2 \in \mathbf{Z}$. In the massless case the zero-mode of Δ, $n_1 = n_2 = 0$, has to be excluded. We see that the zeta function corresponding to the Laplace-Beltrami operator in the massive (resp. massless) case is

$$\zeta_{\Delta+m^2}(s) \;=\; m^{-2s} + \left(\frac{4\pi^2}{\tau_2}\right)^{-s} E\left(s; |\tau|^2, -2\tau_1, 1; \frac{\tau_2 m^2}{4\pi^2}\right), \tag{35}$$

$$\zeta_{\Delta}(s) \;=\; \left(\frac{4\pi^2}{\tau_2}\right)^{-s} E\left(s; |\tau|^2, -2\tau_1, 1; 0\right), \tag{36}$$

respectively. The values at $s = -1/2$ are finite and define the corresponding Casimir energy [14]. After performing the necessary calculations, we obtain as result the quite simple expressions

$$\begin{aligned}
\zeta_{\Delta+m^2}(-1/2) \;=\;& -\frac{m^3}{6\pi} - \frac{2m}{\pi} \sum_{n=1}^{\infty} n^{-1} K_1(nmx_2) \\
& -\sqrt{2}\left(\frac{mx_2}{\pi}\right)^{3/2} \sum_{n=1}^{\infty} n^{-3/2} K_{3/2}(nm/x_2)
\end{aligned} \tag{37}$$

$$-8x_2 \sum_{n=1}^{\infty} n^{-1} \cos(2\pi n x_1^2) \sum_{d|n} d^2 \sqrt{1+\frac{m^2}{(2\pi x_2 d)^2}}\; K_1\left(2\pi n x_2^2 \sqrt{1+\frac{m^2}{(2\pi x_2 d)^2}}\right),$$

and

$$\zeta_{\Delta}(-1/2) = -\frac{\pi}{3x_2} + 4\pi\zeta'(-2)x_2^3 - 8x_2 \sum_{n=1}^{\infty} n^{-1} \cos(2\pi n x_1^2)\sigma_2(n)\, K_1\left(2\pi n x_2^2\right), \tag{38}$$

with

$$x_1 = \sqrt{\frac{\tau_1}{\tau_1^2+\tau_2^2}}, \qquad x_2 = \sqrt{\frac{\tau_2}{\tau_1^2+\tau_2^2}}. \tag{39}$$

The extrema of the corresponding Casimir energy [14], both for the massive and for the massless case, in terms of the original Teichmüller coefficients τ_1 and τ_2, are to be read from these equations. In a three-dimensional plot over the plane τ_1, τ_2, the maximal Casimir energy is seen to be localized on the section $\tau_2 = 1$. On any section $\tau_2 =$ const., a periodic structure is seen to appear, associated with the value of τ_1 along the section. This behavior is easy to recognize from the form of the function $\zeta_\Delta(-1/2)$, (38), and is common to *any* section $\tau_2 =$ const [15]. A plot of the dependence of the Casimir energy in terms of the mass of the field, for fixed values of the Teichmüller parameters, shows the remarkable existence of a preferred non-zero mass in the sense that the Casimir energy density corresponding to this value of the mass has a local maximum. This will be reported in more detail elsewhere.

This work has been supported by DGICYT (Spain) and by CIRIT (Generalitat de Catalunya).

References

[1] S.W. Hawking, Commun. Math Phys. **55**, 133 (1977); J.S. Dowker and R. Critchley, Phys. Rev. **D13**, 3224 (1976); L.S. Brown and G.J. MacLay, Phys. Rev. **184**, 1272 (1969).

[2] E. Elizalde, S.D. Odintsov, A. Romeo, A.A. Bytsenko and S. Zerbini, *Zeta regularization techniques with applications* (World Sci., Singapore, 1994).

[3] E. Elizalde, *Application in cosmology of the zeta function procedure*, Proceedings of the Third Alexander Friedmann International Seminar on Gravitation and Cosmology, Friedmann Laboratory Pub. Ltd. (Saint Petersburg, 1995); *On two complementary approaches aiming at the definition of the determinant of an elliptic partial differential operator*, Trento Univ. preprint UTF 359, hep-th/9508167 (1995).

[4] S. Chowla and A. Selberg, Proc. Nat. Acad. Sci. US **35**, 317 (1949).

[5] E. Elizalde, *Explicit analytical continuation of the inhomogeneous epstein zeta function*, Hiroshima Univ. preprint (1994).

[6] E. Elizalde, J. Math. Phys. **35**, 6100 and 3308 (1994).

[7] E.T. Wittaker and G.N. Watson, *A course of modern analysis* (Cambridge University Press, Cambridge, 4th Ed., 1965).

[8] E. Elizalde, J. Phys. **A22**, 931 (1989); E. Elizalde and A. Romeo, Phys. Rev. **D40**, 436 (1989); E. Elizalde, J. Math. Phys. **31**, 170 (1990).

[9] S. Iyanaga and Y. Kawada, Eds., *Encyclopedic dictionary of mathematics*, Vol. II (The MIT press, Cambridge, 1977), p. 1372 ff.

[10] T. Inagaki, T. Kouno and T. Muta, *Phase structure of four fermion theories at finite temperature and chemical potential in arbitrary dimensions*, Hiroshima preprint HUPD-9402, hep-ph/9409431 (1994).

[11] M. Seriu, *Back-reaction on the topological degrees of freedom in (1+2)-dimensional spacetime*, preprint (1995).

[12] B. Hatfield, *Quantum Field Theory of Point Particles and Strings* (Addison-Wesley, New York, 1992).

[13] D. Lüst and S. Theisen, *Lectures on String Theory* (Springer-Verlag, Berlin, 1989).

[14] H.B.G. Casimir, Proc. Kon. Ned. Akad. Wetenschap **B51**, 793 (1948); G. Plunien, B. Müller, and W. Greiner, Phys. Rep. **134**, 88 (1986); V.M. Mostepanenko and N.N. Trunov, *Casimir effect and its applications* (in russian), (Energoatomizdat, Moscow, 1990).

[15] K. Kirsten and E. Elizalde, *A precise definition of the Casimir energy*, Trento preprint U.F.T. 356, hep-th/9508086 (1995).

Ground State Energy in Smooth Background Fields

Michael Bordag

Abstract

We consider the backreaction problem for a smooth background field depending on one coordinate. We give a reformulation so that the scattering data of the corresponding Schrödinger equation become the independent variables. Using knowledge from the inverse scattering method it is shown that the problem can be reduced to an algebraic one within an arbitrary good approximation. For the simplest case an explicit example is given.

1   Introduction

The backreaction problem arises when considering some quantum field in the background of a classical field. The quantum field by its very existence even when there are no real particles (or excitations) present contributes its *groundstate energy* to the dynamics of the classical background field. So one can consider the gravitational field as classical background and matter fields as quantum fields. In this case one usually deals with the vacuum expectation value of the energy-momentum tensor in the right hand side of the Einstein equation and tries to take it into account as a source of gravitation. A different example is the quantization around a classical solution of the field equations, a Skyrmion for instance. In that case both, the classical field and its quantum fluctuations are of the same nature and the groundstate energy of the quantum fluctuations modifies the classical solution. One of the first examples of this kind were the quantum corrections to the mass of the kink in [1].

In general, the solution of the backreaction problem is a difficult task. Explicit solutions are known only in very special cases (e.g., a massless field with conformal coupling to a isotropic homogeneous gravitational field [2]) as well as several approximative approaches.

A different example is the Casimir effect. Here one considers the ground state energy of some quantum field in a background which is idealized by boundary conditions (the surface of a conductor, for instance). Usually this idealization is considered as an external influence and does not allow the study of the backreaction of the Casimir force on the surfaces. The attempt to do this nevertheless leads one to softened boundary conditions, delta-shaped potentials for instance [4]. However, these potential have an infinite selfenergy. The next step then is the consideration of smooth background fields. This has been done in [5] in the case of a background field depending on one coordinate only. There a formalism has been developed which allows one to express the ground state energy of a quantum field completely in terms of the scattering data of a related Schrödinger equation whose potential is given by the background field.

In the present paper we use that results and study the backreaction problem by expressing it in terms of the scattering data using knowledge from the inverse scattering problem. We show that in a dense subset of the scattering data explicit formulas are possible and solve the backreaction problem in the simplest case explicitly.

2 Ground State Energy Expressed in Terms of Scattering Data

We consider the model given by the Lagrange density

$$L = \frac{1}{2}\Phi\left(\Delta - M^2 - \lambda\Phi^2\right)\Phi + \frac{1}{2}\varphi\left(\Delta - m^2 - \lambda'\Phi^2\right)\varphi \,. \tag{1}$$

Here, Φ is a classical field serving as background field for the quantum field φ. The classical field is assumed to depend on one coordinate only: i.e., $\Phi(x_1)$. Because of this the problem is essentially a $(1+1)$ dimensional one but we prefer to write all formulas in the $(3+1)$ dimensional case having in mind later generalizations to the spherical symmetric case. The energy of the whole system is given by

$$E = \frac{1}{2}V_g + \frac{1}{2}M^2V_1 + \lambda V_2 + V_{\text{eff}} \,, \tag{2}$$

where $V_g \equiv \int_{-\infty}^{\infty} dx_1 \left(\frac{\partial}{\partial x_1}\Phi(x_1)\right)^2$, $V_1 \equiv \int_{-\infty}^{\infty} dx_1\,(\Phi(x_1))^2$ and $V_2 \equiv \int_{-\infty}^{\infty} dx_1\,(\Phi(x_1))^4$ are the contributions resulting from the classical background field and

$$V_{\text{eff}} = \frac{1}{2}\frac{\partial}{\partial s}\text{Tr}(\Delta - m^2 - V(x_1))^{-s}\mu^{2s}\Big|_{s=0} \tag{3}$$

is the ground state energy (or effective potential) of the quantum field φ. Here, we used the zeta functional regularization, μ is an arbitrary parameter with the dimension of a mass.

The scattering problem associated with the background field $\Phi(x)$ is given by the equation

$$\left(-\frac{\partial^2}{\partial x^2} + V(x)\right)\psi(x) = k^2\,\psi(x) \,, \tag{4}$$

which can be considered as a Schrödinger equation with the potential

$$V(x) \equiv \lambda'\,\Phi(x)^2 \,.$$

Under the assumption of a sufficiently fast decreasing of the potential $V(x)$ there exist scattering solutions ψ_1 and ψ_2

$$\psi_1 \underset{x\to-\infty}{\sim} e^{ikx} + s_{12}\,e^{-ikx} \,, \qquad \psi_1 \underset{x\to\infty}{\sim} s_{11}\,e^{ikx} \,,$$

$$\psi_2 \underset{x\to-\infty}{\sim} s_{22}\,e^{-ikx} \,, \qquad \psi_2 \underset{x\to\infty}{\sim} s_{21}\,e^{ikx} + e^{-ikx} \,.$$

which define the one dimensional **S**-matrix $\mathbf{S} = (s_{ij})$ [1]. As it was shown in [5] the effective potential can be expressed in terms of the coefficient s_{11} of the scattering matrix by

$$V_{\text{eff}}^{\text{sub}} = \frac{1}{12\pi^2} \int_m^\infty dk \, (k^2 - m^2)^{3/2} \frac{\partial}{\partial k} \left[\log s_{11}(ik) + \frac{\lambda' V_1}{2k} - \frac{\lambda'^2 V_2}{8k^3} \right]. \tag{5}$$

Here, the renormalization is already carried out by means of a redefinition of the parameters M^2 and λ in the classical part of the Lagrange density. Let us remark that the effective potential $V_{\text{eff}}^{\text{sub}}$ (3) (although it finite by itself [2]) needs a renormalization at last in order to handle the ambiguity resulting from μ. Note, that this renormalization is a finite one. In a different formulation, using a different regularization or the ground state energy instead of the effective potential a infinite renormalization is required yielding the same final answer.

By means of formulas (2) and (5) the complete energy of the system is given in terms of the background field Φ. In order to calculate E for a given Φ one has to calculate V_g, V_1 , V_2 simple by integration and to solve the scattering problem associated with the background field by means of the Schrödinger equation (4) to obtain the scattering coefficient $s_{11}(ik)$ which allows to calculate $V_{\text{eff}}^{\text{sub}}$ (3) simple by integration.

3 Expression Through Scattering Data

The scattering data associated with the Schrödinger equation (4) is the set

$$\{R(q), \, \beta_n, \, \kappa_n \}, \tag{6}$$

where $R(q)$ is the reflection coefficient [3] (q real, $0 \leq q \leq \infty$), the binding energies κ_n of the bound states and the β_n are numbers associated with the normalization of the bound state wave functions. The label $n = 1, \ldots, N$ counts the bound states. It is finite for a sufficiently fast decreasing potential $V(x)$.

Using the dispersion relation

$$\log s_{11}(ik) = \frac{k}{2\pi} \int_0^\infty \frac{\log \left(1 - R(q)\right)}{q^2 + k^2} \, dq + \sum_{n=1}^N \log \frac{k + \kappa_n}{k - \kappa_n} \tag{7}$$

and the relations [6]

$$\lambda' V_1 \equiv \int_{-\infty}^\infty V(x) dx = \frac{-1}{\pi} \int_{-\infty}^\infty \log(1 - R(q)) \, dq - 4 \sum_{n=1}^N \kappa_n, \tag{8}$$

---

[1] We refer to the book of Chadan and Sabatier [3] as the standard monograph concerning the inverse scattering problem. It contains a concise representation of the one dimensional scattering problem too as well as the references to the original papers, especially to the papers by Faddeev and others.

[2] This is because $V_{\text{eff}}^{\text{sub}}$ (5) is in fact a zeta function.

[3] It is connected with the scattering coefficient by the relation $R(q) = 1 - |s_{11}(q)|^2$.

$$\lambda'^2 V_2 \equiv \int_{-\infty}^{\infty} V(x)^2 \mathrm{d}x \;=\; \frac{-4}{\pi} \int_{-\infty}^{\infty} q^2 \log(1 - R(q)) \, \mathrm{d}q + \frac{16}{3} \sum_{n=1}^{N} \kappa_n^3 \,,$$

which express the integrals over the background field which appear in the classical part of the energy by the scattering data we obtain the representation

$$V_{\mathrm{eff}}^{sub} = \tag{9}$$

$$\frac{1}{16\pi^3} \int_0^{\infty} \mathrm{d}q \; q \; \left\{ -2q + \sqrt{q^2 + m^2} \log \frac{\sqrt{q^2 + m^2} + q}{\sqrt{q^2 + m^2} - q} \right\} \; \log \frac{1}{1 - R(q)}$$

$$- \frac{1}{6\pi^2} \sum_{n=1}^{N} \left[(m^2 - \kappa_n^2)^{3/2} \arcsin \frac{\kappa_n}{m} + \frac{4}{3} \kappa_n^3 - \kappa_n m^2 \right] \,,$$

which is a expression of V_{eff}^{sub} in terms of the scattering data.

Now, in order to calculate the complete energy for a given set of scattering data (6), besides V_{eff}^{sub} which is explicit by means of (9), one has to restore the background field Φ. It enters the classical part of the energy through V_1, V_2 which by means of (9) are explicit and through V_g which cannot be expressed by formulas similar to (9). In order to calculate V_g, the inverse scattering problem has to be solved. Omitting a number of details it looks as follows [3]. One has to solve the Gelfand-Levitan-Marchenko equation

$$K(x, y) + M(x, y) + \int_x^{\infty} K(x, z) M(z + y) \, \mathrm{d}y = 0 \tag{10}$$

- a integral equation of Volterra type. Its kernel

$$M(x) = \int_{-\infty}^{\infty} \frac{\mathrm{d}q}{2\pi} R(q) \mathrm{e}^{iqx} + \sum_{n=1}^{N} \beta_n \mathrm{e}^{-\kappa_n x} \tag{11}$$

is determined by the scattering data. Then the restored potential is given by the formula

$$V(x) = -2 \frac{\mathrm{d}^2}{\mathrm{d}x^2} \log \det K(x, x) \tag{12}$$

and the background field correspondingly by $\Phi(x) = \sqrt{V(x)/\lambda'}$. The existence and the uniqueness of the solution of (10) are well known.

By these formulas the backreaction problem is reexpressed in the sense that now the scattering data can be viewed as the independent parameters instead of the initial background field Φ which have to be varied in order to find a minimum of the complete energy (2).

Now, there are two cases when this program can be carried out explicitly. The first one is the case of a reflectionless potential given by a vanishing reflection coefficient

$R(q) = 0$ for all q. In this case the Gelfand-Levitan-Marchenko equation (10) can be solved explicitly yielding

$$V(x) = -2\frac{\mathrm{d}^2}{\mathrm{d}x^2}\log\det\mathbf{A} \tag{13}$$

where the $N \times N$ matrix $\mathbf{A} = (A_{nm})$ is given by

$$\mathbf{A}_{nm} = \delta_{nm} + \frac{\beta_n}{\kappa_n + \kappa_m}\mathrm{e}^{-2\kappa_n x}. \tag{14}$$

An explicit example for $N = 1$ is given in the next section.

A second, more general case possessing an explicit solution arises in the case of a rational reflection coefficient

$$R(q) = \frac{P_m(q)}{Q_n(q)}, \tag{15}$$

where $P_m(q)$ and $Q_n(q)$ are polynomials of order $m \leq n - 2$. They have to satisfy certain algebraic relations in order to ensure the unitarity of the corresponding S-matrix. Now, the Faddeev-Marchenko equation (10) can be solved explicitly again and the potential $V(x)$ can be expressed by formulas similar to that in the case of a reflectionless potential.

Let us remark that the set of rational scattering coefficients is dense in the set of all scattering data [3]. This allows the conclusion that for an approximate calculation of the backreaction problem with any given precision it is sufficient to use the above formulas.

4 An Explicit Example for $N = 1$

We consider a reflectionless potential with one bound state, i.e., we consider the scattering data

$$\{0, \beta, \kappa\},$$

where κ is the binding energy. In this case the matrix A_{nm} (14) is simply a number

$$\mathbf{A} = 1 + \frac{\beta}{2\kappa}\mathrm{e}^{-2\kappa x}$$

and the potential by means of (13) reads

$$V(x) = \frac{-4\kappa\beta\mathrm{e}^{-2\kappa x}}{(\beta/2\kappa + \exp(2\kappa x))^2}$$

- a special case of the known Bargmann potentials. In this simple case the kinetic energy V_g of the background field can be calculated explicitly:

$$V_g = \frac{16}{3}\kappa^3.$$

Finally, taking all contributions together, the complete energy reads

$$E = \frac{4}{3}\frac{\kappa^3}{|\lambda'|} + \frac{2M^2}{|\lambda'|}\kappa + \frac{16\lambda}{3\lambda'}\kappa^3 - \frac{1}{6\pi^2}\left((m^2 - \kappa^2)^{3/2}\arcsin\frac{\kappa}{m} - m\kappa^2 + \frac{4}{3}\kappa^3\right). \quad (16)$$

This formula represents the complete energy through the characteristics of the model – the mass M of the background field Φ, the mass m of the quantum field φ and the couplings λ and λ' – and through the boundstate energy κ (the dependence on β, the other parameter from the scattering data, dropped out in this simple case).

Let us discuss this result in some detail. The parameter κ as being the binding energy cannot be negative by its definition (when viewing it as a pole in the scattering amplitude it can be well in the other half of the momentum plane; this is then a special case of the rational reflection coefficient below). Further, for $\kappa \geq m$ the ground state energy acquires a imaginary part. In that case the background field is strong enough to produce particles and the vacuum state is not stable. In that case the energy is given by

$$E = \frac{4}{3}\frac{\kappa^3}{|\lambda'|} + \frac{2M^2}{|\lambda'|}\kappa + \frac{16\lambda}{3\lambda'}\kappa^3 + \quad (17)$$
$$\frac{1}{6\pi^2}\left((\kappa^2 - m^2)^{3/2}\log\left(\frac{\kappa}{m} + \sqrt{\frac{\kappa^2}{m^2} - 1}\right) + m^2\kappa + \frac{4}{3}\kappa^3\right) -$$
$$\frac{i}{12\pi}(\kappa^2 - m^2)^{3/2}.$$

The energy is plotted in Fig. 1 as a function of κ. There are four cases which are essentially distinguished by the strength λ' of the coupling of the quantized field φ to the background field Φ. For $\lambda' \approx 1$ the quantum corrections are small and do not change the classical energy essentially. This case is shown in the first picture in Fig.1. For moderate κ this is the classical energy of Φ, for large κ it is modified by the logarithmic contribution. But this region corresponds to a very strong external field. For a stronger coupling, $\lambda' = 500$ as shown in the second picture for example, already for $\kappa \approx m$ the influence of the quantum corrections will be essential. For stronger couplings a local minimum of the energy appears and for more stronger ($\lambda' = 1000$ as shown in the last picture) this minimum becomes global and would correspond to a phase transition. However, in the considered model, the local as well as the global minima are at $\kappa > m$, i.e., they are in the region where particle production takes place, so that they do not describe a stable state of the considered model.

## 5	Conclusions

We reformulated the backreaction problem for a simple model (1) with a background field depending on one coordinate only in a way that the scattering data (6) can be varied as independent parameters while scanning for a minimum of the complete

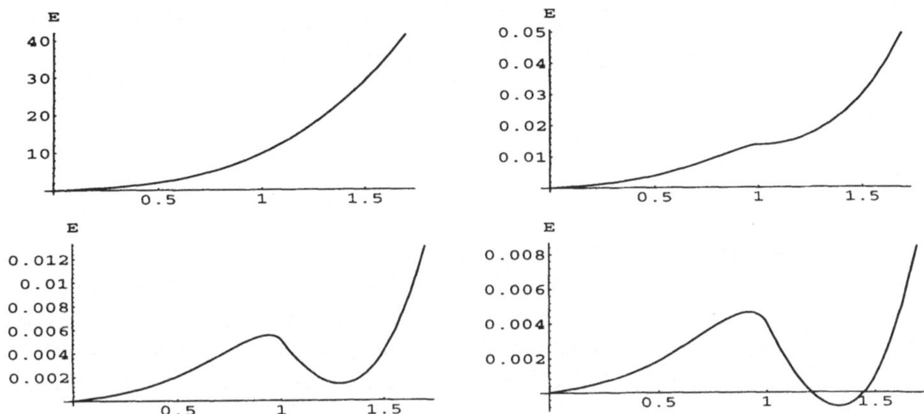

Fig. 1   The complete energy as function of the bound state energy κ for different
values of λ', all other parameters being set to unity

energy of the system consisting of the classical energy of the background field Φ and
the ground state energy of the quantum field φ. In this formulation the solution seems
to be easier by two reasons. In the general case on has to solve the inverse scattering
problem which is given by a well behaved integral equation. On the other hand side,
in the more special case of rational reflection coefficients (which form a dense subset
in the set of all scattering data) an explicit solution of the integral equation is known
and one is left with algebraic expressions and simple integrations. In the simplest case
of scattering data specified by one parameter this is done explicitly. Note that in the
considered example an essential influence of the quantum field is reached only at a
coupling with a remarkably high value ($\lambda' \sim 10^3$) of purely numerical origin.

References

[1] R.F. Dashen, B. Hasslacher, A. Neveu, Phys. Rev. D **10** (1974) 4130.

[2] S.G. Mamaev, V.M. Mostepanenko, *Sov. Phys. - JETP (USA)* **51** (1980) 9; A.A. Starobinsky, Phys. Lett., **91B** (1980) 99.

[3] K. Chadan, P.C. Sabatier, *Inverse Problems in Quantum Scattering Theory*, Springer-Verlag, 2nd. ed., 1989.

[4] M. Bordag, D. Hennig, D. Robaschik, J. Phys. A: Math. Gen. **25** (1992) 4483.

[5] M. Bordag, J. Phys. A: Math. Gen. **28** (1995) 755.

[6] V.E. Zakharov, L.D. Faddeev, Functional Analysis Appl. **5** (1972) 280.

Quantum Field Theory at Finite Temperature and Cosmological Perturbations

Anton K. Rebhan[1]

Abstract

It is shown how quantum field theory at finite temperature can be used to set up self-consistent and gauge invariant equations for cosmological perturbations sustained by an ultrarelativistic plasma. While in the collisionless case, the results are equivalent to those obtained from the Einstein-Vlasov equations, weak self-interactions in the plasma turn out to require the full machinery of perturbative thermal field theories such as resummation of hard thermal loops. Nevertheless it is still possible to use the same methods that yielded exact solutions in the collisionless case.

In order to account for the present large-scale structure of the universe such as galaxies, clusters, superclusters, voids, etc., [1] a cosmological model built on homogeneous and isotropic geometries requires certain imperfections in its symmetries. Through the universally attractive nature of gravitations, initially small perturbations can grow, in particular once that the universe becomes matter dominated and the (maximal) pressure provided by radiation has become inoperative. This picture, which is based on a big-bang scenario, has found dramatic support by the search for and discovery of tiny anisotropies in the cosmic micro-wave background, which in a Friedmann-Robertson-Walker (FRW) universe have wavelengths far exceeding the size of the Hubble horizon at the time when this radiation decoupled from the primordial matter [2].

Whatever the origin of these small deviations from homogeneity and isotropy, there is a rather long epoch of radiation domination which is thought to be well described by a nearly perfect FRW model with metric perturbations evolving in a linear regime. The basic equations for these "cosmological perturbations" are nothing else than the perturbed Einstein equations,

$$\delta G^{\mu\nu} \equiv \frac{\delta(R^{\mu\nu} - \frac{1}{2}g^{\mu\nu}R)}{\delta g^{\alpha\beta}}\delta g^{\alpha\beta} = -8\pi G\,\delta T^{\mu\nu}. \tag{1}$$

In order to have a close set of equations, these have to be supplied with information on the response $\delta T^{\mu\nu}$ of the energy-momentum tensor to metric perturbations

[1]On leave of absence from Institut für Theoretische Physik der Technischen Universität Wien, Wiedner Hauptstr. 8–10, A-1040 Wien, Austria; email: rebhana@email.tuwien.ac.at

$\delta g^{\alpha\beta}$. In a hydrodynamic approach, this is done by sufficiently restricting the form of $\delta T^{\mu\nu}$, specifying the equations of state, and imposing covariant conversation of the full energy-momentum tensor in the perturbed geometry. The simplest case is the one of a perfect (radiation) fluid, which has been studied in the pioneering work of Lifshitz [3]. Many generalizations have since been worked out, and have been cast into a gauge invariant form by Bardeen [4]. A modern geometrical justification and generalization has been given recently by Ellis and co-authors [5].

A more fundamental description of the behaviour of the primordial matter, which in the early universe is mostly a hot plasma of elementary particles, is usually implemented through kinetic theory [6]. However, a truly fundamental description eventually has to take into account quantum field theory. In the following I shall show that an interesting part of the theory of cosmological perturbations can be investigated through the techniques developed for quantum field theory at finite temperature [7], namely the case of a weakly interacting ultrarelativistic particle plasma. In the limiting case of a collisionless ultrarelativistic plasma it turns out to be even possible to obtain exact analytic results [8] where only numerical ones where known before; in the case of weak self-interactions one can still find analytic results [9] which involve such issues as resummation of hard thermal loops that would be very difficult to include in a (quantum) kinetic approach.

For temperatures $T \ll m_{Planck}$ it is sufficient to treat the gravitational field as a classical background field. The energy-momentum tensor can then be defined by the one-point function

$$T_{\mu\nu}(x) = \frac{2}{\sqrt{-g}} \frac{\delta\Gamma[g]}{\delta g^{\mu\nu}}, \tag{2}$$

where $\Gamma[g]$ is the effective action functional that contains all the contributions besides the classical Einstein-Hilbert action. When derived from this effective action, covariant conservation of the energy-momentum tensor is automatic and need not be imposed as a constraint.

The response under perturbations in the metric field is given by

$$\delta T_{\mu\nu}(x) = \int d^4 y \frac{\delta T_{\mu\nu}(x)}{\delta g^{\alpha\beta}(y)} \delta g^{\alpha\beta}(y). \tag{3}$$

Hence, $\delta T_{\mu\nu}$ is determined by the gravitational polarization tensor (or "thermal graviton self-energy")

$$\Pi_{\mu\nu\alpha\beta}(x, y) \equiv \frac{\delta^2 \Gamma}{\delta g^{\mu\nu}(x) \delta g^{\alpha\beta}(y)} = \frac{1}{2} \frac{\delta(\sqrt{-g} T_{\mu\nu}(x))}{\delta g^{\alpha\beta}(y)}. \tag{4}$$

In particle physics terminology, 2 and 4 are the sets of one-particle irreducible diagrams with one and two external graviton line(s) in the background field $g_{\mu\nu}$ given by the cosmological model on which one wants to study the dynamics of cosmological perturbations. The concept of thermal equilibrium makes rigorous sense in conformally trivial situations where $g_{\mu\nu}(x) = \sigma(x)\eta_{\mu\nu}$. This is indeed the case with almost all of the cosmological models of interest. If one knows how the effective action transforms under conformal rescalings of the metric, then the entire problem of determining the highly nonlocal function 4 (and thus the response of the plasma) can be reduced to the evaluation of Feynman diagrams in flat space, where momentum-space techniques can be used. In flat space, temperature can be introduced through periodicity in imaginary time, and retarded Green functions in real time are obtained by analytic continuation.

In the high-temperature limit, where all the momenta and masses of the internal particles are assumed to be much smaller than temperature, the effective action in fact turns out to be invariant under conformal rescalings, so 4 on a curved space with vanishing conformal Weyl tensor can be reconstructed by the simple transformation

$$\Pi_{\mu\nu\rho\sigma}(x,y)\big|_{g_{\mu\nu}=\sigma\eta_{\mu\nu}}$$
$$= \sigma(x) \int \frac{d^4k}{(2\pi)^4} \quad e^{ik(x-y)}\, \tilde{\Pi}_{\mu\nu\rho\sigma}(k)\Big|_{\eta}\, \sigma(y)\ . \tag{5}$$

The Planck mass, which we assumed to be much larger than temperature, does not explicitly appear in $\tilde{\Pi}$ since we are treating the metric field as classical and no higher loop diagrams with graviton self-interactions are involved. So we only need to assume that the (zero-temperature) masses of the thermal matter are small compared to temperature (i.e., the plasma is ultrarelativistic), and that the relevant momentum scales are likewise so. Fortunately, this is just the case of interest with cosmological perturbations. If the latter have typical wavelengths of the order of the Hubble horizon, then $k/T \sim \sqrt{GT^2} \propto T/m_{Planck} \ll 1$.

One-loop diagrams correspond to collisionless thermal matter which has only gravitational interactions. The leading temperature contributions to 4 have been first calculated in [10] (see also Ref. [11]) and turn out to have a universal structure, where only the overall factor varies among the various forms of thermal matter according to their energy density. It is highly nonlocal and comes with a complicated tensor structure, since with $\eta^{\mu\nu}$, $u_\mu = \delta^0_\mu$, and $K^\mu = (K^0, \mathbf{k})$, one can build 14 tensors to form a basis for $\tilde{\Pi}^{\mu\nu\alpha\beta}(K) = \rho \sum_{i=1}^{14} c_i(K) T_i^{\mu\nu\alpha\beta}(K)$, see Table 1.

However, $\tilde{\Pi}$ satisfies the Ward identities corresponding to diffeomorphism invariance and conformal invariance, and this reduces the number of independent structure functions to 3. They can be chosen as

$$A(K) \equiv \tilde{\Pi}_{0000}(K)/\rho, \quad B(K) \equiv \tilde{\Pi}_{0\mu}{}^\mu{}_0(K)/\rho, \quad C(K) \equiv \tilde{\Pi}_{\mu\nu}{}^{\mu\nu}(K)/\rho \tag{6}$$

($\rho = T_{00}$), and the $c_{1\ldots14}$ are determined by the linear combinations given in Table 2. The universal result for ultrarelativistic collisionless thermal matter then reads

$$A^{(1)}(K) = \omega\,\mathrm{artanh}\frac{1}{\omega} - \frac{5}{4}, \qquad B^{(1)} = -1, \qquad C^{(1)} = 0, \tag{7}$$

with $\omega \equiv K_0/k$.

Cosmological perturbations can be classified according to their transformation behaviour under spatial coordinate transformations [4] as scalar, vector, or tensor, which corresponds to compressional, rotational, or radiative perturbations in the plasma. The above 3 independent components of $\tilde{\Pi}$ determine, in certain combinations, the connection between the respective perturbations in the energy-momentum tensor and in the metric field.

In the radiation-dominated epoch the standard choice is that of a spatially flat Einstein-de Sitter model with line element

$$ds^2 = \sigma(\tau)(d\tau^2 - d\mathbf{x}^2), \qquad \sigma(\tau) = \frac{8\pi G\rho_0}{3}\tau^2, \tag{8}$$

(ρ_0 is the energy density when $\sigma = 1$), and, given 8, it is moreover natural to decompose all perturbations in plane waves, since in linear perturbation theory the different modes evolve independently. The problem is thus reduced to a one-dimensional one, and it is convenient to introduce a dimensionless time variable

$$x \equiv k\tau = \frac{R_H}{\lambda/(2\pi)}, \tag{9}$$

which measures the (growing) size of the Hubble horizon over the wavelength of a given mode (which is constant in comoving coordinates).

Of all the numerous components of 1, only a few are independent by virtue of general covariance and turn out to involve only those gauge-invariant combinations of the components of the metric perturbations $\delta g_{\mu\nu}$ that have been studied by Bardeen [4]. For instance, the scalar part of metric perturbations can be parametrized in terms of four scalar functions

$$\delta g_{\mu\nu}^{(S)} = \sigma(\tau) \begin{pmatrix} C & D_{,i} \\ D_{,j} & A\delta_{ij} + B_{,ij} \end{pmatrix} \tag{10}$$

of which always two can be gauged away. Instead of fixing a gauge, we can also use the gauge-invariant combinations

$$\Phi = A + \frac{\dot\sigma}{\sigma}\left(D - \frac{1}{2}\dot B\right) \tag{11}$$

$$\Pi = \frac{1}{2}\left(\ddot B + \frac{\dot\sigma}{\sigma}\dot B + C - A\right) - \dot D - \frac{\dot\sigma}{\sigma}D, \tag{12}$$

where a dot denotes differentiation with respect to the conformal time variable τ.

Each spatial Fourier mode with wave vector $\mathbf{k}$ is related to perturbations in the energy density and anisotropic pressure according to

$$\delta = \frac{1}{3}x^2\Phi, \qquad \pi_{anis.} = \frac{1}{3}x^2\Pi. \tag{13}$$

Here energy density perturbations δ are defined with respect to space-like hypersurfaces representing everywhere the local rest frame of the full energy-momentum tensor, whereas $\pi_{anis.}$ is an unambiguous quantity, since there is no anisotropic pressure in the background.

Correspondingly, when specifying to scalar perturbations, there are just two independent equations contained in 1. Because of conformal invariance, the trace of 1 is particularly simple and yields a finite-order differential equation in x,

$$\Phi'' + \frac{4}{x}\Phi' + \frac{1}{3}\Phi = \frac{2}{3}\Pi - \frac{2}{x}\Pi' \tag{14}$$

(a prime denotes differentiation with respect to the dimensionless time variable x). The other components, however, involve the nonlocalities of the gravitational polarization tensor. These lead to an integro-differential equation, which upon imposing retarded boundary conditions reads [8]

$$(x^2 - 3)\Phi + 3x\Phi' = 6\Pi - 12\int_{x_0}^{x} dx'\, j_0(x - x')(\Phi'(x') + \Pi'(x')) + \varphi(x - x_0) \tag{15}$$

where $j_0(x) = \sin(x)/x$ arises as Fourier transform of $A(\omega)$ in 7. $\varphi(x - x_0)$ encodes the initial conditions, the simplest choice of which corresponds to $\varphi(x - x_0) \propto j_0(x - x_0)$.

Similar integro-differential equations have been obtained from coupled Einstein-Vlasov equations in particular gauges, and the above one can be shown to arise from a gauge-invariant reformulation of classical kinetic theory [12]. Usually, these equations were studied numerically, but in fact they can be solved analytically [8]. If initial conditions are formulated for $x_0 \to 0$, a power series ansatz for Φ and Π leads to recursion relations that can be solved and lead to an alternating series that converges faster than trigonometric functions.

This also holds true for the vector and tensor perturbations and when the more realistic case of a two-component system of perfect radiation fluid and ultrarelativistic plasma is considered [13].

In Fig. 1, the solution for the energy-density contrast is given in a doubly-logarithmic plot (full line) and compared with the perfect-fluid case (dotted line). In the latter, one has growth of the energy-density contrast as long as the wavelength of the perturbation exceeds the size of the Hubble radius ($x \ll 1$). After the Hubble horizon has grown such as to encompass about one half wavelength ($x = \pi$), further growth of the perturbation is stopped by the strong radiation pressure, turning it into an (undamped) acoustic wave propagating with the speed of sound in radiation, $v = 1/\sqrt{3}$. The collisionless case is similar as concerns the superhorizon-sized perturbations, but after horizon crossing, there is strong damping $\sim 1/x$, and the phase velocity is about 1. This indeed reproduces the findings of the numerical studies of Ref. [14]. They can be understood as follows: a energy-density perturbation consisting of collisionless particles propagates with the speed of their constituents, which in the ultrarelativistic

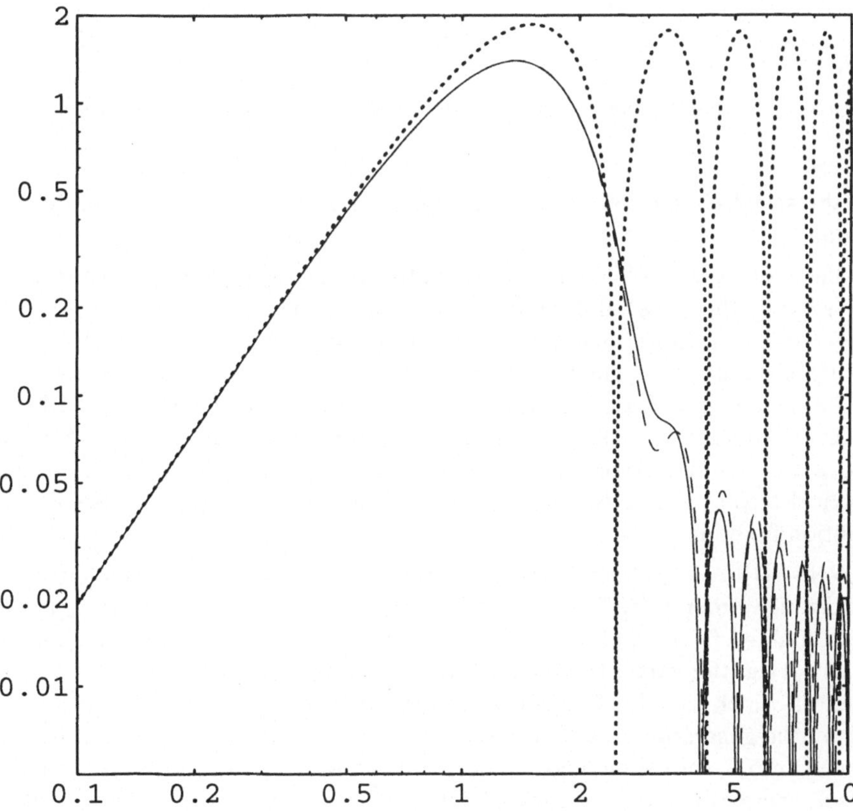

Figure 1: The energy-density contrast (arbitrary normalization) as a function of x/π for a collisionless ultrarelativistic plasma (full line), a scalar plasma with quartic self-interactions $\lambda\phi^4$ and $\lambda = 1$ (dashed line), and a perfect radiation fluid (dotted line).

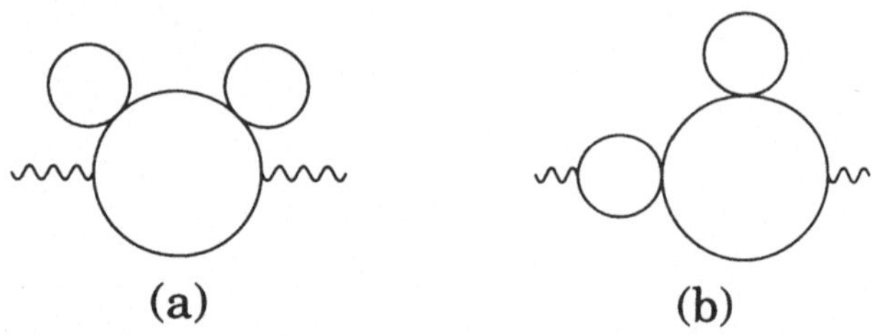

Figure 2: Two examples of infrared divergent graphs beyond two-loop order.

case is the speed of light, and there is collisionless damping in the form of directional dispersion.

While with purely collisionless ultrarelativistic matter, all results are equivalent [12] to solving the classical Einstein-Vlasov equations, a quantum-field-theoretical treatment comes into its own when self-interactions within the thermal matter are taken into account. In a kinetic treatment one could add in a collision term to the coupled Einstein-Boltzmann equations, but eventually one would have to abandon the classical concept of a distribution function for the thermal matter. A virtue of the above thermal-field-theoretical approach is that everything is formulated in purely geometrical terms, without explicit recourse to perturbations in the (gauge variant) distribution function.

In Ref. [15], the gravitational polarization tensor has been calculated in a $\lambda\phi^4$ theory through order $\lambda^{3/2}$. The next-to-leading order contributions to $\Pi_{\mu\nu\alpha\beta}$ at order λ^1 are contained in the high-temperature limit of two-loop diagrams and their evaluation is straightforward. However, starting at three-loop order, there are infrared divergences which signal a breakdown of the convential perturbative series. This is caused by the generation of a thermal mass $\propto \sqrt{\lambda}T$ for the hot scalars. If this is not resummed into a correspondingly massive scalar propagator, repeated insertions of scalar self-energy diagrams in a scalar line produces arbitrarily high powers of massless scalar propagators all with the same momentum, and thus increasingly singular infrared behaviour (Fig. 2a).

However, it is not sufficient to resum this thermal mass for the hot scalars. After all, this would break conformal invariance. Indeed, there are also vertex subdiagrams $\propto \lambda T^2$ that have a similar effect as a self-energy insertion, see. Fig. 2b. As in the hard-thermal-loop resummation program developed for high-temperature quantum chromodynamics [16], one has to resum also nonlocal vertex contributions. Doing so, the result turns out to satisfy both the diffeomorphism and conformal Ward identities.

In the low-momentum limit that is of interest in our application to cosmological perturbations, the function A in 7 that governs the evolution of scalar perturbations reads through order $\lambda^{3/2}$

$$
\begin{aligned}
A \;=\;& \omega \operatorname{artanh}\frac{1}{\omega} - \frac{5}{4} \\[4pt]
&+ \frac{5\lambda}{8\pi^2}\left[2\left(\omega\operatorname{artanh}\frac{1}{\omega}\right)^2 - \omega\operatorname{artanh}\frac{1}{\omega} - \frac{\omega^2}{\omega^2-1} \right] \\[4pt]
&+ \frac{5\lambda^{3/2}}{8\pi^3}\left[3\left(\omega^2-1-\omega\sqrt{\omega^2-1}\right)\left(\omega\operatorname{artanh}\frac{1}{\omega}\right)^2 \right.\\[4pt]
&\qquad\quad +6\left(\omega\sqrt{\omega^2-1}-\omega^2-\frac{\omega}{\sqrt{\omega^2-1}}\right)\omega\operatorname{artanh}\frac{1}{\omega} \\[4pt]
&\qquad\quad \left. +\frac{\omega}{(\omega^2-1)^{3/2}} + 3\frac{\omega^2}{\omega^2-1} + 6\frac{\omega}{\sqrt{\omega^2-1}} - 3\omega\sqrt{\omega^2-1} + 3\omega^2 \right]
\end{aligned}
\tag{16}
$$

and similarly complicated expressions arise for B and C, which in the collisionless limit were pure numbers.

The Fourier transform of this expression determines the kernel in the convolution integral of 15. At order λ^1, it can still be expressed in terms of well-known special functions [9], whereas at order $\lambda^{3/2}$ this would involve rather intractable integrals over Lommel functions. However, all that is needed for finding analytical solutions is their power series representations which are comparatively simple. Given them, it is as easy as before to solve the perturbation equations, however one finds that the asymptotic behaviour $x \gg 1$ eventually becomes sensitive to higher and higher loop orders. The reason for this is that higher loop orders come with increasingly singular contributions at $\omega = \pm 1$ to $A(\omega)$, and the large-x behaviour is dominated by the latter. This could be cured by a further resummation similar to the one introduced for hot quantum chromodynamics in Ref. [17], but it turns out that a particular Padé-approximant based on the perturbative result reflects the effects of this further resummation quite well [18]. The result for the density perturbations in a scalar plasma with $\lambda\phi^4$-interactions and $\lambda = 1$ are shown in Fig. 1 by the dashed line, where it is compared with the collisionless case (full line) and the one of a perfect radiation fluid (dotted line). The effects of the self-interactions within the ultrarelativistic plasma become important only for $x \gtrsim \pi$, where the strong collisionless damping is somewhat reduced and the phase velocity is smaller than 1.

A full analysis of scalar, vector, and tensor perturbations in the general case of a two-component system containing also a perfect radiation fluid is given in Ref. [18]. Let us just mention one of the more spectacular results, which arise in the case of vector (rotational) perturbations. This case has not been investigated much previously, presumably because in the perfect-fluid case there are no regular solutions — rotational perturbations necessarily lead to strongly anisotropic initial singularities. This can be explained by the Helmholtz-Kelvin circulation theorem [19] which states that in a perfect fluid the circulation around a closed curve following the motion of matter is conserved. However, this theorem does not apply generally. Indeed, in a (nearly)

$$T_1^{\alpha\beta\mu\nu} = \eta^{\alpha\nu}\,\eta^{\beta\mu} + \eta^{\alpha\mu}\,\eta^{\beta\nu}$$

$$T_2^{\alpha\beta\mu\nu} = u^\mu\left(u^\beta\,\eta^{\alpha\nu} + u^\alpha\,\eta^{\beta\nu}\right) + u^\nu\left(u^\beta\,\eta^{\alpha\mu} + u^\alpha\,\eta^{\beta\mu}\right)$$

$$T_3^{\alpha\beta\mu\nu} = u^\alpha\,u^\beta\,u^\mu\,u^\nu$$

$$T_4^{\alpha\beta\mu\nu} = \eta^{\alpha\beta}\,\eta^{\mu\nu}$$

$$T_5^{\alpha\beta\mu\nu} = u^\mu\,u^\nu\,\eta^{\alpha\beta} + u^\alpha\,u^\beta\,\eta^{\mu\nu}$$

$$T_6^{\alpha\beta\mu\nu} = u^\beta\left(\bar{K}^\nu\,\eta^{\alpha\mu} + \bar{K}^\mu\,\eta^{\alpha\nu}\right) + \bar{K}^\beta\left(u^\nu\,\eta^{\alpha\mu} + u^\mu\,\eta^{\alpha\nu}\right)$$
$$+ u^\alpha\left(\bar{K}^\nu\,\eta^{\beta\mu} + \bar{K}^\mu\,\eta^{\beta\nu}\right) + \bar{K}^\alpha\left(u^\nu\,\eta^{\beta\mu} + u^\mu\,\eta^{\beta\nu}\right)$$

$$T_7^{\alpha\beta\mu\nu} = \bar{K}^\nu\,u^\alpha\,u^\beta\,u^\mu + \bar{K}^\mu\,u^\alpha\,u^\beta\,u^\nu + \bar{K}^\beta\,u^\alpha\,u^\mu\,u^\nu + \bar{K}^\alpha\,u^\beta\,u^\mu\,u^\nu$$

$$T_8^{\alpha\beta\mu\nu} = \bar{K}^\beta\,\bar{K}^\nu\,\eta^{\alpha\mu} + \bar{K}^\beta\,\bar{K}^\mu\,\eta^{\alpha\nu} + \bar{K}^\alpha\,\bar{K}^\nu\,\eta^{\beta\mu} + \bar{K}^\alpha\,\bar{K}^\mu\,\eta^{\beta\nu}$$

$$T_9^{\alpha\beta\mu\nu} = \bar{K}^\mu\,\bar{K}^\nu\,u^\alpha\,u^\beta + \bar{K}^\alpha\,\bar{K}^\beta\,u^\mu\,u^\nu$$

$$T_{10}^{\alpha\beta\mu\nu} = \left(\bar{K}^\beta\,u^\alpha + \bar{K}^\alpha\,u^\beta\right)\left(\bar{K}^\nu\,u^\mu + \bar{K}^\mu\,u^\nu\right)$$

$$T_{11}^{\alpha\beta\mu\nu} = \bar{K}^\beta\,\bar{K}^\mu\,\bar{K}^\nu\,u^\alpha + \bar{K}^\alpha\,\bar{K}^\mu\,\bar{K}^\nu\,u^\beta + \bar{K}^\alpha\,\bar{K}^\beta\,\bar{K}^\nu\,u^\mu + \bar{K}^\alpha\,\bar{K}^\beta\,\bar{K}^\mu\,u^\nu$$

$$T_{12}^{\alpha\beta\mu\nu} = \bar{K}^\alpha\,\bar{K}^\beta\,\bar{K}^\mu\,\bar{K}^\nu$$

$$T_{13}^{\alpha\beta\mu\nu} = \bar{K}^\mu\,\bar{K}^\nu\,\eta^{\alpha\beta} + \bar{K}^\alpha\,\bar{K}^\beta\,\eta^{\mu\nu}$$

$$T_{14}^{\alpha\beta\mu\nu} = \left(\bar{K}^\nu\,u^\mu + \bar{K}^\mu\,u^\nu\right)\eta^{\alpha\beta} + \left(\bar{K}^\beta\,u^\alpha + \bar{K}^\alpha\,u^\beta\right)\eta^{\mu\nu}$$

Table 1: A basis of 14 independent tensors $T_i^{\alpha\beta\mu\nu}$ built from $\eta^{\mu\nu}$, $u^\mu = \delta_0^\mu$, and $\bar{K}^\mu \equiv K^\mu/k = (\omega, \mathbf{k}/k)$.

collisionless medium one can have small initial anisotropies in the distribution function that gives rise to a growing vorticity on superhorizon scales [13], which decays after horizon crossing through directional dispersion. In a two-component system one can even have such perturbations which do not decay by arranging for vorticity in a perfect-fluid component that is compensated by an initially matching one with reversed sign in the nearly collisionless plasma component. Then net vorticity is generated by the decay of the vector perturbation in the plasma component, see Fig. 3. This is particularly interesting in that vector perturbations generally lead to the generation of primordial magnetic fields at the time when the universe changes from radiation to matter domination [20]. Because tiny primordial magnetic fields can act as seed fields for galactic dynamos, such rotational perturbations may therefore be of interest with respect to the still unsolved problem of the origin of galactic and intergalactic magnetic fields.

Acknowledgements

I am grateful to Ulli Kraemmer, Herbert Nachbagauer, and Dominik Schwarz, for their most enjoyable collaboration in various stages of the presented work.

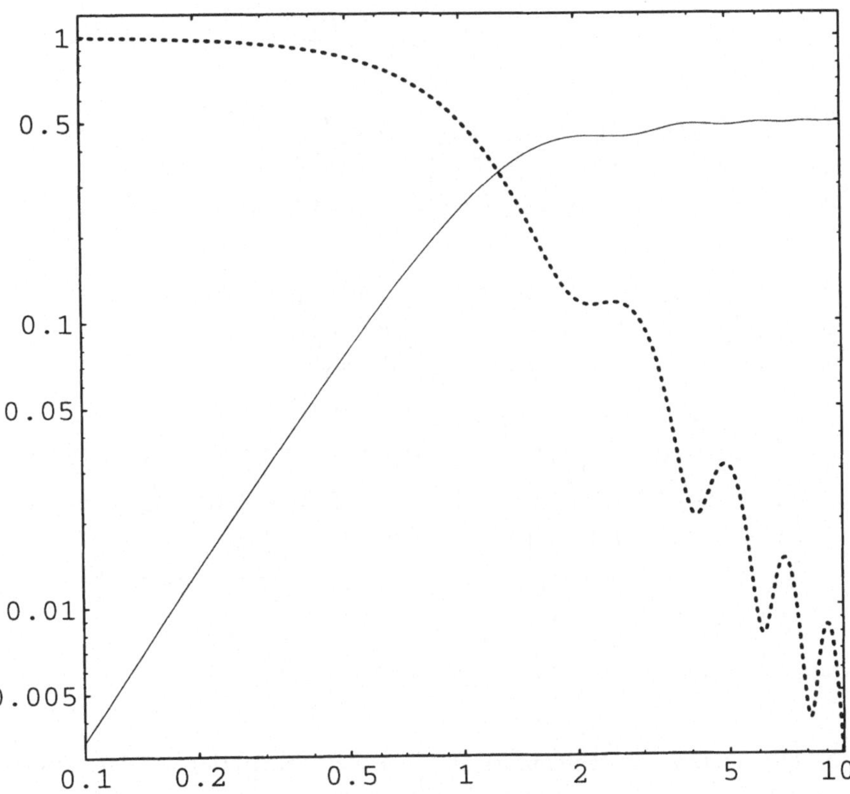

Figure 3: Rotational perturbations in a two-component system consisting of 50 % perfect radiation fluid and 50 % ultrarelativistic scalar plasma ($\lambda = 1$). Given is the velocity amplitude for the total system (full line) and the weakly interacting plasma (dashed line) in arbitary units as a function of x/π.

$$c_1 = \tfrac{1}{8}\bar{K}^4 A + \tfrac{1}{2}\bar{K}^2 B + \tfrac{1}{4}C + \tfrac{1}{32}\bar{K}^4 + \tfrac{11}{24}\bar{K}^2 + \tfrac{1}{6}$$

$$c_2 = \tfrac{5}{8}\bar{K}^6 A + \bar{K}^4 B + \tfrac{1}{4}\bar{K}^2 C + \tfrac{5}{32}\bar{K}^6 + \tfrac{19}{24}\bar{K}^4 + \tfrac{1}{12}\bar{K}^2 - \tfrac{1}{3}$$

$$c_3 = \bar{K}^2 \left\{ \tfrac{35}{8}\bar{K}^6 A + \tfrac{5}{2}\bar{K}^4 B + \tfrac{1}{4}\bar{K}^2 C + \tfrac{35}{32}\bar{K}^6 + \tfrac{25}{24}\bar{K}^4 + \tfrac{7}{12}\bar{K}^2 - \tfrac{1}{3} \right\}$$

$$c_4 = \bar{K}^2 \left\{ \tfrac{1}{8}\bar{K}^2 A - \tfrac{1}{2}B - \tfrac{1}{4}C + \tfrac{1}{32}\bar{K}^2 - \tfrac{13}{24} \right\}$$

$$c_5 = \bar{K}^2 \left\{ \tfrac{5}{8}\bar{K}^4 A - \tfrac{1}{2}\bar{K}^2 B - \tfrac{1}{4}C + \tfrac{5}{32}\bar{K}^4 - \tfrac{17}{24}\bar{K}^2 + \tfrac{1}{12} \right\}$$

$$c_6 = \omega \left\{ \tfrac{-5}{8}\bar{K}^4 A - \bar{K}^2 B - \tfrac{1}{4}C - \tfrac{5}{32}\bar{K}^4 - \tfrac{19}{24}\bar{K}^2 - \tfrac{1}{12} \right\}$$

$$c_7 = \omega \left\{ \tfrac{-35}{8}\bar{K}^6 A - \tfrac{5}{2}\bar{K}^4 B - \tfrac{1}{4}\bar{K}^2 C - \tfrac{35}{32}\bar{K}^6 - \tfrac{25}{24}\bar{K}^4 - \tfrac{7}{12}\bar{K}^2 + \tfrac{1}{3} \right\}$$

$$c_8 = \left(\tfrac{5}{8}\bar{K}^2 + \tfrac{1}{2} \right) \bar{K}^2 A + \left(\bar{K}^2 + \tfrac{1}{2} \right) B + \tfrac{1}{4}C + \tfrac{5}{32}\bar{K}^4 + \tfrac{11}{12}\bar{K}^2 + \tfrac{5}{12}$$

$$c_9 = \left(\tfrac{35}{8}\bar{K}^2 + \tfrac{15}{4} \right) \bar{K}^4 A + \left(\tfrac{5}{2}\bar{K}^2 + 3 \right) \bar{K}^2 B + \left(\tfrac{1}{4}\bar{K}^2 + \tfrac{1}{2} \right) C + \tfrac{35}{32}\bar{K}^6 + \tfrac{95}{48}\bar{K}^4 + \tfrac{7}{3}\bar{K}^2 + \tfrac{1}{6}$$

$$c_{10} = \left(\tfrac{35}{8}\bar{K}^2 + \tfrac{15}{4} \right) \bar{K}^4 A + \left(\tfrac{5}{2}\bar{K}^2 + \tfrac{3}{2} \right) \bar{K}^2 B + \tfrac{1}{4}\bar{K}^2 C + \tfrac{35}{32}\bar{K}^6 + \tfrac{95}{48}\bar{K}^4 + \tfrac{5}{6}\bar{K}^2 + \tfrac{1}{6}$$

$$c_{11} = \omega \left\{ \left(\tfrac{-35}{8}\bar{K}^2 - \tfrac{5}{2} \right) \bar{K}^2 A - \left(\tfrac{5}{2}\bar{K}^2 + 1 \right) B - \tfrac{1}{4}C - \tfrac{35}{32}\bar{K}^4 - \tfrac{5}{3}\bar{K}^2 - \tfrac{3}{4} \right\}$$

$$c_{12} = \left(\tfrac{35}{8}\bar{K}^4 + 5\bar{K}^2 + 1 \right) A + \left(\tfrac{5}{2}\bar{K}^2 + 2 \right) B + \tfrac{1}{4}C + \tfrac{35}{32}\bar{K}^4 + \tfrac{55}{24}\bar{K}^2 + \tfrac{7}{6}$$

$$c_{13} = \left(\tfrac{5}{8}\bar{K}^2 + \tfrac{1}{2} \right) \bar{K}^2 A - \tfrac{1}{2}\bar{K}^2 B - \tfrac{1}{4}C + \tfrac{5}{32}\bar{K}^4 - \tfrac{7}{12}\bar{K}^2 - \tfrac{1}{12}$$

$$c_{14} = \omega \left\{ \tfrac{-5}{8}\bar{K}^4 A + \tfrac{1}{2}\bar{K}^2 B + \tfrac{1}{4}C - \tfrac{5}{32}\bar{K}^4 + \tfrac{17}{24}\bar{K}^2 - \tfrac{1}{12} \right\}$$

Table 2: The structure of the conformally covariant gravitational polarization tensor $\tilde{\Pi}^{\mu\nu\alpha\beta}/\rho = \sum_{i=1}^{14} c_i T_i^{\mu\nu\alpha\beta}$ in terms of $A \equiv \tilde{\Pi}_{0000}/\rho$, $B \equiv \tilde{\Pi}_{0\mu}{}^{\mu}{}_{0}/\rho$, and $C \equiv \tilde{\Pi}_{\mu\nu}{}^{\mu\nu}/\rho$.

References

[1] G. Börner, *The Early Universe: Facts and Fiction* (Springer-Verlag, Berlin, 1993).

[2] G.F. Smoot et al., Astrophys. J. **396**, L1 (1992).

[3] E. Lifshitz, Zh. Eksp. Teor. Fiz. **16**, 587 (1946);
E. Lifshitz and I. Khalatnikov, Adv. Phys. **12**, 185 (1963).

[4] J.M. Bardeen, Phys. Rev. D **22**, 1882 (1980).

[5] G.F.R. Ellis and M. Bruni, Phys. Rev. D **40**, 1804 (1989);
G.F.R. Ellis, J. Hwang and M. Bruni, Phys. Rev. D **40**, 1819 (1989).

[6] J. Ehlers, in *General Relativity and Gravitation*, edited by R.K. Sachs (Academic Press, New York, 1971);
J.M. Stewart, *Non-equilibrium Relativistic Kinetic Theory*, (Springer-Verlag, New York, 1971).

[7] J.I. Kapusta, *Finite-temperature field theory*, (Cambridge University Press, Cambridge, 1989);
M. Le Bellac, *Thermal Field Theory* to appear in Cambridge University Press.

[8] U. Kraemmer and A. Rebhan, Phys. Rev. Lett. **67**, 793 (1991).

[9] H. Nachbagauer, A.K. Rebhan, and D.J. Schwarz, Phys. Rev. D **51**, R2504 (1995).

[10] A. Rebhan, Nucl. Phys. **B351**, 706 (1991).

[11] J. Frenkel and J.C. Taylor, Z. Phys. C **49**, 515 (1991);
F.T. Brandt, J. Frenkel and J.C. Taylor, Nucl. Phys. **B374**, 169 (1992);
A.P. de Almeida, F.T. Brandt and J. Frenkel, Phys. Rev. D **49**, 4196 (1994);
J. Frenkel, E.A. Gaffney and J.C. Taylor, Nucl. Phys. **B439**, 131 (1995).

[12] A.K. Rebhan and D.J. Schwarz, Phys. Rev. D **50**, 2541 (1994).

[13] A. Rebhan, Nucl. Phys. **B368**, 479 (1992).

[14] J.R. Bond and A.S. Szalay, Astrophys. J. **274**, 443 (1983).

[15] H. Nachbagauer, A.K. Rebhan, and D.J. Schwarz, preprint DESY 95-136, to appear in Phys. Rev. D (1995).

[16] E. Braaten and R.D. Pisarski, Nucl. Phys. **B337**, 569 (1990).

[17] U. Kraemmer, A.K. Rebhan and H. Schulz, Ann. Phys. (NY) bf 238, 286 (1995);
F. Flechsig and A.K. Rebhan, preprint DESY 95-170.

[18] H. Nachbagauer, A.K. Rebhan, and D.J. Schwarz, preprint DESY 95-191.

[19] B.J.T. Jones, Rev. Mod. Phys. **48**, 107 (1976).

[20] A. Rebhan, Astrophys. J. **392**, 385 (1992).

Algorithms for the Calculation of the Heat Kernel Coefficients

Ivan G. Avramidi and <u>Rainer Schimming</u>

Abstract

We present a brief overview of several approaches for calculating the local asymptotic expansion of the heat kernel for Laplace-type operators. The different methods developed in the papers of both authors some time ago are described in more detail.

1 Laplace-type differential operators

The convential Laplacian $\Delta = div\,grad$ on the Euclidean space $\mathbb{R}^n$ admits a natural generalization to a Riemannian manifold (M, g):

$$\Delta = g^{ab}\nabla_a\nabla_b.$$

Here

$$g = g_{ab}dx^a dx^b, \qquad (g^{ab}) := (g_{ab})^{-1}$$

with respect to local coordinates $x^a = x^1, x^2, \ldots, x^d$, and ∇ is the Levi-Civita derivative to g acting on scalar or tensor fields. Another far-reaching generalization is a *Laplace-type* linear differential operator of the form

$$L = g^{ab}D_a D_b + W \tag{1}$$

acting on the smooth sections of a vector bundle E over the manifold M. A covariant derivative D of these sections and a section W of the endomorphism bundle $End\,E$ to E enter the expression for L.

An intrinsic definition, alternative to the coordinate-dependent definition (1), can be given [54]. Let the map ad_f for a smooth scalar field $f \in C^\infty(M)$ act on any linear operator L according to

$$ad_f L := [L, f] \equiv Lf - fL.$$

(Here and in the following a scalar field is identified with the operator of multiplication by this scalar field.) A linear operator L acting on $\varphi \in C^\infty(E)$ is called a Laplace-type differential operator if

$$(ad_f)^2 L = 2|df|^2.$$

It can be shown then that

$$2 < df, D\varphi >:= (ad_f L)\varphi - (\Delta f)\varphi$$

defines a covariant derivative D on $C^\infty(E)$ and that

$$W\varphi := L\varphi - g^{ab}D_aD_b\varphi$$

defines an endomorphism $W \in C^\infty(End\,E)$. We use the notation

$$< v, w >:= g^{ab}v_aw_b, \qquad |v|^2 :=< v, v >$$

for bundle-valued one-forms $v = v_a dx^a$, $w = w_a dx^a$.

Clearly, the class of Laplace-type operators L is form-invariant under diffeomorphisms Φ, when the objects g, D, W are carried along with Φ. This class is also form-invariant under *gauge transformations*

$$\bar{L} := \Lambda^{-1}L\Lambda,$$

where $\Lambda \in C^\infty(Aut\,E)$ is a section of the automorphism bundle to E, and under *conformal transformations*

$$\bar{L} := e^{-(m+2)f}Le^{mf},$$

where $f \in C^\infty(M)$ and $m := (d-2)/2$, d being the dimension of M.

We have

$$\bar{g} = g, \qquad \bar{D} = \Lambda^{-1}D\Lambda, \qquad \bar{W} = \Lambda^{-1}W\Lambda$$

under a gauge transformation with Λ, and

$$\bar{g} = e^{2f}g, \qquad \bar{W} = e^{-2f}\left(W + I \cdot e^{-mf}\Delta e^{mf}\right)$$

under a conformal transformation with f. Multiple covariant derivatives D of sections of E behave like multiple Levi-Civita derivatives of a scalar field under conformal transformations. In particular,

$$\bar{D}\varphi = D\varphi \text{ for } \varphi \in C^\infty(E).$$

For details cf. [54,33].

2 The heat kernel coefficients

Some sequence of locally defined two-point quantities $H_k = H_k(x, x')$ $(k = 0, 1, 2, \ldots)$ is associated to any Laplace-type operator L. Proposition 1. A formal series expression

$$K = K(t; x, x') = (4\pi t)^{-d/2}\exp\left(-\frac{\sigma(x, x')}{2t}\right)\sum_{k=0}^{\infty}H_k(x, x')t^k \tag{2}$$

solves the heat equation

$$\frac{\partial K}{\partial t} = LK$$

with the initial condition

$$K(+0; x, x') = \delta(x, x')$$

if and only if

$$(\mathcal{D} + \mu)\, H_0 = 0, \qquad H_0(x, x) = I, \tag{3}$$

and

$$(\mathcal{D} + k + \mu)\, H_k = L H_{k-1}, \qquad \text{for } k = 1, 2, \ldots. \tag{4}$$

Here $\sigma(x, x')$ is the so called geodetic interval defined as one half the square of the distance along the geodesic between the points x and x', $\mathcal{D}$ is a first-order linear differential operator given by

$$\mathcal{D}\varphi = <d\sigma, D\varphi>,$$

I is the identity endomorphism, and $\mu = \frac{1}{2}(\Delta\sigma - d)$.

All the analysis in this paper is purely local. This means that we fix some small regular region of the manifold M and consider the points x and x' to lie inside this region. *Proposition 2.* There is a neighbourhood N of the diagonal of $M \times M$ and a unique smooth system of solutions $H_k = H_k(x, x')$ in N of the differential recursion system (4). These $H_k = H_k(x, x')$ are called the *Hadamard coefficients* to L.

Note that the quantities $H_k = H_k(x, x')$ are also widely known under the name Hadamard-Minakshisundaram-De Witt-Seeley (HMDS or HAMIDEW coefficients), according to the papers of these authors [36,42,43,21,22,51,52], or heat kernel coefficients, the last being used in the title of this paper. Hadamard [36] introduced them already in 1923 for scalar operators L and established the essential properties.

3 Methods for effective calculations

We want to classify the methods for an effective calculation of the Taylor coefficients $[D^n H_k]$ $(p, k = 0, 1, 2, \ldots)$ of the Hadamard coefficients or, which means a little more, of the diagonal values of their multiple covariant derivatives with respect to x, $[D_{a_1} D_{a_2} \cdots D_{a_n} H_k]$ $(n, k = 0, 1, 2, \ldots)$. The square brackets [ ] applied to a two-point quantity means the restriction to the diagonal of $M \times M$. Actually, we arrange the methods to five groups.

I. Taylor expansion in normal coordinates. Naturally, this is a method which comes first to the mind. It has been used with great skill by Günther, Wünsch, McLenaghan and others [31-33,65,41].

II. Manifestly covariant methods. The Taylor expansion in normal coordinates can be reinterpreted in a coordinate-independent manner. So, if one works with covariant derivatives instead of partial derivatives it should be possible to replace the method I. by manifestly covariant formulas. In fact, Synge [60,61] and De Witt and Brehme [21,22] made the first steps in this direction. Their work can be continued in two ways to an algorithm for the Taylor coefficients $[D^n H_k]$, which are given by the diagonal values of the *symmetrized* covariant derivatives of H_k, or to an algorithm for the diagonal values of the (*non-symmetrized*) covariant derivatives. We will present such algorithms, developed by the authors, in the next sections.

III. INVARIANT-THEORETICAL METHODS. Let R_{abcd} denote the components of the Riemannian curvature of (M, g) and F_{ab} be the components of the curvature of the bundle connection D,

$$[D_a, D_b]\varphi = F_{ab}\varphi \qquad \text{for } \varphi \in C^\infty(E).$$

The components of $[D^n H_k]$ or of the diagonal values of the multiple covariant derivatives of the H_k, $[D_{a_1} D_{a_2} \cdots D_{a_n} H_k]$, exhibit a controlled behavior under diffeomorphisms and gauge transformations. This implies that they are just polynomials in g_{ab}, g^{ab}, R_{abcd}, F_{ab}, $\nabla_{a_1} R_{abcd}$, $D_{a_1} F_{ab}$, $\nabla_{a_1} \nabla_{a_2} R_{abcd}$, $D_{a_1} D_{a_2} F_{ab}$, etc. Some careful analysis shows that the coefficients of these polynomials are determined by the functorial properties under several decompositions: Riemannian product $(M, g){=}(M_1, g_1) \times (M_2, g_2)$, direct vector bundle sum $E = E_1 \oplus E_2$, tensor product of vector bundles $E = E_1 \otimes E_2$, and splitting of the potential term $W = W_1 + W_2$. Gilkey [26-29] extensively applied the invariant-theoretical method as we call this analysis. Let us mention also the papers [65,33,25].

IV. HEAT SEMIGROUP METHODS. If L is elliptic, i.e. the metric is positive definite, and if the manifold M is compact, then the linear operator $\exp(tL)$ is well defined and forms a semigroup, the so-called heat semigroup. (Let us mention that there is also a construction of the heat semigroup for non-compact M under certain additional conditions.) The linear operator $\exp(tL)$ has a kernel $K = K(t; x, x')$. The latter admits the formal series (2) as an asymptotic series as $t \to +0$, cf. [42,43,18,28].

In quantum field theory there have been developed several perturbational techniques for handling a semigroup of the form $\exp(tL)$. Let us mention the Feynman path integrals, graph methods, the Dyson formula etc. Such methods can produce results for the kernel $K = K(t; x, x')$ of $\exp(tL)$ and, hence, for the Hadamard coefficients. Osborn, Zuk, Nepomechie and others used this approach [44-48,69-72,23]. In recent papers of one of the authors (I.G.A.) the whole heat kernel diagonal $K(t; x, x)$ (not only the $H_k(x, x)$) in low-energy approximation was constructed using the semigroup operator approach [8-14].

The semigroup $\exp(tL)$ has also a probabilistic interpretation: it describes some stochastic process. In particular, $\exp(tL)$ on scalar fields describes Brownian motion; the general $\exp(tL)$ belongs to some more complicated Markov process. The probabilistic approach gives further insights; let us only mention the Feynman-Kac formula and let us quote the papers [17,39,19,20].

V. PSEUDODIFFERENTIAL OPERATOR METHODS. Hadamard in his book [36] constructed an asymptotic expansion for the Green function to L; he called it elementary solution. Let L be elliptic and replace it by $L - \lambda I$, where the parameter λ belongs to the resolvent set of L, i.e. λ is not a spectral value of L. Hadamard's construction for $L - \lambda I$ can be reinterpreted in terms of pseudodifferential operators as a parametrix construction. The coefficients of the asymptotic series are just the Hadamard coefficients, up to numerical factors. Knowing this, one can produce results for the H_k ($k = 0, 1, 2, \ldots$) by means of refined pseudodifferential operator methods. Gilkey, Fulling and Kennedy, Gusynin, Obukhov and others did this succesfully [27,28,24,35,46,16,64].

Note that our systematics considers only *general* methods. Additionally to them, there are *special* methods for restricted classes of Riemannian manifolds, vector bundles and/or differential operators. Let us only mention the harmonic analysis on Lie groups or homogeneous manifolds: a spectral analysis of group-invariant differential operators by means of representation theory.

4 An algorithm for the Taylor coefficients of the H_k

The exposition in this section is due to the papers of one of the authors (I.G.A.) [2,3,6,7]. It is not difficult to show that

$$H_0(x, x') = \Delta^{1/2}(x, x')\mathcal{P}(x, x').$$

where $\Delta(x, x')$ is the Van-Vleck-Morette determinant and $\mathcal{P}(x, x')$ is the parallel displacement operator with respect to D along the geodesic. The modified coefficients

$$a_k := \frac{(-1)^k}{k!} H_0^{-1} H_k$$

fulfill the simplified differential recursion system [22]

$$a_0 = I, \qquad \left(1 + \frac{1}{k}\mathcal{D}\right) a_k = M a_{k-1} \tag{5}$$

where

$$M := -H_0^{-1} L H_0. \tag{6}$$

We assume that there is a unique geodesic connecting the arguments x and x' of our two-point functions and consider all two-point quantities to be single-valued analytic functions.

The formal solution of the recursion relations (5) has the form

$$a_k = \left(1 + \frac{1}{k}\mathcal{D}\right)^{-1} M \left(1 + \frac{1}{k-1}\mathcal{D}\right)^{-1} M \cdots (1 + \mathcal{D})^{-1} M$$

Let us expand a_k in a covariant Taylor series

$$a_k = \sum_{n \geq 0} |n><n|a_k>,$$

where some compact notation is used. Namely, let $grad'\sigma$ denote the vector field with components

$$\sigma^{\mu'} = g^{\mu'\nu'}\sigma_{\nu'} \equiv g^{\mu'\nu'}\nabla_{\nu'}\sigma,$$

$$(grad'\sigma)^n := grad'\sigma \otimes \cdots \otimes grad'\sigma$$

be the n-fold (symmetric) tensor product and

$$|n> := \frac{(-1)^n}{n!}(grad'\sigma)^n.$$

Let further the linear functionals $<n|$ (dual to the system of functions $|n>$) be defined by

$$<n|\varphi> := [D^n\varphi]$$

for a two-point field φ, where $D^n = DD\cdots D$ is the multiple symmetric covariant derivative with respect to x, and the square brackets $[\,]$ denote, as before, the restriction to the diagonal. Then $|n><n|\varphi>$ denotes the inner product

$$|n><n|\varphi> = (|n>)^{\mu_1\ldots\mu_n}(<n|\varphi>)_{\mu_1\ldots\mu_n}.$$

From the remarkable eigenvalue formula

$$\mathcal{D}|n> = n|n>$$

one can obtain an expression for the Taylor coefficients of the a_k, namely

$$<n|a_k> = \sum_{n_1,\cdots,n_{k-1}\geq 0}\left(1+\frac{n}{k}\right)^{-1}\left(1+\frac{n_{k-1}}{k-1}\right)^{-1}\cdots(1+n_1)^{-1}$$

$$\times\; <n|M|n_{k-1}><n_{k-1}|M|n_{k-2}>\cdots<n_1|M|0>,$$

where $<m|M|n>$ are the 'matrix elements' of the operator M (6)

$$<m|M|n> = \left[D^m M\frac{(-1)^n}{n!}(grad'\sigma)^n\right]. \tag{7}$$

The matrix elements $<m|M|n>$ are tensors with components $M^{\nu_1\ldots\nu_n}{}_{\mu_1\ldots\mu_m}$ which are symmetric both in upper and lower indices. The inner product of the matrix elements is again such a tensor with components

$$(<n|M|k><k|M|m>)^{\nu_1\ldots\nu_n}{}_{\mu_1\ldots\mu_m} = M^{\nu_1\ldots\nu_n}{}_{\lambda_1\ldots\lambda_k}M^{\lambda_1\ldots\lambda_k}{}_{\mu_1\ldots\mu_m}.$$

These tensors are expressible in terms of diagonal values of derivatives of basic two-point quantities. The components of these matrix elements were found in [2,3,6,7].

In order to present explicit formulas for these matrix elements we use a calculus of matrix-valued, vector-valued and $EndE$-valued symmetric differential forms. So we introduce first matrix-valued symmetric forms $K_{(n)} = (K^\nu{}_{\mu(n)})$:

$$K^\nu{}_{\mu(n)} = \nabla_{(\mu_1} \cdots \nabla_{\mu_{n-2}} R^\nu{}_{\mu_{n-1}|\mu|\mu_n)} dx^{\mu_1} \vee \cdots \vee dx^{\mu_n},$$

where $\vee$ denotes the symmetric tensor product of symmetric forms. Then we define the matrix-valued symmetric forms $\gamma_{(n)} = (\gamma^\alpha{}_{\beta(n)})$, $\eta_{(n)} = (\eta^\alpha{}_{\beta(n)})$ and $(X^{\mu\nu}{}_{(n)})$ by

$$\gamma_{(n)} = \sum_{1 \le k \le [\frac{n}{2}]} (-1)^{k+1} \sum_{\substack{n_1,\cdots,n_k \ge 2 \\ n_1 + \cdots + n_k = n}} N^{-1}(n_1, \cdots, n_k) K_{(n_k)} \vee \cdots \vee K_{(n_2)} \vee K_{(n_1)},$$

where

$$N(n_1, \cdots, n_k) = \tfrac{(n+1)}{(n-1)!}(n_1 - 2)! \cdots (n_k - 2)!$$
$$\times n_1(n_1 + 1)(n_1 + n_2)(n_1 + n_2 + 1) \cdots (n_1 + \cdots + n_{k-1})(n_1 + \cdots + n_{k-1} + 1),$$

and

$$\eta_{(n)} = - \sum_{1 \le k \le [\frac{n}{2}]} \sum_{\substack{n_1,\cdots,n_k \ge 2 \\ n_1 + \cdots + n_k = n}} \frac{n!}{n_1! \cdots n_k!} \ \gamma_{(n_k)} \vee \cdots \vee \gamma_{(n_2)} \vee \gamma_{(n_1)},$$

$$X^{\mu\nu}{}_{(n)} = \sum_{0 \le k \le n} \binom{n}{k} \eta^{(\mu}{}_{\alpha(k)} \vee \eta^{\nu)\alpha}{}_{(n-k)},$$

and scalar symmetric forms $\zeta_{(n)}$

$$\zeta_{(n)} = \sum_{1 \le k \le [\frac{n}{2}]} \frac{1}{2k} \sum_{\substack{n_1,\cdots,n_k \ge 2 \\ n_1 + \cdots + n_k = n}} \frac{n!}{n_1! \cdots n_k!} \mathrm{tr} \left(\gamma_{(n_k)} \vee \cdots \vee \gamma_{(n_2)} \vee \gamma_{(n_1)} \right).$$

Next, let us introduce the vector-valued (and $EndE$-valued) symmetric forms $F_{(n)} = (F_{\mu(n)})$:

$$F_{\mu(n)} = D_{(\mu_1} \cdots D_{\mu_{n-1}} F_{|\mu|\mu_n)} dx^{\mu_1} \vee \cdots \vee dx^{\mu_n}$$

and $A_{(n)} = (A_{\mu(n)})$

$$A_{(n)} = \frac{n}{n+1} \left\{ F_{(n)} - \sum_{1 \le k \le n-2} \binom{n-1}{k-1} F_{(k)} \vee \gamma_{(n-k)} \right\},$$

and $EndE$-valued symmetric forms $W_{(n)}$

$$W_{(n)} = W_{\mu_1 \cdots \mu_n} dx^{\mu_1} \vee \cdots \vee dx^{\mu_n} = D_{(\mu_1} \cdots D_{\mu_n)} W dx^{\mu_1} \vee \cdots \vee dx^{\mu_n}.$$

Using the introduced quantities we define finally some more symmetric forms

$$X^{\nu\alpha}{}_{\alpha(n)} = X^{\nu\alpha}{}_{\alpha\mu_1 \ldots \mu_n} dx^{\mu_1} \vee \cdots \vee dx^{\mu_n},$$

and

$$A_{\alpha\beta(n)} = A_{\alpha\beta\mu_1\ldots\mu_n} dx^{\mu_1} \vee \cdots \vee dx^{\mu_n}, \qquad \zeta_{\alpha\beta(n)} = \zeta_{\alpha\beta\mu_1\ldots\mu_n} dx^{\mu_1} \vee \cdots \vee dx^{\mu_n}.$$

where $X^{\mu\nu}{}_{\mu_1\ldots\mu_n}$, $A_{\mu\,\mu_1\ldots\mu_n}$ and $\zeta_{\mu_1\ldots\mu_n}$ are the components of the forms $X^{\mu\nu}{}_{(n)}$, $A_{\mu(n)}$ and $\zeta_{(n)}$. The components of the matrix elements $< m|M|n >$ are given then by

$$< m|M|n >= 0 \qquad \text{for} \quad n > m+2 \quad \text{and} \quad n = m+1,$$

$$M^{\nu_1\ldots\nu_n}{}_{\mu_1\ldots\mu_m} = \binom{m}{n}\delta^{\nu_1\cdots\nu_n}_{(\mu_1\cdots\mu_n} Z_{\mu_{n+1}\cdots\mu_m)} + \binom{m}{n-1}\delta^{(\nu_1\cdots\nu_{n-1}}_{(\mu_1\cdots\mu_{n-1}} Y^{\nu_n)}{}_{\mu_n\cdots\mu_m)}$$
$$- \binom{m}{n-2} I\, \delta^{(\nu_1\cdots\nu_{n-2}}_{(\mu_1\cdots\mu_{n-2}} X^{\nu_{n-1}\nu_n)}{}_{\mu_{n-1}\cdots\mu_m)},$$

where $X^{\mu\nu}{}_{\mu_1\ldots\mu_n}$, $Y^{\nu}{}_{\mu_1\cdots\mu_n}$ and $Z_{\mu_1\cdots\mu_n}$ are the components of the forms $X^{\mu\nu}{}_{(n)}$, $Y^{\nu}{}_{(n)}$ and $Z_{(n)}$:

$$Y^{\nu}{}_{(n)} = -I\, X^{\nu\alpha}{}_{\alpha(n)} + 2 \sum_{0\leq k\leq n} \binom{n}{k} X^{\nu\mu}{}_{(k)} \vee A_{\mu(n-k)},$$

$$Z_{(n)} = -W_{(n)}$$
$$+ \sum_{0\leq k\leq n} \binom{n}{k}\left\{ X^{\alpha\beta}{}_{(k)} \vee \left(-A_{\alpha\beta(n-k)} + I\,\zeta_{\alpha\beta(n-k)}\right) + X^{\beta\alpha}{}_{\alpha(k)} \vee \left(-A_{\beta(m)} + I\,\zeta_{\beta(m)}\right)\right\}$$
$$+ \sum_{\substack{m,k\geq 0 \\ m+k\leq n}} \frac{n!}{k!m!(n-k-m)!} X^{\alpha\beta}{}_{(k)} \vee \left(A_{\alpha(m)} \vee A_{\beta(n-k-m)} - I\,\zeta_{\alpha(m)} \vee \zeta_{\beta(n-k-m)}\right).$$

5 An algorithm for the derivatives of the H_k.

The following is due to the papers of the other author (R.Sch.) [53-58]. Let us begin with a new tensor notation: indices at a tensor symbol shall not denote the components but the valences, that means the entries of the tensor taken as a multilinear functional. Positive integers $1, 2, 3 \ldots$ are preferred indices for the valences. Thus, $u_{12\ldots p}$ denotes a covariant p-tensor, $u^{12\ldots p}$ denotes a contravariant p-tensor, $u_{12\ldots p} + v_{12\ldots p}$ is the sum of two tensors, $u_{12\ldots p} v_{p+1,\ldots,p+q}$ is a tensor product, the natural action of a permutation π of $1, 2, \ldots, p$ on a $p-$tensor is

$$\pi u_{12\ldots p} = u_{\pi(1)\pi(2)\ldots\pi(p)}.$$

More generally, if Π is a subset of the symmetric group S_p (of permutations of $1, 2, \ldots, p$), then

$$\Pi u_{12\ldots p} := \sum_{\pi\in\Pi} \pi u_{12\ldots p}.$$

We are interested here in special permutations: a q-shuffle of $1, 2, \ldots, p$ is a $\pi \in S_p$ such that

$$\pi(1) < \pi(2) < \cdots < \pi(q) \qquad \text{and} \qquad \pi(q+1) < \pi(q+2) < \cdots < \pi(p).$$

The action of the set of all q-shuffles on a covariant tensor $u_{12\ldots p}$ is denoted by

$$u_{\underline{12\ldots q}\ \underline{q+1\ldots p}}.$$

For example:

$$u_{\underline{1}\ \underline{23}} = u_{123} + u_{213} + u_{312},$$

$$u_{\underline{12}\ \underline{34}} = u_{1234} + u_{1324} + u_{1423} + u_{2314} + u_{2413} + u_{3412}.$$

Let us define tensors $S_{a12\ldots p}$ $(p \geq 3)$ and $End\,E$-valued tensors $M_{12\ldots p}$ $(p \geq 2)$ by

$$S_{a123} := R_{a123}, \qquad S_{a1234} := R_{a123;4}$$

$$S_{a12\ldots p} := R_{a1\underline{23};\underline{45\cdots p}} + \sum_{q=2}^{p-3} R_{a1b2;\underline{3\cdots q}} S^{b}{}_{\underline{q+1\ldots p}},$$

$$M_{12} := F_{12}, \qquad M_{123} := F_{1\underline{2};\underline{3}},$$

$$M_{12\ldots p} := F_{1\underline{2};\underline{34\ldots p}} + \sum_{q=1}^{p-3} F_{a1;\underline{23\ldots q}} S^{a}{}_{\underline{q+1\ldots p}},$$

where indices after a semicolon express covariant derivatives. We define also

$$s_{a12\ldots p} := [\sigma_{;a12\ldots p}] + S_{a12\ldots p}$$

and

$$m_{12\ldots p} := [\mu_{;12\ldots p}] - M_{12\ldots p}.$$

The diagonal values of the derivatives of the Hadamard coefficients are then recursively given by

$$(p + k)[H_{k;12\ldots p}] = [(L\,H_{k-1})_{;12\ldots p})]$$

$$- \sum_{q=2}^{p} m_{12\ldots p}[H_{k;q+1,\ldots p}] - \sum_{q=3}^{p} s^{a}{}_{12\ldots q}[H_{k;a\underline{q+1\ldots p}}],$$

where

$$(LH_k)_{;12\ldots p} = H_{k;}{}^{a}{}_{a12\ldots p} + \sum_{q=0}^{p} \binom{p}{q} W_{;12\ldots q} H_{k;\underline{q+1\ldots p}}.$$

6 Discussion

Let us add some historical remarks on the explicit calculation of the diagonal values of the Hadamard coefficients and their Taylor coefficients. Already in the thirties Heisenberg and Euler [37] and Mathisson [40] have calculated $[H_1]$, $[DH_1]$, $[D^2H_1]$ for Laplace-type operators in flat space, $L = g^{ab}D_aD_b + W$, $g^{ab} = $ const. For scalar operators in curved space let us mention Hölder [38] for $[H_1]$, Günther [31] for $[H_1]$, $[H_2]$, Sakai [50] for $[H_3]$, Amsterdamski, Berkin and O'Connor [1] for $[H_4]$, Günther [32] for $[DH_1]$ and $[D^2H_1]$ and Wünsch [66-68] for $[DH_2]$ and $[D^2H_2]$. For non-scalar operators in flat space we mention [62] for $[H_4]$ and [63] for $[H_5]$.

Results for the general Laplace-type operator have been obtained by De Witt [22]: $[H_1]$, $[H_2]$, Gilkey [26]: $[H_1],[H_2],[H_3]$, one of the authors (I.G.A.) [2,3,6,7]: $[H_1],[H_2],[H_3],[H_4]$, the other author (R.Sch.) [53,54]: $[DH_1],[D^2H_1],[D^3H_1]$, Wünsch [65]: $[D^4H_1]$, and others.

The Hadamard coefficients H_k or rather the one-point quantities $[D^nH_k]$ or $[D_{a_1}D_{a_2} \cdots D_{a_n}H_k]$ derived from them have many appearances or applications:

- — Huygens' principle for hyperbolic L [40,31-33,41,53-55,65-68],

- — Spectral geometry for elliptic L and compact Riemannian (M,g) with positive definite metric g [42,43,18],

- — "Heat kernel proofs" of the index theorems [34,28],

- — Zeta-function regularization,

- — Regularized energy-momentum tensor,

- — Korteveg-De Vries hierarchy [56,15].

  It should be noted that there are some other problems in mathematical physics that require a similar technique for investigation of transport equations. These are:

- — expansion of the metric and other quantities in normal coordinates [33],

- — harmonic manifolds [49] and harmonic differential operators [58],

- — volume problems in the sense of [30] (to read geometric information from the volume of geodesic balls, truncated light cones, ...) [30,57],

- — Brownian motion.

Acknowledgements

The work of I.G.A. was supported by the Alexander von Humboldt Foundation. He is grateful to J. Eichhorn for his hospitality at the University of Greifswald.

References

[1] P. Amsterdamski, A.L. Berkin and D.J. O'Connor, Class. Quant. Grav. **6** (1989) 1981

[2] I.G. Avramidi, *The covariant methods for calculation of the effective action in quantum field theory and the investigation of higher derivative quantum gravity*, PhD Thesis, Moscow State University, Moscow, 1986; hep-th/9510140

[3] I.G. Avramidi, Teor. Mat. Fiz. **79** (1989) 219; (Theor. Math. Phys. **79** (1989) 494)

[4] I.G. Avramidi, Yad. Fiz. **49** (1989) 1185; (Sov. J. Nucl. Phys., **49** (1989) 735)

[5] I.G. Avramidi, Phys. Lett. B **236** (1990) 443

[6] I.G. Avramidi, Phys. Lett. B **238** (1990) 92

[7] I.G. Avramidi, Nucl. Phys. B **355** (1991) 712

[8] I.G. Avramidi, Phys. Lett. B **305** (1993) 27

[9] I.G. Avramidi, Phys. Lett. B **336** (1994) 171

[10] I.G. Avramidi, *Covariant methods for calculating the low-energy effective action in quantum field theory and quantum gravity*, University of Greifswald (1994), gr-qc/9403036

[11] I.G. Avramidi, *A new algebraic approach for calculating the heat kernel in quantum gravity*, University of Greifswald (1994), hep-th/9406047, J. Math. Phys. **37** (1) (1996), to appear

[12] I.G. Avramidi, *New algebraic methods for calculating the heat kernel and the effective action in quantum gravity and gauge theories*, gr-qc/9408028, in: 'Heat Kernel Techniques and Quantum Gravity', Discourses in Mathematics and Its Applications, No. 4, Ed. S.A. Fulling, Texas A&M University, (College Station, Texas, 1995), to appear

[13] I.G. Avramidi, J. Math. Phys. **36** (1995) 5055

[14] I.G. Avramidi, *Covariant approximation schemes for calculation of the heat kernel in quantum field theory*, University of Greifswald (1995), hep-th/9509075, Proc. Int. Sem. "Quantum Gravity", Moscow, June 12-19, 1995, to appear

[15] I.G. Avramidi and R. Schimming, J. Math. Phys. **36** (1995) 5042

[16] A.O. Barvinsky and G.A. Vilkovisky, Phys. Rep. **119** (1985) 1

[17] C. Bellaiche, Asterisque, **84/85** (1981) 151

[18] M.P. Berger, P. Gauduchon and E. Mazet, *Le spectre d'une variete riemannienne*, Lecture Notes in Math. **194**, Berlin 1971

[19] J.M. Bismut, J. Funct. Anal. **57** (1984) 56

[20] J.M. Bismut, Comm. Math. Phys. **98** (1985) 213

[21] B.S. De Witt and R.W. Brehme, Ann. Phys. **9** (1960) 220

[22] B.S. De Witt, *Dynamical theory of groups and fields*, (Gordon and Breach , N.Y., 1965)

[23] Y. Fujiwara, T.A. Osborn and S.F. Wilk, Phys. Rev. A **25** (1982) 14

[24] S. Fulling and G. Kennedy, Transac. Am. Math. Soc. **310** (1988) 583

[25] S. Fulling, R.C. King, B.G. Wybourne and C.C. Cummins, Class. Quant. Grav. **9** (1992) 1151

[26] P.B. Gilkey, J. Diff. Geom. **10** (1975) 601

[27] P.B. Gilkey, Compositio Math. **38** (1979) 201

[28] P.B. Gilkey, *Invariance theory , the heat equation and the Atiyah-Singer index theorem*, (Publish or Perish, Wilmington, 1984)

[29] P.B. Gilkey, Contemp. Math. **73** (1988) 79

[30] A. Gray and L. Vanhecke, Acta Math. **142** (1979) 157

[31] P. Günther, Ber. Verhand. Sächs. Akad. d. Wiss. Leipzig **100** (1952) Heft 2

[32] P. Günther, Math. Nachr. **22** (1960) 285

[33] P. Günther, *Huygens' Principle and Hyperbolic Equations*, (Academic Press, San Diego, 1988)

[34] P. Günther and R. Schimming, J. Diff. Geom. **12** (1977) 599

[35] V. P. Gusynin, Nucl Phys. B **333** (1990) 296

[36] J. Hadamard, *Lectures on Cauchy's problem*, (Yale Univ. Press, New Haven, 1923)

[37] W. Heisenberg and H. Euler, Z. Phys. **98** (1936) 714

[38] E. Hölder, Ber. Verh. Sächs. Akad. Wiss. Leipzig, **99** (1938) 55

[39] T. Jacobson, J. Math. Phys. **26** (1985) 1600

[40] M. Matthisson, Acta Math. **71** (1939) 249

[41] R. G. McLenaghan, Proc. Camb. Phil. Soc., **65** (1969) 139

[42] S. Minakshisundaram and A. Plejel, Canad. J. Math. **1** (1949) 242

[43] S. Minakshisundaram, J. Indian Math. Soc. **17** (1953) 158

[44] F.H. Molzahn and T.A. Osborn, J. Math. Phys. **27** (1986) 88

[45] R.I. Nepomechie, Phys. Rev. D **31** (1985) 3291

[46] Yu.N. Obukhov, Nucl. Phys. B **212** (1983) 237

[47] T.A. Osborn and F.H. Molzahn, Phys. Rev. A **34** (1986) 1696

[48] T.A. Osborn and R.A. Corns, J. Math. Phys. **26** (1985) 453

[49] H.S. Ruse, A.G. Walker, T.J. Willmore, *Harmonic spaces*, (Edizioni Cremonese, Roma, 1961)

[50] T. Sakai, Tohoku Math. J. **23** (1971) 589

[51] R.T. Seeley, AMS, Proc. Symp. Pure Math. 10 (1967) 288

[52] R.T. Seeley, Am. J. Math. **91** (1969) 889

[53] R. Schimming, Ukrainsk. Mat. Z., **29** (1977) 351

[54] R. Schimming, Beiträge zur Analysis, **11** (1978) 45

[55] R. Schimming, Beiträge zur Analysis, **15** (1981) 77

[56] R. Schimming, Z. f. Analysis u. ihre Anwend., **7** (1988) 263

[57] R. Schimming, Archivum Math. Brno, **24** (1988) 5

[58] R. Schimming, Forum Math., **3** (1991) 177

[59] R. Schimming, *Calculation of the heat kernel coefficients*, in: *Analysis, Geometry and Groups. A Riemann Legacy Volume*, Eds. H. M. Srivastava and Th. M. Rassias, (Hadronic Press, Palm Harbour, 1993), p. 627

[60] J.L. Synge, Proc. London Math. Soc. **32** (1931) 241

[61] J.L. Synge, *Relativity. The general theory*, (North Holland, Amsterdam, 1960)

[62] A.I. Vainstein, V.I. Zakharov, V.A. Novikov and M.A. Shifman, Yad. Fiz. 39 (1984) 124

[63] A.E.M. Van de Ven, Nucl. Phys. B250 (1985) 593

[64] H. Widom, Bull. Sci. Math. 104 (1980) 19

[65] V. Wünsch, Math. Nachr. **47** (1970) 131

[66] V. Wünsch, Math. Nachr. **73** (1976) 37

[67] V. Wünsch, Beiträge zur Analysis **13** (1979) 147

[68] V. Wünsch, Math. Nachr. **120** (1985) 175

[69] J.A. Zuk, Phys. Rev. D **32** (1985) 2650

[70] J.A. Zuk, Phys. Rev. D **33** (1986) 3645

[71] J.A. Zuk, Phys. Rev. D **34** (1986) 1791

[72] J.A. Zuk, Nucl. Phys. B **280** (1987) 125

Finite Temperature Field Theory: Physical Effects of Nontrivial Spectral Functions

Peter A. Henning

At finite temperature particles are subject to collisions, therefore stable asymptotic states, i.e., with a sharp dispersion law, do not exist. In this talk I report on three different physical topics connected with an approach to finite temperature field theory in terms of a continuous mass spectrum [1]:

- **Equilibrium Systems:**
  The self consistent determination of appropriately parameterized spectral functions is an entirely non-perturbative method [2]. In hot plasma systems, a nonzero spectral width it leads to a suppression of long wavelength photon emission [3].

- **Non-Equilibrium Systems:**
  Nontrivial spectral functions lead to memory effects in the relaxation of many-body systems: The relaxation process is essentially non-Markovian [4]. For a toy model of a quark–gluon plasma it is found, that this may substantially hinder the thermalization of the plasma over long time-scales [5].

- **Formal:**
  If spectral functions are determined by approximate methods, one may be subject to serious errors: The four dimensional Fourier transform of the spectral function is the expectation value of the (anti-) commutator of two field operators, hence should vanish for spacelike coordinate differences. Commonly used models for the spectral function violate this requirement [6].

References

[1] P.A. Henning, Phys.Rep. **253** (1995) 235

[2] P.A. Henning, R. Sollacher und H. Weigert, *Fermion damping rate in a hot medium*, GSI Preprint 94-56 (1994), hep-ph 9409280

[3] P.A. Henning und E. Quack, Phys.Rev.Lett. **75** (1995) 2811

[4] P.A. Henning, Nucl.Phys. **A582** (1995) 633 (Erratum: **A586** (1995) 777)

[5] P.A. Henning, *Quantum Transport Theory*, *in:* Proceedings of the 4th Workshop on Thermal Field Theories, eds. Y.X. Gui et.al. (World Scientific, Singapore); hep-ph 9510315

[6] P.A. Henning, E. Poliatchenko und T. Schilling, *Approximate spectral functions in thermal field theory*, hep-ph 9510322

Bose-Einstein Condensation under External Conditions

Klaus Kirsten[1]

We examine the occurrence of Bose-Einstein condensation in both nonrelativistic and relativistic systems with no self-interactions in a general setting. A simple condition for the occurrence of Bose-Einstein condensation can be given if we adopt generalized ζ-functions to define the quantum theory. We show that the crucial feature governing Bose-Einstein condensation is the dimension q associated with the continuous part of the eigenvalue spectrum of the Hamiltonian for nonrelativistic systems or the spatial part of the Klein-Gordon operator for relativistic systems. In either case Bose-Einstein condensation can only occur if $q \geq 3$ [1]. Several examples, some of them not completely solved before, may be treated very easily using our simple criterion [2]. These examples are charged free gases confined in finite volumes as torus, boxes, spheres, cylinders (see also [3, 4, 5, 6]) and in the presence of general constant external magnetic fields ([7, 8, 9, 10]).

References

[1] K. Kirsten and D.J. Toms, Simple criterion for the occurrence of Bose-Einstein condensation, to appear in Phys. Lett. B

[2] K. Kirsten and D.J. Toms, in preparation

[3] C.S. Zasada and R.K. Pathria, Phys. Rev. A **15** (1977) 2439

[4] M.B. Al'taie, J. Phys. A **11** (1978) 1603

[5] J.S. Dowker and J.P. Schofield, Nucl. Phys. B **327** (1989) 267

[6] D.J. Toms, Phys. Rev. Lett. **8** (1992) 1152

[7] M.R. Schafroth, Phys. Rev. **100** (1955) 463

[8] J. Daicic, N.E. Frankel, R.M. Gailis and V. Kowalenko, Phys. Rep. **237** (1994) 63

[9] D.J. Toms, Phys. Rev. D **50** (1994) 6457

[10] P. Elmfors, P. Liljenberg, D. Persson and B.-S. Skagerstam, Phys. Lett. B **348** (1995)462

---

[1]The presented talk is a result of a collaboration with David J. Toms, Department of Physics, University of Newcastle Upon Tyne, Newcastle Upon Tyne, U. K. NE1 7RU

Speed of Light in Nontrivial Vacua

José I. Latorre[1]

The usual vacuum in a Quantum Field Theory is a state characterized by the absence of real particles and classical fields and by its Minkowskian geometry. Electromagnetic and gravitational fields as well as massless particles propagate though it with the same constant, Lorentz invariant speed, c. When the vacuum is modified, so is the speed of propagation of particles and fields. This quantum field theoretical effect has been analyzed within QED for low–energy photons in several cases, resulting in a universal formula

$$v = 1 - \frac{44}{135}\, \alpha^2\, \frac{\rho}{m_e^4}$$

where ρ is the energy density relative to the standard vacuum and where, if the vacuum is a gravitational one, one α has to be substituted by $m_e^2 G_N$. It follows automatically that if the vacuum has a lower energy density than the standard vacuum, $\rho < 0$ and $v > 1$, and viceversa. Therefore, whether photons move faster or slower than c depends on the lower or higher energy density of the modified vacuum, respectively. Physically, a higher energy density is characterized by the presence of real particles in the vacuum whereas a lower one stems from the absence of some virtual modes.

We have also studied modifications of the speed of light in other theories up to two loops. The most concise summary of our results states that only low–energy massless particles corresponding to a massive theory show genuine modifications of their speed while remaining massless. All other modifications are mass-related, or running mass-related.

---

[1]based on: J. I. Latorre, P. Pascual and R. Tarrach, Nucl.Phys. **B437**(1995) 60; hep-th/9408016.

Influence of Condensates and External Fields on Physical Effects in Gauge Theories

Vladimir Ch. Zhukovsky

Some recent results of investigations of the influence of external conditions such as condensate and external fields, finite temperature and matter density on various physical effects in gauge theories, obtained at the Department of Theoretical Physics of the Moscow State University with contributions of A.V.Borisov, P.A.Eminov and A.S.Vshivtsev as well as of some other collaborators are reviewed in this talk.

The outline of the talk is as follows.

1. QCD excitations at short and large distances. Condensate fields. Effective models for the low momenta regions [1,2].

2. Thermodynamic potential for the SU(2) gauge theory in the presence of condensate fields [3,4].

3. The role of condensate fields in the high energy scattering processes.

    3.1. Quark condensate formation [5,6].

    3.2. Gluon condensate fields and their influence on the deep inelastic scattering processes [5].

    3.3. Gluon condensate and its influence on the lepton pairs annihilation process [7].

4. Radiative effects.

    4.1. Electron mass operator in a magnetic field [8.9], the quark mass shift in the gluon condensate field [10] and the energy shift and anomalous magnetic moment of a neutrino in a constant field [11] at finite temperature and matter density.

    4.2. Two–loop thermodynamic potential of QED [12] and polarization operator in 2+1 dimensional QED [13] in a constant magnetic field.

    4.3. Production of photinos [14] and radiative neutrino decay [15] in an intense magnetic field.

References

[1] D. Ebert, Yu.L. Kalinovsky, L. Münchow and M.K. Volkov, Int. Journal of Mod. Phys. A, vol.8, no.7., 1295, 1993.

[2] J. Ellis and K. Geiger, Real–time description of parton–hadron conversion and confinement dynamics, CERN–TH., 95–35, March 1995.

[3] A.S. Vshivtsev, V.Ch. Zhukovsky and A.O. Starinets, Z.Phys. C, vol.61, no.2, 285, 1994.

[4] A.O. Starinents, A.S. Vshivtsev and V.Ch. Zhukovsky, Phys. Lett., vol.B 322, 403, 1994.

[5] A.V. Averin, A.V. Borisov and V.Ch. Zhukovsky, Z.Phys. C, vol.48, 457, 1990.

[6] A.S. Vshivtsev, V.Ch. Zhukovsky and A.V. Tatarintsev, Izv.Vissh.Uchebn.Zaved., Fiz., no.1, 39, 1994.

[7] V.Ch. Zhukovsky, A. Grigoruk, Z.Phys. C (to be published).

[8] V.Ch. Zhukovskii, P.G. Midodashvili and P.A. Eminov, Zh.Eksp.Teor.Fiz., vol.106, 929, 1994.

[9] V.Ch. Zhukovskii, T.L. Shoniya and P.A. Eminov, Physics of Atomic Nuclei, vol.57, no.8, 1365 (Yad.Fiz., vol.57, no.8, 1437, 1994).

[10] V.Ch. Zhukovsky, K.G. Levtchenko, T.L. Shoniya and P.A. Eminov, Vest.Mosk.Univ., Fiz.Astr. (to be published).

[11] V.Ch. Zhukovskii, T.L. Shoniya and P.A. Eminov, Zh.Eksp.Teor.Fiz., vol.104, 3269, 1993.

[12] V.Ch. Zhukovsky, T.L. Shoniya and P.A. Eminov, Zh.Eksp.Teor.Fiz., vol. 107, no. 2, 299, 1995.

[13] P.A. Eminov, K.V. Zhukovsky, Yad.Fiz. (to be published).

[14] V.Ch. Zhukovsky, P.A. Eminov, Yad.Fiz., vol.52, no.5(11), 1473, 1990.

[15] P.A. Eminov, A. Grigoruk, V.Ch. Zhukovsky, Phys.Lett. (to be published).

Effective Potential in Yang-Mills Theory and Stability of Chromomagnetic Vacuum

Ivan G. Avramidi[1]

In non-Abelian gauge theories of general form

$$S = \int dx \left\{ -\frac{1}{2g^2}\mathrm{tr}(\mathcal{F}_{\mu\nu}^2) + i\bar{\psi}(\gamma^\mu \nabla_\mu + M(\varphi))\psi + \frac{1}{2}\varphi^T(-\Box)\varphi + V(\varphi) \right\}.$$

the effective interaction coupling becomes strong at long distances and leads to the confinement of color. The main problem in studying the confinement is the investigation of the structure of the vacuum that turns out to be far more complicated than the perturbative one. A possible approach to study the vacuum structure is to put forward some explicit simple model, which allows the analytical investigation. This was initiated first by Savvidy [1] who showed that in Yang-Mills $SU(2)$ model the perturbative empty vacuum is unstable under the creation of a constant chromomagnetic field that leads to the negative vacuum energy. The Savvidy's result was significantly specified in further investigations, where it was shown, in particular, that the Savvidy's magneto-vacuum is still unstable (for the refs. see [2]). Thus the space filled with a constant homogeneous chromomagnetic field can serve as a simple and visual model of the nonperturbative vacuum.

The natural tool for investigating the vacuum structure is the effective potential, i.e. the low-energy limit of the effective action, which is determined by covariantly constant background fields. To define the low-energy limit in gauge theories one has to factorize out the gauge degrees of freedom. That is why it is the *covariantly* constant background that should be used in defining the effective potential.

In his pioneering work Savvidy studied the simplest non-Abelian $SU(2)$ gauge group and the only possible covariantly constant chromomagnetic field with only one nonvanishing color and space-time component. However, when investigating more complicated groups and space-times of higher dimension the covariantly constant background can have far more complex structure with *many independent color and space-time invariants*. We investigate the effective action for this *general* case and analyze some new opportunities to ensure the stability of the vacuum that it provides.

On the covariantly constant background the one-loop effective action does not depend on the gauge fixing parameters [2] and is determined by the ζ-functions

$$\Gamma_{(1)} = -\frac{1}{2}\zeta_\Delta'(0) + \zeta_F'(0) - \frac{1}{2}\zeta_N'(0) + \frac{1}{2}\zeta_K'(0),$$

---

[1]Alexander von Humboldt Fellow, on leave of absence from Research Institute for Physics, Rostov State University, Stachki 194, 344104 Rostov-on-Don, Russia.

where

$$\Delta^\mu_{\ \nu} = -\Box\,\delta^\mu_{\ \nu} - 2\mathcal{F}^\mu_{\ \nu}, \quad F = -\Box, \quad N = -\Box + Q, \quad K = -\Box - \frac{1}{2}\gamma^{\mu\nu}\mathcal{R}_{\mu\nu} + M^2,$$

Q is the mass matrix of the scalar fields, $\mathcal{R}_{\mu\nu}$ is the field strength in representation of spinor fields, and $\zeta_L(p)$ is defined via the heat kernel

$$\zeta_L(p) = \mu^{2p}\mathrm{Tr}\,L^{-p} = \frac{\mu^{2p}}{\Gamma(p)}\int_0^\infty dt\ t^{p-1}\mathrm{Tr}\exp(-tL).$$

The heat kernels for operators Δ, F, N and K are explicitly calculated in [2] by means of covariant algebraic method of [3].

The vacuum is described by the spectrum of the operators Δ, F, N and K determining the dynamics of the quantum fields on the given background. It is stable if these operators are positive, i.e. do not have negative modes. The presence of the negative modes leads to the imaginary part of the effective action and, as a consequence, to the instability of the vacuum. To study the problem of stability it is not needed to know the whole spectrum. The minimal eigenvalue of an operator can be found by analyzing the asymptotic behavior of the heat kernels at $t \to \infty$. Using the explicit results for the heat kernels we have found the minimal eigenvalues of all operators and analyzed therefrom the conditions of absence of negative modes.

The ghost and matter fields operators F, N and K are found to be positive provided some positive singlet contributions in mass matrices M^2 and Q are present. The real problem is the gauge operator Δ. It is positive only in the case when the background satisfies a *condition of stability of the vacuum* that is explicitly formulated in [2] in form of an inequality that must be satisfied by the invariants of the gauge field strength $\mathcal{F}_{\mu\nu}$. Roughly speaking, to ensure the stability there must be more than one spacetime invariant of the gauge field strength and *the maximal invariant must be smaller than the sum of all other ones*. In other words, the Savvidy-like vacuum with constant chromomagnetic fields can be stable only in the case when more than one constant chromomagnetic fields are present and the values of these fields differ little from each other. This is possible only in the spaces of dimension greater or equal than four.

In four-dimensional space, but only with *Euclidean* signature, there is a special stable background field configuration (with two equal field invariants). Only beginning with $d \geq 5$ there exists such a background that, on the one hand, ensures the operator Δ to be positive definite and, on the other hand, assumes the analytic continuation on the pseudo-Euclidean space-time of Lorentzian signature.

References

[1] G.K. Savvidy, Phys. Lett. **B71** (1977) 133

[2] I.G. Avramidi, J. Math. Phys. **36** (1995) 1557

[3] I.G. Avramidi, Phys. Lett. **B305** (1993) 27

Ground State Energy Density in Smooth Background Fields

Joachim Lindig and Michael Bordag

The concept of vacuum energy has found important applications in several fields of physics since its discovery in 1948 by H.B.G. Casimir [1]. A typical situation is encountered in cosmology when quantizing matter fields in classical gravitational background. The presence of the gravitational background results in a non-trivial vacuum expectation value of the energy-momentum-tensor (EMT) which, in turn, acts as a source of gravitational field [2, 3]. As implied by this example, the EMT is one of the fundamental quantities when describing quantum fields in interaction with an external system.

In the following, a real scalar field in the presence of an arbitrary background field $V(x)$ will be investigated [4]. The problem of renormalization is solved. In the particular case of a background field depending on one coordinate only, an integral representation for the renormalized expectation value of the EMT is given. It is accessible to numerical analysis. More specifically, the system under consideration is described by the Lagrange density

$$L = \frac{1}{2}\Phi(-\partial_\mu\partial^\mu - M^2 - \lambda\Phi^2)\Phi + \frac{1}{2}\varphi(-\partial_\mu\partial^\mu - m^2 - \lambda'\Phi^2)\varphi \tag{1}$$

where φ denotes the quantized scalar field. Φ is a classical Φ^4-self-interacting field modelling the background interaction via the identification $V(x) = \lambda'\Phi^2$. The vacuum expectation value of the canonical EMT is now expressed in terms of the causal propagator. The heat-kernel-expansion [2] of the propagator provides the appropriate technique to isolate the counterterms. After adding the total divergence term $\frac{\kappa}{2}\left(-\partial^\mu\partial^\nu\Phi^2 + g^{\mu\nu}\partial_\rho\partial^\rho\Phi^2\right)$ to the canonical EMT, we are led to the following renormalization of the constants of the classical system Φ

$$M^2 \longrightarrow M^2 - \frac{m^2\lambda'}{16\pi^2}\left(\frac{1}{s} - 2C + \ln\frac{\mu^2}{m^2}\right)$$

$$\lambda \longrightarrow \lambda - \frac{\lambda'^2}{32\pi^2}\left(\frac{1}{s} - 2C - 1 + \ln\frac{\mu^2}{m^2}\right)$$

$$\kappa \longrightarrow \kappa + \frac{\lambda'}{48\pi^2}\left(\frac{1}{s} - 2C + \ln\frac{\mu^2}{m^2}\right). \tag{2}$$

The normalisation condition has been chosen so that the EMT of the quantum system does not contain contributions proportional to the classical terms. It is evident that the renormalization requires the embedding in the classical system Φ. As expected, the direct calculation of the vacuum energy using the mode summation method does not produce the total divergence term [5].

If the background field depends on one coordinate only $V(x) = V(x_3)$ (and decreases sufficiently fast at infinity), the causal propagator can be represented in terms of the one-dimensional scattering basis $\varphi_1(x_3), \varphi_2(x_3)$. The counterterms can be recovered by asymptotically expanding the integrand. Then, the renormalization procedure as described above provides us with the following integral representation for the renormalized energy density

$$
\begin{aligned}
< T^{00}(x) >_0^{ren} \;=\; & \int_0^\infty \frac{dr\, r^2}{(2\pi)^2}\left[\frac{\left(\gamma^2 - \frac{2}{3}r^2 + V\right)\varphi_1\varphi_2 + \varphi_1'\varphi_2'}{W(\varphi_1,\varphi_2)}\right. \\
& + \left.\frac{1}{2\gamma}\left\{\frac{2}{3}r^2 - \frac{1}{3}\frac{r^2}{\gamma^2}V + \frac{1}{\gamma^2}\frac{1}{4}V'' - \frac{1}{3}\frac{r^2}{\gamma^4}\left(\frac{1}{4}V'' - \frac{3}{4}V^2\right)\right\}\right].
\end{aligned}
\tag{3}
$$

The 33-component is treated likewise. The energy density vanishes with the background potential tending to zero. It has been calulated explicitly for the square well potential, the piecewise linear and the piecewise oscillatory potential where the scattering basis can be expressed by special functions. These examples reflect the fact that. in general, the energy density may exhibit a singularity at points of discontinuity of the potential or its derivatives. A detailed analysis shows terms diverging as ε^{-2}, ε^{-1} and $\ln\varepsilon$ where ε measures the distance to the point of discontinuity. The coefficients of these terms are proportional to the jump of the potential, its first and second derivative, respectively. Hence, a twice continuously differentiable background potential results in a continuous energy density. The singularities arise as a consequence of the boundary conditions the scattering basis has to obey at points of discontinuity. Thus. they reflect the idealized nature of discontinuous background potentials.

References

[1] H.B.G. Casimir, Proc. Kon. Ned. Akad. Wetenschap **B 51** (1948) 793, Physica **19** (1953) 846

[2] N.D. Birrell & P.C.W.Davies, Quantum Fields in Curved Space. Cambridge Univ. Press.(1982)

[3] A.A. Grib, S.G. Mamayev & V.M. Mostepanenko, Vacuum Quantum Effects in Strong Fields, Friedmann Lab. Publ. (1994)

[4] J. Lindig, Diploma Thesis "The Calculation of the Ground State Energy Density in Finite External Potentials" (in German), University of Leipzig, 1995

[5] M. Bordag, J. Phys. A **28**, (1995) 755

Part IV

Quantum Fields in Black Hole Background

Black Hole Entropy

Valeri Frolov

1 Dynamical Degrees of Freedom and No-Boundary Wave Function of a Black Hole

According to the thermodynamical analogy in black hole physics, the Bekenstein-Hawking entropy of a black hole in the Einstein theory of gravity is $S^{BH} = A_H/(4l_{\rm P}^2)$, where A_H is the area of a black hole surface and $l_{\rm P} = (\hbar G/c^3)^{1/2}$ is the Planck length.

The success of the thermodynamical analogy in black hole physics allows one to hope that this analogy may be even deeper and it is possible to develop statistical-mechanical foundation of black hole thermodynamics. The thermodynamical and statistical-mechanical definitions of the entropy are logically different. *Thermodynamical entropy* S^{TD} is defined by the response of the free energy F of a system to the change of its temperature: $dF = -S^{TD}dT$. This definition applied to a black hole determines its Bekenstein-Hawking entropy. *Statistical-mechanical entropy* S^{SM} is defined as $S^{SM} = -\mathrm{Tr}(\hat{\rho}\ln\hat{\rho})$, where $\hat{\rho}$ is the density matrix describing the internal state of the system under consideration.

Is the analogy between black holes thermodynamics and the 'standard' thermodynamics complete? Are there internal degrees of freedom of a black hole responsible for its entropy? Is it possible to apply the statistical-mechanical definition of the entropy to black holes and how is it related with the Bekenstein-Hawking entropy? In the present talk I discuss these questions.

The problem of the dynamical origin of the black hole entropy was intensively discussed recently. The proposed basic idea is to relate the dynamical degrees of freedom of a black hole with its quantum excitations. This idea has different realizations [1, 2, 3, 4, 5, 6] (see also review [7]). In the *dynamical-black-hole-interior* model proposed in Ref. [5] the dynamical degrees of freedom of a black hole are identified with the internal modes of all physical fields that might exist in the black hole interior. The set of the fields must include the gravitational one. It was shown (see [5]) that the contribution of a field to the statistical-mechanical entropy of a black hole calculated by counting the internal modes of a black hole is formally divergent: $S^{SM} = \alpha A/l^2$, where A is the surface area of a black hole, l is a proper distance cut-off parameter and a dimensionless parameter α depends on the type of the field.

The calculation of the black hole entropy can be simplified by using the following trick. Consider an eternal version of a black hole, i.e. an eternal black hole with the same mass M as the original black hole formed as the result of collapse. At late time the geometry of both holes are identical. One can trace back in time all the perturbations, propagating at late time in the geometry of the eternal version of a black hole . As the result one can relate perturbations at late time in a spacetime of a real black hole with initial data on the Einstein-Rosen bridge (spatial slice of $t =$const)

176   V. Frolov

of the eternal black hole geometry.

We denote the 3-surface of the Einstein-Rosen bridge by Σ. This surface has the topology $S^2 \times R^1$. The 2-surface S of the horizon $r = 2M$ splits it into two isometric parts: 'external' Σ_+ and 'internal' Σ_-. In a spacetime of the eternal black hole the Killing vector ξ which is used to define energy is future directed on Σ_+ and past directed on Σ_-. For this reason initial date having a support located on Σ_+ correspond to the field configurations having positive energy, while the energy of the field configurations with the initial data on Σ_- possess negative energy. The former describes external degrees of freedom of a black hole, while the latter describes the internal ones. The set of fields representing the degrees of freedom of a black hole contains the gravitational perturbations. For given initial values of fields and gravitational perturbations on Σ the gravitational constraint equations determine the deformation of the 3-geometry of the Einstein-Rosen bridge. We shall use the notion 'deformation' in order to describe not only deformed geometry of the Einstein-Rosen bridge, but also the physical fields on it. By using this terminology we can say that the states of a black hole at late time are uniquely related with deformations of the Einstein-Rosen bridge.

It was proposed in [6] to introduce a wave function of a black hole as the functional over the space of deformations of the Einstein-Rosen bridge. In this approach the wave function of a black hole depends on data located on both parts of the Einstein-Rosen bridge: an external Σ_+ (external degrees of freedom) and an inner Σ_- one (internal degrees of freedom).

Certainly there exist infinite number of different wavefunctions of a black hole. Our aim is to get a useful tool for the description of the canonical ensemble of black holes inside the cavity restricted by a spherical boundary of the radius r_B and with fixed inverse temperature β on it. For this reason only very special wavefunctions will be important for us.

Here we present a modified version of the no-boundary approach of Ref. [6] which is analogous to the 'no-boundary ansatz' in quantum cosmology [8]. This ansatz singles out a set of no-boundary wavefunctions which is convenient for our purpose.

Instead of the complete Einstein-Rosen bridge we consider its part Σ' lying between two spherical 2-dimensional boundaries $S_\pm$ located from both sides of S at $r = r_B$. Σ' has the topology $S^2 \times I$, where I is the unit interval $[0, 1]$. We denote by M_β a Euclidean manifold with a boundary ∂M_β, which consists of two parts: Σ' and another 3-surface Σ^B with the same topology $S^2 \times I$, which intersects Σ' at S_+ and S_-, and which represents the Euclidean evolution of the external boundary B. We define the no-boundary wavefunction depending on one parameter β by the following path integral

$$\Psi_\beta({}^3g(x), \varphi(x)) = \int \mathcal{D}\,{}^4g\,\mathcal{D}\phi\,\exp\left(-I[{}^4g, \phi]\right). \tag{1}$$

Here $I[{}^4g, \phi]$ is the Euclidean gravitational action. The integral is taken over Euclidean 4-geometries and matter-field configurations on a spacetime M_β with a boundary $\partial M_\beta \equiv \Sigma' \cup \Sigma^B$. The integration variables are subject to the conditions $({}^3g(\boldsymbol{x})$, $\varphi(\boldsymbol{x}))$, $\boldsymbol{x} \in \partial M_\beta$, – the collection of 3-geometry and boundary matter fields on ∂M_β, which are just the argument of the wavefunction (1).

The no-boundary wavefunction $\boldsymbol{\Psi}_\beta[M, \varphi_+, \varphi_-]$ of a black hole in the semiclassical approximation is

$$\boldsymbol{\Psi}_\beta = N \, \exp\left(-\frac{1}{2\beta r_B}[1 - (1 - \frac{2M}{r_B})^{\frac{1}{2}}] + 2\pi M^2 - \sum I_2[\varphi_+, \varphi_-] \right). \qquad (2)$$

Here N is a normalization constant, and M is a mass of a black hole. The symbol of summation in the exponent indicates that the additional a summation over all physical fields must be done. $I_2[\varphi_+, \varphi_-]$ is the following quadratic form:

$$I_2 = \sum_\lambda \left\{ \frac{\omega_\lambda \cosh(\beta_\infty \omega_\lambda/2)}{2 \sinh(\beta_\infty \omega_\lambda/2)} \, (\varphi_{\lambda,+}^2 + \varphi_{\lambda,-}^2) - \frac{\omega_\lambda}{\sinh(\beta_\infty \omega_\lambda/2)} \, \varphi_{\lambda,+} \varphi_{\lambda,-} \right\}, \qquad (3)$$

and $\varphi_{\lambda,\pm}$ are coefficients of the decomposition of the boundary values φ on $\Sigma_\pm$ in the basis of spatial harmonics R_λ on M_β: $\varphi_\pm(\boldsymbol{x}) = \sum_\lambda \varphi_{\lambda,\pm} R_\lambda(\boldsymbol{x})$. This action is a sum of Euclidean actions for quantum oscillators of frequency ω_λ for the interval $\beta_\infty \equiv \beta(1 - 2M/r_B)^{1/2}$ of the Euclidean time with the initial value of its amplitude φ_- and the final value φ_+.

The square of this wavefunction gives the probability to find a given configuration in the state determined by the parameter β. For large M ($M \gg m_{\rm P}$) this probability is a sharp peak with width $\approx m_{\rm P}$ located near the value of $M = M_\beta \equiv \beta_\infty/8\pi$. For $\beta_\infty = 8\pi M$ and $r_B \to \infty$ this wavefunction coincides with a no-boundary wavefunction obtained in [6].

For fixed M the density matrix for internal variables φ_- of a black hole is

$$\hat{\rho}_\beta[\varphi_-, \varphi_-'] = \int D\varphi_+ \boldsymbol{\Psi}_\beta[\varphi_+, \varphi_-] \boldsymbol{\Psi}_\beta[\varphi_+, \varphi_-'] = \hat{\rho}[\varphi_-, \varphi_-'] = P' \, e^{-\tilde{I}_2[\varphi_-, \varphi_-']}, \qquad (4)$$

where $\tilde{I}_2$ is given by the expression (3) with β changed by 2β. It is easy to show that: $\hat{\rho}[\varphi_-, \varphi_-'] = P''\langle\varphi_-|e^{-\beta\hat{H}}|\varphi_-'\rangle$, where P'' is a normalization constant and $\hat{H}$ is the Hamiltonian of free fields φ propagating on the Schwarzschild background. The statistical-mechanical entropy S^{SM} of a black hole obtained by using this density matrix coincides with the expression obtained in Ref. [5].

2   Renormalized Effective Action and Free Energy

The partition function $Z(\beta)$ for the canonical ensemble of black holes with a given inverse temperature β at the boundary is given by the Euclidean path integral [9]

$$Z(\beta) = \int D[\phi] \exp(-I[\phi]). \qquad (5)$$

Here the integration is taken over all fields including the gravitational one that are real on the Euclidean section and are periodic in the imaginary time coordinate τ with period β. The quantity ϕ is understood as the collective variable describing the fields. In particular it contains the gravitational field. Here $D[\phi]$ is the measure of the space of fields ϕ and I_E is the Euclidean action of the field configuration. The action I_E includes the Euclidean Einstein action. The state of the system is determined by the choice of the boundary conditions on the metrics that one integrates over. For the canonical ensemble for the gravitational fields inside a spherical box of radius r_B at temperature T one must integrate over all the metrics inside r_B which are periodically identified in the imaginary time direction with period $\beta = T^{-1}$. The partition function Z is related with the effective action $\Gamma = -\ln Z$ and with the free energy $F = \beta^{-1}\Gamma = -\beta^{-1}\ln Z$.

By using the stationary-phase approximation one gets $\beta F \equiv -\ln Z = I[\phi_0] - \ln Z_1 + \dots$. Here ϕ_0 is the (generally speaking, complex) solution of classical field equations for action $I[\phi]$ obeying the required periodicity and boundary conditions. Besides the tree-level contribution $I[\phi_0]$, the free energy includes also one-loop corrections $\ln Z_1$, connected with the contributions of the fields perturbations on the background ϕ_0, as well as higher order terms in loops expansion, denoted by $(\dots)$. The one-loop contribution of a field ϕ can be written as follows $\ln Z_1 = -\frac{1}{2}\mathrm{Tr}\ln(-D)$, where D is the field operator for the field ϕ inside the box r_B. The one-loop contribution contains divergences and required the renormalization. In order to be able to absorb these divergences in the renormalization of the coefficients of the bare classical action we chose the latter in the form

$$I_{cl} = \int d^4 x \sqrt{g}\left[-\frac{\Lambda_B}{8\pi G_B} - \frac{R}{16\pi G_B} + c_B^1 R^2 + c_B^2 R_{\mu\nu}^2 + c_B^3 R_{\alpha\beta\mu\nu}^2 \right]. \tag{6}$$

The divergent part of the one-loop effective action has the same structure as the initial classical action (6) and hence one can write $\Gamma = \Gamma_{cl}^{ren} + \Gamma_1^{ren}$, $\Gamma_1^{ren} = \Gamma_1 - \Gamma_1^{div} = \Gamma_1^{fin}$. Here Γ_{cl}^{ren} is identical to the initial classical action with the only change that all the bare coefficients Λ_B, G_B, and c_B^i are substituted by their renormalized versions Λ_{ren}, G_{ren}, and c_{ren}^i. After multiplying the the renormalized effective action by β^{-1} we get the expansion for the renormalized free energy.

The effective action Γ contains the complete information about the system under consideration. In particular the variation of Γ with respect to the metric provides one with the equations for the quantum average metric $\bar{g} = \langle g \rangle$: $\delta\Gamma/\delta\bar{g} = 0$. One usually assumes that quantum corrections are small and solves this equation perturbatively: $\bar{g} = g_{cl} + \delta g$, where g_{cl} is a solution of the classical equations. One usually assumes that the renormalized values of Λ_{ren} and c_{ren}^i vanish $\Lambda_{ren} = c_{ren}^i = 0$. It means that in general case their bare values were not vanishing unless one is dealing with some special type of theory (e.g. assuming supersymmetry).To provide the condition that δg is small, one is to begin with the metric g_{cl} that is an extremum of Γ_{cl}^{ren} : $\delta\Gamma_{cl}^{ren}/\delta g_{cl} = 0$.We assume that the renormalization of the coupling constants in the classical action is made from the very beginning. In this case g_{cl} is the Euclidean black hole metric, while the metric $\bar{g}$ describes the Gibbons-Hawking instanton deformed due to the presence of quantum corrections to the metric. The quantity $\Gamma[\bar{g}]$ being expressed as the function of boundary conditions (β and r_B) specifies the thermodynamical properties of a black hole.

3   Relation Between S^{SM} and S^{TD}

The leading tree-level contribution to the renormalized effective action given by the Euclidean gravitational action for the Euclidean black hole solution (the so-called Gibbons-Hawking instanton) is [11, 12]:

$$F_0^{ren} \equiv \beta^{-1}\Gamma_{cl}^{ren}[g_{cl}] = r_B \left(1 - \sqrt{1 - r_+/r_B}\right) - \pi r_+^2 \beta^{-1}.$$

Here $r_+ = 2G_{ren}M$ is the gravitational radius of a black hole of mass M, which for a given temperature β^{-1} at the boundary r_B is defined by the relation $\beta = 4\pi r_+(1 - r_+/r_B)^{1/2}$. One can easily verify that

$$S_0^{TD} = -dF_0^{ren}/dT \equiv \beta^2 dF_0^{ren}/d\beta = A_H/4l_P^2, \tag{7}$$

and hence S_0^{TD} coincides with the Bekenstein-Hawking expression S^{BH}. (It is assumed that r_+ in F_0^{ren} is expressed in terms of β and r_B before differentiation with respect to β.) One-loop contribution describes quantum correction to the entropy of a black hole as well as the entropy of thermal radiation in its exterior. The latter evidently depends on the radius r_B of the boundary. Since the Euclidean black hole background is regular the corresponding contribution F_1^{ren} is finite. For this reason the quantum corrections to the Bekenstein-Hawking entropy $4\pi M^2$ are small unless the mass of a black hole M is comparable with the Planckian mass [13]. Due to the presence of the conformal anomalies one might expect that the leading corrections are of the order $\ln M$ (see, e.g. [14]).

The derivation of the thermodynamical entropy of a black hole requires the *on-shell* calculations. It means that one uses only a regular Euclidean metrics that are the solutions of the field equations. The discussion of the relation of the thermodynamical and statistical-mechanical entropy of a black hole requires *off-shell* calculations.

To discuss the statistical-mechanical entropy and its relation to the thermodynamical entropy one must generalize the calculation of the one-loop contribution to the renormalized free energy to the case of spaces with cone-like singularity. The new feature which arises is that the corresponding renormalized one-loop corrections would contain new type of divergence, which is directly connected with the presence of cone singularity. To make the answer finite one might introduce spatial cut-off in the volume integrals near the cone-singularity. It is convenient to restrict the integration by some proper distance l from singularity. The renormalized one-loop contribution F_1^{ren} to the free energy is of the form $F_1^{ren} = F_1^{ren}[\beta, \beta_H, \varepsilon]$, where $\varepsilon = (l/2G_{ren}M)^2$ is the dimensionless cut-off parameter. For $\varepsilon \to 0$ and $\beta \neq \beta_H$ the one-loop free energy F_1^{ren} is divergent $F_1^{ren} \sim \varepsilon^{-1} f(\beta, \beta_H)$. For $\beta = \beta_H$ the divergence disappears, so that $F_1^{ren}[\beta_H, \beta_H, 0]$ is finite. This quantity is directly related with quantum (one-loop) corrections to the thermodynamical entropy of a black hole

$$S_1^{TD} = \beta_H^2 \frac{d}{d\beta_H} F_1^{ren}[\beta_H, \beta_H, 0] \equiv \left[\beta^2 \frac{\partial F_1^{ren}}{\partial \beta} + \beta_H^2 \frac{\partial F_1^{ren}}{\partial \beta_H} \right]_{\beta = \beta_H}. \tag{8}$$

Dowker and Kennedy [15] and Allen [16] made an important observation that $F_1^{ren} = F_{vac}^{ren} + F_{therm}^{ren}$, $\partial F_{vac}^{ren}/\partial \beta = 0$,

$$F_{therm}^{ren} = -\beta^{-1} \ln \mathrm{Tr} \left[\mathrm{e}^{-\beta \hat{H}} \right] = \ln[\sum \exp(-\beta E_n)], \tag{9}$$

where E_n is the energy (eigenvalue of the Hamiltonian $\hat{H}$ of the field φ).

By using the expansion in eigenfunctions one can obtain

$$F_{therm}^{ren} = \sum_\lambda f(\beta \omega_\lambda) = \int d\omega N(\omega | \beta_H, \varepsilon) f(\beta \omega). \tag{10}$$

$f(\beta\omega) = \beta^{-1} \ln[1 - \exp(-\beta\omega)]$ is free energy of an oscillator of frequency ω at inverse temperature β, and $N(\omega | \beta_H)$ is the density of number of states at the given energy ω in a spacetime of a black hole of mass M. This density of number of states diverges. In order to make it finite we introduced the cut-off ε. We include ε as the argument of N in order to remind about this. The expression (10) is usually a starting point for 'brick wall' model [2]. The statistical-mechanical entropy S^{SM} is

$$S^{SM} = \left[\frac{\partial F_1^{ren}}{\partial \beta} \right]_{\beta_H} = \left[\frac{\partial F_{therm}^{ren}}{\partial \beta} \right]_{\beta_H} = \int d\omega N(\omega | \beta_H, \varepsilon) s(\beta \omega) \tag{11}$$

Here $s(\beta\omega) = \beta\omega/(e^{\beta\omega} - 1) - \ln(1 - e^{-\beta\omega})$ is the entropy of a quantum oscillator of frequency ω with inverse temperature β. S^{SM} is divergent in the limit $\varepsilon \to 0$. The divergence is directly related with the divergency of the density of number of states located in the narrow region in the vicinity of the horizon. By comparing the expressions (8) and (11) we can conclude that S^{TD} and S^{SM} differs from one another. It happens for the following two reasons: (1) Vacuum polarization (F_{vac}^{ren}) depends on M and hence on β_H; (2) $d/d\beta$ does not commute with Tr-operation. In general case one gets $S_1^{TD} = S^{SM} + \Delta S$. In the limit $\varepsilon \to 0$ $\Delta S \neq 0$ is also divergent, but S_1^{TD} remains finite and (for $M \gg m_{\mathrm{P}}$) small. To summarize, in order to obtain one-loop contribution to the observable (thermodynamical) entropy of a black hole one need to renormalize the statistical-mechanical entropy. As the result of this renormalization all the cut-off dependent terms will disappear.

References

[1] W.H. Zurek and K.S. Thorne, Phys.Rev.Lett. **54** (1985) 2171.

[2] G. 't Hooft, Nucl.Phys. **B256** (1985) 727.

[3] L. Bombelli, R.K. Koul, J. Lee, and R. Sorkin, Phys.Rev. **D34** 373 (1986).

[4] M. Srednicki, Phys.Rev.Lett. **71**, 666 (1993).

[5] V. Frolov and I. Novikov, Phys.Rev. **D48** (1993) 4545.

[6] A.I. Barvinsky, V.P. Frolov, and A.I. Zelnikov, Phys.Rev.**D51** (1995) 1741.

[7] J.D. Bekenstein, preprint gr-qc/9409015 (1994).

[8] J.B. Hartle and S.W. Hawking, Phys.Rev. **D28**, 2960 (1983).

[9] S.W. Hawking, In: *General Relativity: An Einstein Centenary Survey.* (eds. S.W. Hawking and W. Israel), Cambridge Univ.Press, Cambridge, 1979.

[10] V. Frolov, Phys.Rev.Lett. **74** (1995) 3319.

[11] J.W. York, Phys.Rev. **D33** (1986) 2092.

[12] H.W. Braden, J.D. Brown, B.F. Whiting, and J.W. Jork, Phys.Rev. **D42** (1990) 3376.

[13] P.R. Anderson, W.A. Hiscock, J. Whitesell, and J.W. York, Phys.Rev. **D50** (1994) 6427.

[14] D.V. Fursaev, Phys.Rev.**D51** (1995) 5352.

[15] J.S. Dowker and G. Kennedy, J.Phys. **A 11** (1978) 895.

[16] B. Allen, Phys.Rev. **D33** (1986) 3640.

Thermodynamic Features of Black Holes Dressed with Quantum Fields

David Hochberg

Abstract

The thermal properties of black holes in the presence of quantum fields can be revealed through solutions of the semi-classical Einstein equation. We present a brief but self-contained review of the main features of the semi-classical back reaction problem for a black hole in the microcanonical ensemble. The solutions, obtained for conformal scalars, massless spinors and U(1) gauge bosons, are used to calculate the $O(\hbar)$ corrections to the temperature and thermodynamical entropy of a Schwarzschild black hole. In each spin case considered, the entropy corrections $\Delta S(r)$, are positive definite and monotone increasing with increasing distance r from the hole, and are of the same order as the naive flat space radiation entropy.

1 Introduction

The physics of black holes provides a fertile ground in which the confluence of gravitation, quantum mechanics and thermodynamics takes place. Progress in our understanding of the thermal features of black holes demands a deeper understanding of the relationships among the state functions of black holes in thermodynamic equilibrium with quantized matter. A black hole can exist in thermodynamical equilibrium provided it is surrounded by radiation with a suitable distribution of stress-energy. In effect, one studies the consequences of coupling the black hole with its own Hawking radiation. In the semi-classical approach, such radiation is characterized by the expectation value of a stress energy tensor obtained by renormalization of a quantum field on the classical spacetime geometry of a black hole. Using such a stress tensor as a source in the semi-classical Einstein equation defines the back reaction problem.

In this talk, we shall review briefly the steps involved in the 1-loop back reaction problem for black holes placing special emphasis on the need to impose boundary conditions and the issue of the perturbative validity of the modified black hole metric. When then go on to use the solutions so obtained to calculate the corrections to the temperature at spatial infinity, T_∞, and to the entropy ΔS by which quantum fields of spin $0, 1/2, 1$ augment the usual Bekenstein-Hawking entropy. One finds that $T_\infty \neq T_H = \frac{\hbar}{8\pi M}$, while $\Delta S \geq 0$ has a global minimum at the renormalized event horizon and is monotonically increasing for increasing distance from the black hole. The topics sketched here can be found in more detail in [1, 2, 3, 4].

2 Mechanics of the 1-loop Back-reaction

We begin by choosing the background spacetime. We therefore suppose we have solved the classical vacuum Einstein equation

$$\hat{G}_{\mu\nu}(\hat{g}) = 0, \tag{1}$$

for $\hat{g}_{\mu\nu}$, the classical metric. We now inject, or superimpose, a collection of quantum fields on this classical background and solve the semi-classical field equation

$$G_{\mu\nu}(\hat{g} + \delta g) = 8\pi < T_{\mu\nu}(\phi, \psi, A_\mu, \cdots) >_{\hat{g}}, \tag{2}$$

where $< T_{\mu\nu} >_{\hat{g}}$ represents the 1-loop quantum stress-energy tensor for the fields $\phi, \psi, A_\mu, ...$ (scalars, spinors, vector bosons, etc.) renormalized over the background spacetime whose metric is $\hat{g}$. The effect, or back-reaction, of the quantum fields on the geometry described by the metric $\hat{g}$ induces a modification of this metric, δg, so that

$\hat{g} \to \hat{g} + \delta g$ when we "turn on" the sources. These quantum stress tensors obey the background conservation law

$$\hat{\nabla}_\mu < T^{\mu\nu} >= 0, \tag{3}$$

with respect to the background covariant derivative: $\hat{\nabla} \sim \partial + \hat{\Gamma}(\hat{g})$. As a consequence, it turns out we may only solve for the modified metric to order $O(\hbar)$. This we can demonstrate easily by expanding the Einstein tensor G in powers of the Planck constant and using (1) and (3), thus leading to

$$\hat{\nabla}_\mu (\hat{G} + \delta G + \Delta G)^{\mu\nu} = 8\pi \, \hat{\nabla}_\mu < T^{\mu\nu} >= 0, \tag{4}$$

so that $\hat{\nabla}_\mu(\delta G + \Delta G)^{\mu\nu} = 0$, or to order $\hbar$, $\hat{\nabla}_\mu \delta G^{\mu\nu} = 0$. Here, δG and ΔG denote the order $\hbar$ and higher-order contributions, respectively. Since the operator $\hat{\nabla}$ is $O(1)$ while the stress tensor is $O(\hbar)$, the back-reaction equation (2) reduces to

$$\delta G^{\mu\nu}(\hat{g} + \delta g) = 8\pi < T^{\mu\nu} > . \tag{5}$$

The components of the stress tensors are treated as input. To solve (5), one inserts the most general metric ansatz compatible with the symmetries and coordinate dependence of $< T_{\mu\nu} >$ into the left hand side.

Turning now to the case at hand, we consider the black hole background, whose metric is

$$\hat{g}_{\mu\nu} = \text{diag} \left(-(1 - \frac{2M}{r}), (1 - \frac{2M}{r})^{-1}, r^2, r^2 \sin^2\theta \right), \tag{6}$$

and M is the mass of the black hole. As shown in [1, 3], the modified metric, which is static and spherically symmetric, can be put into the following form:

$$ds^2 = -\left(1 - \frac{2m(r)}{r}\right)(1 + 2\epsilon\bar{\rho}(r)) \, dt^2 + \left(1 - \frac{2m(r)}{r}\right)^{-1} dr^2 + r^2 d\Omega^2, \tag{7}$$

D. Hochberg

where $m(r), \bar{\rho}(r)$ are two functions depending on r and $\epsilon = (M_{Planck}/M_{BH})^2 < 1$ is the expansion parameter. Note that $\epsilon = \hbar/M^2$ in units where $G = c = k_B = 1$. This parametrization of the new equilibrium metric reflects the fact that the back reaction induces static, spherically symmetric metric perturbations. Indeed, the stress tensors renormalized over $\hat{g}$ are static and depend only on the radial coordinate. $d\Omega^2$ is the standard metric of a normal round unit sphere. The mass function has the form $m(r) = M[1 + \epsilon(\mu(r) + CK^{-1})]$ where C is a constant of integration which serves to renormalize the bare Schwarzschild mass M. Indeed, to the order we are working, we may write

$$m(r) = M \left(1 + \epsilon C K^{-1}\right) [1 + \epsilon\mu(r)] \equiv M_{ren}[1 + \epsilon\mu(r)]. \tag{8}$$

We henceforth write M for the black hole mass in what follows, with the understanding that this represents the physical black hole mass. We note from (11) that $M_{rad} = \epsilon M\mu(r)$ is the usual expression for the effective mass of a spherical source. The metric in (7) is completed by the determination of $\bar{\rho}$, where

$$\bar{\rho}(r) = \rho(r) + kK^{-1}, \tag{9}$$

with k another constant of integration; $K = 3840\pi$.

The corresponding semiclassical field equations, valid to $O(\epsilon)$ are $(w = 2M/r)$

$$\begin{aligned}
\epsilon\frac{d\rho}{dw} &= -\frac{16\pi M^2}{w^3}(1 - w)^{-1} < T_r^r - T_t^t > \\
\epsilon\frac{d\mu}{dw} &= \frac{32\pi M^2}{w^4} < T_t^t > .
\end{aligned} \tag{10}$$

These may be obtained substituting (7) into (5) keeping only the $O(\epsilon)$ terms. Naturally, indices are raised/lowered with the background metric.

Once the indicated components of the renormalized stress-energy tensors are known, the solution of the semiclassical back-reaction equation (5) follows immediately from two simple integrations:

$$\mu(r) = \frac{1}{\epsilon M} \int_{2M}^{r} < -T_t^t > 4\pi\tilde{r}^2 \, d\tilde{r}, \tag{11}$$

$$\rho(r) = \frac{1}{\epsilon} \int_{2M}^{r} < T_r^r - T_t^t > 4\pi\tilde{r}^2 \, d\tilde{r}. \tag{12}$$

Actually, the back reaction problem as it stands has no <u>definite</u> solution unless boundary conditions are specified [1]. There are a number of reasons for why this must be so. In the first instance, the constant of integration k, appearing in $\bar{\rho}$ remains undetermined unless a boundary condition is invoked (asymptotic flatness does *not* fix this constant [1]). More importantly, the renormalized stress tensors employed are asymptotically *constant*, thus the radiation in a sufficiently large spatial region would collapse onto the black hole thereby producing a larger one. It is therefore necessary to implant the system consisting of the black hole plus radiation in a finite cavity with wall radius $= r_o > 2M$. As discussed in [5], a very important consequence of considering black holes in spatially bounded regions, quite independent of the back reaction problem, is that the cavity stabilizes the black hole in the thermodynamic sense and yields a *positive* heat capacity for the hole. There are at least two distinct types of physically relevant boundary conditions one may choose to impose. In the case of canonical boundary conditions, we specify the temperature of the cavity wall $T(r_o)$ and immerse the cavity containing the black hole and radiation in an external heat reservoir whose temperature $T = T(r_o)$. To obtain microcanonical boundary conditions, we specify the total energy at the cavity wall $E(r_o)$ and match on an external Schwarzschild spacetime with effective mass $m(r_o)$. In the former case, the integration constant is absorbed by a redefinition of the time coordinate. This is possible as coordinate time has no special meaning unless the metric is asymptotically constant. We can choose the timelike Killing vector to be $\frac{\partial}{\partial t} = \lambda \frac{\partial}{\partial \hat{t}}$ for λ a constant. The choice $\lambda = (1 - \epsilon k K^{-1})$ removes k from expressions for the physical quantities. In the latter case, we fix k by requiring continuity of the metric across the cavity wall: $k = -\rho(r_o)K$.

How does one determine the size of the cavity? The answer to this question is tied up with the perturbative validity of the solutions, which may be maintained provided the cavity wall radius satisfies a certain inequality, to which we now turn. We recall that all the stress tensors renormalized on the black hole background satisfy [6, 8, 9]

$$< T_\nu^\mu > \longrightarrow (\text{spin} - \text{dependent const.}) \times \text{diag}(-3, 1, 1, 1)_\nu^\mu, \tag{13}$$

as $r \to \infty$, which results in asymptotically unbounded metric perturbations $\delta g_{\mu\nu} = (g_{\mu\nu} - \hat{g}_{\mu\nu})$. In fact, one can show [1, 2, 3, 4] that the relative corrections diverge quadratically

$$\left|\frac{\delta g}{\hat{g}}\right| \longrightarrow = \epsilon \, \alpha_s \left(\frac{2}{3K}\right) \left(\frac{r}{2M}\right)^2, \tag{14}$$

(tt and rr components only, since $\delta g_{\theta\theta} = \delta g_{\phi\phi} = 0$) with α_s a spin-dependent constant: $\alpha_s = (1/2, 7/8, 1)$ for the spins $s = (0, 1/2, 1)$, respectively. So, we can obtain solutions that are uniformly small over the entire range $2M < r < r_{domain}$, taking $\left|\frac{\delta g}{\hat{g}}\right| < \delta < 1$, which from (14), and by saturating the previous inequality, defines the radius of the domain of perturbative validity via

$$\left(\frac{r_{domain}}{2M}\right)^2 = \frac{3K}{2\alpha_s}\left(\frac{\delta}{\epsilon}\right). \tag{15}$$

So, we should take the cavity radius $r_o < r_{domain}$. By way of illustration, setting $\epsilon = \delta < 1$ results in rather large perturbatively valid domains, indeed $r_{domain} \approx (380M, 286M, 268M)$ for the spins $s = (0, 1/2, 1)$, respectively.

3   The Quantum Stress-Energy Tensors

These have been obtained in exact form for the conformal scalar [6] and U(1) gauge boson [8], and in an approximate form for the massless spin 1/2 fermion [9]. For the former two cases, they are expressed as

$$< T_\nu^\mu >_{renormalized} = < T_\nu^\mu >_{analytic} + \left(\frac{\hbar}{\pi^2 (4M)^4} \right) \Delta_\nu^\mu, \tag{16}$$

where the analytic piece, in the case of the conformal scalar, was first given by Page [10]. The term Δ_ν^μ is obtained from a numerical mode sum. As this term is small in comparison to the analytic piece, we do not include it here. This affects none of our qualitative results; both pieces separately obey the the required regularity and consistency conditions. The analytic part has the exact trace anomaly in both cases. We display only the analytic part for the U(1) case and direct the interested reader to the original literature for the other cases. Dropping the angular brackets, we have $(w = 2M/r)$

$$T_t^t = -\frac{1}{3} a T_H^4 \left(3 + 6w + 9w^2 + 12w^3 - 315w^4 + 78w^5 - 249w^6 \right),$$

$$T_r^r = \frac{1}{3} a T_H^4 \left(1 + 2w + 3w^2 - 76w^3 + 295w^4 - 54w^5 + 285w^6 \right),$$

$$T_\theta^\theta = T_\phi^\phi = \frac{1}{3} a T_H^4 \left(1 + 2w + 3w^2 + 44w^3 - 305w^4 + 66w^5 - 579w^6 \right),$$

where $a = (\pi^2/15\hbar^3)$ and $T_H = \hbar/8\pi M$ is the (uncorrected) Hawking temperature.

4   Solutions

The explicit solutions to the $O(\epsilon)$ back reaction (5) obtained using the above stress-tensors may be summarized compactly as follows. Denoting with the subscripts S, f, V the conformal scalar, massless fermion and vector boson respectively, the metric functions in (11,12) turn out to be [1, 2, 3, 4]

$$K\mu_S = \frac{1}{2} \left[\frac{2}{3} w^{-3} + 2w^{-2} + 6w^{-1} - 8\ln(w) - 10w - 6w^2 + 22w^3 - \frac{44}{3} \right],$$

$$K\rho_S = \frac{1}{2} \left[\frac{2}{3} w^{-2} + 4w^{-1} - 8\ln(w) - \frac{40}{3} w - 10w^2 - \frac{28}{3} w^3 + \frac{84}{3} \right], \tag{17}$$

for the conformal scalar,

$$K\mu_f = \frac{7}{8} \left[\frac{2}{3} w^{-3} + 2w^{-2} + 6w^{-1} - 8\ln(w) - \frac{90}{7} w - \frac{62}{7} w^2 + \frac{46}{3} w^3 - \frac{16}{7} \right],$$

$$K\rho_f = \frac{7}{8} \left[\frac{2}{3} w^{-2} + 4w^{-1} - 8\ln(w) - \frac{200}{21} w - \frac{50}{7} w^2 - \frac{52}{7} w^3 + \frac{136}{7} \right], \tag{18}$$

for the massless spinor, and

$$K\mu_V = \frac{2}{3}w^{-3} + 2w^{-2} + 6w^{-1} - 8\ln(w) + 210w - 26w^2 + \frac{166}{3}w^3 - 248,$$

$$K\rho_V = \frac{2}{3}w^{-2} + 4w^{-1} - 8\ln(w) + \frac{40}{3}w + 10w^2 + 4w^3 - 32, \tag{19}$$

for the U(1) gauge boson. The various spin cases are distinguished by the spin dependence of the numerical coefficients.

5   Corrected Black Hole Temperature

As is well known, a black hole in empty space radiates quanta possessing a temperature characterized by the mass M of the hole. At large distances from the hole ($r >> M$), the temperature of the radiation approaches $T_\infty = \hbar/8\pi M \equiv T_H$. The presence of this radiation however, leads to modifications of this temperature when the back reaction is properly taken into account. For microcanonical boundary conditions, the corrected temperature at spatial infinity takes the form

$$T_\infty = \frac{\hbar}{8\pi M}\left(1 + \epsilon f(r_o, s, M)\right), \tag{20}$$

where f is a calculable function of the cavity radius r_o, the spin s of the quantum radiation and the renormalized mass M of the hole. Note that generally, T_∞ is *not* equal to the Hawking temperature.

To find the form of f we recall that

$$T_\infty = \frac{\hbar \kappa_H}{2\pi}, \tag{21}$$

where κ_H is the surface gravity of the event horizon; for a "free" black hole, $\kappa_H = (4M)^{-1}$. In the case of back reaction, the surface gravity is given by [1]

$$\kappa_H = \frac{1}{4M}\left(1 + \epsilon(\bar{\rho} - \mu) + 8\pi r^2 < T_t^t >\right)_{r=2M}. \tag{22}$$

With the microcanonical boundary conditions, one then obtains [1, 3]

$$T_\infty^{(s)} = \frac{\hbar}{8\pi M}\left(1 - \epsilon\rho_s(r_o) + \epsilon n_s K^{-1}\right), \tag{23}$$

where the constant $n_s = (12, -4, 304)$ for the spin $s = (0, 1/2, 1)$. The local temperature, T_{loc}, is obtained by blueshifting back from infinity to a finite value of r (by means of the Tolman factor [7]):

$$T_{loc}(r) = T_\infty[-g_{tt}(r)]^{-1/2}. \tag{24}$$

This yields the local temperature valid for all $r > 2M$ (dropping the spin labels)

$$T_{loc}(r) = T_H(1 - w)^{-1/2}\left[1 + \epsilon(\rho(r) - nK^{-1} - \frac{w}{2}(1 - w)^{-1}\mu(w))\right]. \tag{25}$$

6 Thermodynamical Entropy

One way to calculate the entropy is as follows. Consider first the case of the free black hole. From the 1st law of thermodynamics applied to slightly differing equilibrium systems

$$dE = dQ + dW, \tag{26}$$

and as no work is done on the system, $dW = 0$, so that

$$dS = \frac{dQ}{T} = \frac{dE}{T} = \frac{dM}{T_\infty}, \tag{27}$$

where M is the black hole mass and T_∞ the temperature at spatial infinity. Integrating, we have that the entropy

$$S = \int \frac{dM}{T_\infty} = \int (\frac{8\pi M}{\hbar}) dM = \frac{4\pi M^2}{\hbar} + \text{const.} \tag{28}$$

This yields the usual Bekenstein-Hawking expression for the black hole entropy after setting the constant of integration to zero: $S = S_{BH}$.

Now turn on the back reaction and compute S from (28). We must of course use the corrected asymptotic temperature (23) and the effective mass of the combined system of black hole plus radiation: $m(r_o) = M[1 + \epsilon\mu(r_o)]$ in (28). With (remembering that $\epsilon = \epsilon(M)$)

$$dm(r_o) = \left[1 + \epsilon\,(w\frac{\partial\mu}{\partial w} - \mu(w))\right]_{r_o} dM, \tag{29}$$

we obtain ($dr_o = 0$) that

$$S = S_{BH} + \Delta S(r_o) + \text{constant.} \tag{30}$$

The constant of integration is fixed by requiring that $\Delta S(r_o = 2M) = 0$, that is, with no "room" for the fields to contribute anything further, one should obtain just the Bekenstein-Hawking entropy $\frac{1}{4}A_H/\hbar$, as would be expected. With this choice, we find [4] for the conformal scalar field

$$\Delta S_S = \frac{8\pi}{K}(\frac{1}{2})(\frac{8}{9}w^{-3} + \frac{8}{3}w^{-2} + 8w^{-1} + \frac{32}{3}\ln(w) \quad - \quad \frac{40}{3}w - 8w^2$$
$$+ \frac{104}{9}w^3 - \frac{16}{9}), \tag{31}$$

while for the massless spin-$\frac{1}{2}$ fermion

$$\Delta S_f = \frac{8\pi}{K}(\frac{7}{8})(\frac{8}{9}w^{-3} + \frac{8}{3}w^{-2} + 8w^{-1} + \frac{128}{7}\ln(w) \quad - \quad \frac{200}{21}w - 8w^2$$
$$+ \frac{488}{63}w^3 - \frac{16}{9}), \tag{32}$$

and

$$\Delta S_V = \frac{8\pi}{K}\left(\frac{8}{9}w^{-3} + \frac{8}{3}w^{-2} + 8w^{-1} - 96\ln(w) + \frac{40}{3}w - 8w^2 + \frac{344}{9}w^3 - \frac{496}{9}\right), \tag{33}$$

for the U(1) gauge field. Among the features enjoyed by these entropy corrections, it is important to note that they all are positive definite, $\Delta S \geq 0$, and are monotone increasing functions of r_o . They are thus amenable to arguments relating thermodynamical to statistical entropy. Moreover, the corrections are of order $O(1)$ in $\hbar$ and so are of the same order as the naive flat space radiation entropy: $S_{flat} = \frac{4}{3}aT_H^3 V$. These corrections are therefore important.

7   Concluding Remarks

We should like to comment on the following points.

Lowest order solutions of the semi-classical back reaction problem have been used to calculate the $O(\hbar)$ corrections to the temperature and the order $O(1)$ correction to the entropy of a Schwarzschild black hole. It should be emphasized that this is a perturbative calculation. It would be of interest to extend the results to higher order, although a non-perturbative treatment would clearly be preferable. At higher orders, one needs to be careful, however, because the semi-classical field equation may well receive corrections arising from the stress tensor renormalization:

$$G_{\mu\nu} + \Lambda g_{\mu\nu} + aH_{\mu\nu}^{(1)} + bH_{\mu\nu}^{(2)} = 8\pi G < T_{\mu\nu} >, \tag{34}$$

where Λ, a, b are constants and the tensors $H_{\mu\nu}^{(1)}$ and $H_{\mu\nu}^{(2)}$ are linear combinations of quadratic curvature terms [12, 13].

Perhaps of greater importance is the fact that the contribution of quantum metric fluctuations has not been taken into account. Provided the semi-classical back reaction program leads qualitatively in the right direction (that is, towards a correct quantum gravity), one should include the spin-2 graviton contribution to the effective stress-energy tensor. One in fact expects the effects of linear gravitons to contribute a term of the same order to the stress tensor as those coming from ordinary matter and radiation fields. The corresponding calculations could be carried out once a renormalized effective energy-momentum tensor for quantized linear metric perturbations over a black hole background is worked out. However, at present, certain technical obstructions appear to make this a difficult task [11].

Lastly, one can use the results of the back reaction problem to explore the nature of the modified black hole spacetime by means of the effective potential [4], which determines the motion of test particles in the vicinity of the perturbed black hole, and to estimate the black hole free energy, of particular importance for assessing under what conditions the nucleation phase transition of black holes from hot flat space is likely to occur [14].

Acknowledgements

This work is based on collaborations with Thomas Kephart and James W. York, Jr. I thank Dr. Michael Bordag of Leipzig University for providing me with the partial financial support which facilitated my attending this Workshop.

References

[1] J.W. York, Jr., Phys. Rev. **D31**, 775 (1985).

[2] D. Hochberg and T.W. Kephart, Phys. Rev. **D47**, 1465 (1993).

[3] D. Hochberg, T.W. Kephart, and J.W. York, Jr., Phys. Rev. **D48**, 479 (1993).

[4] D. Hochberg, T.W. Kephart, and J.W. York, Jr., Phys. Rev. **D49**, 5257 (1994).

[5] J.W. York, Jr., Phys. Rev. **D33**, 2092 (1986).

[6] K.W. Howard, Phys. Rev. **D30**, 2532 (1984); K.W. Howard and P. Candelas, Phys. Rev. Lett. **53**, 403 (1984).

[7] R.C. Tolman, *Relativity, Thermodynamics and Cosmology* (Dover, New York, 1987).

[8] B.P. Jensen and A. Ottewill, Phys. Rev. **D39**, 1130 (1989).

[9] M.R. Brown, A.C. Ottewill, and D.N. Page, Phys. Rev. **D33**, 2840 (1986).

[10] D.N. Page, Phys. Rev. **D25**, 1499 (1982).

[11] B.P. Jensen, J.G. Mc Laughlin and A.C. Ottewill, Phys. Rev. **D51**, 5676 (1995).

[12] N.D. Birrell and P.C.W. Davies, *Quantum Fields in Curved Space* (Cambridge University Press, Cambridge, 1982).

[13] A.A. Grib, S.G. Mamayev and V.M. Mostepanenko, *Vacuum Quantum Effects in Strong Fields* (Friedmann Laboratory Publishing, St. Petersburg, 1994).

[14] D. Hochberg, Phys. Rev. **D51**, 5742 (1995).

Thermodynamics in D-dimensional Rindler-like Spaces

Andrei A. Bytsenko, Guido Cognola, Luciano Vanzo and Sergio Zerbini

1 Introduction

In this contribution, we would like to address some quantum aspects in the physics of black holes, which recently have been considered in the literature. Several issues like the interpretation of the Bekenstein-Hawking classical formula for the black hole entropy, the puzzle of loss of information in the black hole evaporation and the interpretation of the Hawking temperature have been discussed (see, for example, the review [1]). Furthermore in many papers, it has been pointed out that it should be desirable to have the usual statistical interpretation of the black hole entropy as the number of the gravitational states at the horizon and to try to understand the dynamical origin of the black hole entropy (see the Frolov's contribution and Refs. [2, 3]).

On general grounds, the density of levels as a function of the mass M, of a D-dimensional black hole should read

$$\Omega(M) \simeq C_D(M) \exp\left(\frac{4\pi \hat{G}_D}{D-2} M^{\frac{D-2}{D-3}}\right).$$

where $C_D(M)$ is a quantum prefactor and $\hat{G}_D$ is related to the generalized Newton constant. For a 4-dimensional black hole $\Omega(M) \simeq C_4(M) \exp(4\pi G M^2)$, $4\pi G M^2$ being the well known Bekenstein-Hawking classical entropy. As 't Hooft has pointed out, the prefactor $C_4(M)$ (the first quantum correction to the classical result), which is usually computable for quantum fields or extended objects (such as strings or p-branes) in an ultrastatic space-time background, turns out to be divergent [4]. This prefactor may be regarded as the contribution associated with the first quantum correction to the classical free energy. Several different methods have been used in dealing with such divergences, for example "the brick wall method" [4], the conical singularity method [5].

The physical origin of the horizon divergences may be described by the following simple considerations. In a D-dimensional static space-time with horizons, the equivalence principle implies that a system in thermal equilibrium has a local Tolman temperature given by $T(x) = T/\sqrt{-g_{00}(x)}$, T being the asymptotic temperature. Roughly speaking, near the canonical horizon (this means that the quantity g_{00} has simple zeros), a static space-time may be regarded as a Rindler-like space-time. We will show that, if one denotes by ρ the proper distance from the horizon, one gets for the Tolman temperature $T(\rho) = T/\rho$. As a consequence, the total entropy for a quantum bosonic gas reads (omitting a multiplicative constant)

$$S \equiv \int d\vec{x} \int_0^\infty T(\rho)^{D-1} d\rho \simeq A T^{D-1} \int_0^\infty \rho^{-D+1} d\rho,$$

where A is the integral on the transverse coordinates $\vec{x}$, namely the area of the horizon. The latter integral is clearly divergent. Introducing a horizon cutoff parameter ε we may rewrite it as (Λ being an infrared cutoff)

$$S \simeq AT \int_{\varepsilon}^{\Lambda} \rho^{-1}\, d\rho \simeq AT \ln \frac{\Lambda}{\varepsilon}\,, \qquad \text{for } D = 2\,,$$

$$S \simeq AT^{D-1} \int_{\varepsilon}^{\infty} \rho^{-D+1}\, d\rho \simeq \frac{AT^{D-1}}{(D-2)\varepsilon^{D-2}}\,, \qquad \text{for } D > 2\,.$$

For the sake of generality, we write down the asymptotic high temperature expansion for the entropy of a quantum gas on a $D = N+1$-dimensional static space-time defined by the metric

$$ds^2 = g_{00}(\vec{x})(dx^0)^2 + g_{ij}(\vec{x})dx^i dx^j\,, \qquad \vec{x} = \{x^j\}\,, \qquad i,j = 1,...,N\,,$$

$g(\vec{x})$ denoting its determinant. Again the equivalence principle leads to

$$S \simeq T^N \int \left(\frac{g(\vec{x})}{g_{00}(\vec{x})} \right)^{-N/2} dx^N\,.$$

As a consequence, the horizon divergences depend on the nature of the poles of the integrand $(g/g_{00})^{-N/2}$.

These considerations suggest the use of the optical metric $\bar{g}_{\mu\nu} = g_{\mu\nu}/g_{00}$, conformally related to the original one, in order to investigate these issues. It is our opinion that the conformal transformation techniques are particularly suitable for studying finite temperature effects for fields in space-times with horizons and here we would like to present some examples of computation. This method is well known (see for example the cited References in [6]).

Here we would like to report on the implementation of this idea in the case of massive fields in D-dimensional Rindler-like space-times [6]. To start with, we recall that Rindler-like space-times are static space-times admitting canonical horizons and having the topology of the form $\mathbb{R} \times \mathbb{R}^+ \times \mathcal{M}^{N-1}$. Their metric reads

$$ds^2 = -\frac{b^2 \rho^2}{r_H^2} dx_0^2 + d\rho^2 + d\sigma_{N-1}^2\,, \tag{1}$$

where r_H is a dimensional constant, b a constant factor and $d\sigma_{N-1}^2$ the spatial metric related to the $N-1$-dimensional manifold $\mathcal{M}^{N-1}$. If $\mathcal{M}^{N-1} \equiv \mathbb{R}^{N-1}$, $b = 1$ and $r_H = 1/a$, a being a constant acceleration, one has to deal with the Rindler space-time. If $\mathcal{M}^{N-1} \equiv S^{N-1}$, $b = (D-3)/2$ and r_H is the Schwarzschild radius of a black hole, then we shall show that one is dealing with a space-time which approximates, near the horizon and in the large mass limit, a D-dimensional black hole.

It is well known that space-times with canonical horizons admit a distinguished temperature, the (Unruh) Hawking temperature. For the metric (1), one has

$$\beta_H = \frac{2\pi r_H}{b}\,.$$

It is also important to stress that the variable ρ defined by the metric in Eq. (1) has the meaning of radial proper distance between the horizon and a point outside it and so, the divergences of thermodynamical quantities are automatically expressed in an invariant way.

2 Conformal transformation techniques and optical manifold

Here we briefly summarize the method of conformal transformations using ζ-function regularization [7]. These techniques permit to compute all physical quantities in an ultrastatic manifold (called the optical manifold) and, at the end of calculations, transform them back to a static one. This method is particularly useful in dealing with finite temperature effects for quantum fields, since these effects can be easily investigated in ultrastatic space-times.

To start with, we consider a non self-interacting scalar field on a $D = N + 1$-dimensional static space-time defined by the metric

$$ds^2 = g_{00}(\vec{x})(dx^0)^2 + g_{ij}(\vec{x})dx^i dx^j , \qquad \vec{x} = \{x^j\} , \qquad i,j = 1, ..., N .$$

The one-loop partition function is given by (we perform the Wick rotation $x_0 = -i\tau$, thus all the differential operator one is dealing with will be elliptic)

$$Z = \int d[\phi] \exp\left(-\frac{1}{2}\int \phi L_D \phi d^D x\right) , \tag{2}$$

where $L_D = -\Delta_D^g + m^2 + \xi R^g$, m being the mass and ξ an arbitrary parameter, while Δ_D^g and R^g are respectively the Laplace-Beltrami operator and the scalar curvature of the manifold in the original metric g.

The ultrastatic metric $\bar{g}_{\mu\nu}$ can be related to the static one by the conformal transformation $\bar{g}_{\mu\nu}(\vec{x}) = e^{2\sigma(\vec{x})} g_{\mu\nu}(\vec{x})$ with $\sigma(\vec{x}) = -\frac{1}{2}\ln g_{00}$. In this manner, $\bar{g}_{00} = 1$ and $\bar{g}_{ij} = g_{ij}/g_{00}$ (optical metric). For the operator one obtains

$$\bar{L}_D = -\Delta_D^{\bar{g}} + \xi_D R^{\bar{g}} + e^{-2\sigma}\left[m^2 + (\xi - \xi_D)R^g\right] = -\partial_\tau^2 + \bar{L}_N , \tag{3}$$

and for the one-loop partition function $\bar{Z}_{beta} = J[g,\bar{g}] Z_{beta}$, $\xi_D = (D-2)/4(D-1)$ being the conformal factor and $J[g,\bar{g}]$ the Jacobian of the conformal transformation. Since we have to deal with finite temperature theories, it is understood that the fields have to be periodic in τ, with period equal to the inverse temperature $\beta = 1/T$ and $0 \leq \tau \leq \beta$.

By means of the ζ-function regularization one easily obtain

$$\ln \bar{Z}_\beta = -\frac{\beta}{2}\left[\mathrm{PP}\,\zeta(-\tfrac{1}{2}|\bar{L}_N) + (2 - 2\ln 2\ell)\,\mathrm{Res}\,\zeta(-\tfrac{1}{2}|\bar{L}_N)\right]$$

$$+ \lim_{s\to 0}\frac{d}{ds}\frac{\beta}{\sqrt{4\pi}\Gamma(s)}\sum_{n=1}^{\infty}\int_0^\infty t^{s-3/2}\,e^{-n^2\beta^2/4t}\,\mathrm{Tr}\,e^{-t\bar{L}_N}\,dt . \tag{4}$$

where PP and Res stand for the principal part and the residue of the function. The free energy is related to the canonical partition function by means of equation

$$F(\beta) = -\frac{1}{\beta}\ln Z_\beta = -\frac{1}{\beta}\left(\ln \bar{Z}_\beta - \ln J[g,\bar{g}]\right) = F_0 + F_\beta\,, \tag{5}$$

where F_0 represents the vacuum energy. It is important to note that the Jacobian of the conformal transformation gives contributions only to F_0. The entropy and the internal energy of the system are given by the usual thermodynamical formulae

$$S_\beta = \beta^2\partial_\beta F(\beta) = \beta^2\partial_\beta F_\beta \qquad\qquad U_\beta = \beta\partial_\beta F(\beta) + F_\beta = -\partial_\beta \ln Z_\beta\,. \tag{6}$$

3   The heat kernel for massive fields in the Optical manifold

As we have mentioned, in many interesting physical cases, the Euclidean optical metric my be written in the form (see Eq. (1))

$$d\bar{s}^2 = d\tau^2 + \frac{r_H^2}{\rho^2}\left(d\rho^2 + d\sigma_{N-1}^2\right)\,, \tag{7}$$

where $d\tau = b\,dx^0$, r_H being a characteristic length (for example the horizon radius), $(r_H/\rho)^2$ the conformal factor and $d\sigma_{N-1}^2$ the metric of a $N-1$-dimensional manifold.

First we write down the spectral measure of the operator $\bar{L}_N$, as defined by Eq. (3), acting on scalars in the spatial section of the manifold defined by the metric Eq. (7) (for more details see Ref. [6]). Using such equations (for convenience now we put $r_H = 1$; in this way all quantities are dimensionless; the dimensions will be easily restored at the end of calculations) we easily obtain

$$\begin{aligned}
dV &= \rho^{-N}\,d\rho\,dV_{N-1}\,, \\
\bar{L}_N &= -\Delta_N^{\bar{g}} - \varrho_N^2 + C\rho^2\,, \\
\Delta_N^{\bar{g}} &= \rho^2\partial_\rho^2 - (N-2)\rho\,\partial_\rho + \rho^2\Delta_{N-1}\,,
\end{aligned}$$

where Δ_{N-1} is the Laplace-Beltrami operator on the manifold $\mathcal{M}^{N-1}$ and dV_{N-1} its invariant measure. We have also set $\varrho_N = (N-1)/2$ and $C = m^2 + \xi R^g$. It should be noted the appearance of an effective "tachionic" mass $-\varrho_N^2$, which has important consequences on the structure of the ζ-function related to the operator $\bar{L}_N$.

The eigenfunctions of the elliptic operator $\bar{L}_N$ can be written in the form $\Psi_{r\alpha}(x) = f_\alpha(\vec{x})\phi_{r\alpha}(\rho)$ and the corresponding eigenvalues $\lambda_r^2 = r^2$ ($r \geq 0$), where $f_\alpha(\vec{x})$ are the (normalized) eigenfunctions of the reduced operator $L_{N-1} = -\Delta_{N-1} + C$ with eigenvalues λ_α^2, C being a constant. For $\phi_{r\alpha}(\rho)$ one obtains

$$\phi_{r\alpha}(\rho) = \rho^{\frac{N-1}{2}}K_{ir}(\rho\lambda_\alpha)\,, \qquad\qquad (\Psi_{r\alpha},\Psi_{r'\alpha'}) = \delta_{\alpha\alpha'}\frac{\delta(r-r')}{\mu(r)}\,,$$

$$\mu(r) = \frac{2}{\pi|\Gamma(ir)|^2} = \frac{2}{\pi^2}r\sinh\pi r\,,$$

where $\mu(r)$ is the spectral measure associated with the continuum spectrum.

It is convenient to make the sum over α and integrate over the manifold defining the total spectral measure

$$\mu_I(r) = \mu(r) \int_0^\infty \sum_\alpha |K_{ir}(\rho\lambda_\alpha)|^2 \frac{d\rho}{\rho} .$$

Thus, for any suitable function $f(\bar{L}_N)$, we have

$$\mathrm{Tr}\, f(\bar{L}_N) = \int_0^\infty f(r^2)\, \mu_I(r)\, dr . \tag{8}$$

In order to derive $\mu_I(r)$, we make use of the Mellin-Barnes representation

$$K_{ir}^2(\rho\lambda_\alpha) = \frac{1}{4i\sqrt{\pi}} \int_{\mathrm{Re}\, z>1} \frac{\Gamma(z+ir)\Gamma(z-ir)\Gamma(z)}{\Gamma(z+1/2)} \rho^{-2z}\lambda_\alpha^{-2z}\, dz$$

and observe that, for $\mathrm{Re}\, z > (N-1)/2$, the sum over α can be performed and gives $\zeta(z|L_{N-1})$. In making the integration over ρ, we introduce a horizon cutoff parameter ε and, when possible, we take the limit $\varepsilon \to 0$. As a result

$$\mu_I(r) = \frac{\mu(r)}{8i\sqrt{\pi}} \int_{\mathrm{Re}\, z=c>(N-1)/2} \frac{\Gamma(z+ir)\Gamma(z-ir)\Gamma(z)\zeta(z|L_{N-1})}{z\Gamma(z+1/2)\varepsilon^{2z}}\, dz .$$

Finally, performing the integral over z, we obtain

$$\begin{aligned}
\mu_I(r) \;=\; & \sum_{n=0}^{\left[\frac{N-2}{2}\right]} \frac{K_{2n}(L_{N-1})\, \Phi_{N-2n}(r)}{N-1-2n} \left(\frac{4\pi}{\varepsilon^2}\right)^{\frac{N-1-2n}{2}} \\
& + \frac{\zeta'(0|L_{N-1})}{2\pi} + \frac{\zeta(0|L_{N-1})}{2\pi}\left[\psi(ir) + \psi(-ir) - 2\ln\frac{\varepsilon}{2} - \pi\delta(r)\right] ,
\end{aligned}$$

where $\left[\frac{N-2}{2}\right]$ is the integer part of the number $\frac{N-2}{2}$, $\psi(z)$ the logarithmic derivative of Γ, $K_n(L_{N-1})$ the heat kernel coefficients and $\delta(r)$ the usual Dirac δ-function. Note that for even N, $\zeta(0|L_{N-1})$ is vanishing and so the last term in the latter equation disappears.

The trace of the heat kernel can be computed by using Eq. (8) with $f(r^2) = \exp(-tr^2)$. We write it in the form

$$\begin{aligned}
\mathrm{Tr}\, e^{-t\bar{L}_N} \;=\; & \sum_{n=0}^{\frac{N-3}{2}} \frac{K_{2n}(L_{N-1})\, K(t| - \Delta_H^{N-2n} - \varrho_{N-2n}^2)}{N-1-2n} \left(\frac{4\pi}{\varepsilon^2}\right)^{\frac{N-1-2n}{2}} \\
& + \frac{1}{4\sqrt{\pi t}}\left[\zeta'(0|L_{N-1}) - 2\zeta(0|L_{N-1})\ln\frac{\varepsilon}{2}\right] \\
& - \frac{\zeta(0|L_{N-1})}{4} + \frac{\zeta(0|L_{N-1})}{2\pi}\int_{-\infty}^\infty \psi(ir)e^{-tr^2}\, dr , \tag{9}
\end{aligned}$$

$$\mathrm{Tr}\, e^{-t\bar{L}_N} = \sum_{n=0}^{\frac{N-2}{2}} \frac{K_{2n}(L_{N-1})\, K(t|-\Delta_{H^{N-2n}} - \varrho^2_{N-2n})}{N-1-2n} \left(\frac{4\pi}{\varepsilon^2}\right)^{\frac{N-1-2n}{2}}$$

$$+ \frac{1}{4\sqrt{\pi t}} \zeta'(0|L_{N-1}), \quad (10)$$

valid for odd and even N respectively. Here by $K(t|-\Delta_{H^{N-2n}} - \varrho^2_{N-2n})$ we indicate the diagonal heat kernel of a Laplace-like operator on H^{N-2n}. Of course, it does not depend on the coordinates since hyperbolic manifolds are homogeneous. Such a kernel is known in any dimension [8].

4 Physical applications

Now it is quite straightforward to obtain the partition function and then all the other thermodynamical quantities by means of Eqs. (4) and (5). Since the vacuum energy has been extensively studied in many papers, here we concentrate our attention on the temperature dependent part of the free energy (statistical sum) $F_\beta = \bar{F}_\beta = -\ln \bar{Z}_\beta/\beta$. Using Eqs. (9) and (10) we get

$$F_\beta^{even\, D} = \sum_{n=0}^{\frac{N-3}{2}} \frac{K_{2n}(L_{N-1})\, \mathcal{F}_{N-2n}^\beta}{N-1-2n} \left(\frac{4\pi}{\varepsilon^2}\right)^{\frac{N-1-2n}{2}}$$

$$-\frac{\pi}{12\beta^2} \left[\zeta'(0|L_{N-1}) - 2\zeta(0|L_{N-1})\ln\frac{\varepsilon}{2}\right] - \frac{\zeta(0|L_{N-1})}{4}\frac{\ln\beta}{\beta}$$

$$+\frac{\zeta(0|L_{N-1})}{2\pi\beta} \int_0^\infty [\psi(ir)+\psi(-ir)][1-e^{-\beta r}]\, dr \quad (11)$$

$$F_\beta^{odd\, D} = \sum_{n=0}^{\frac{N-2}{2}} \frac{K_{2n}(L_{N-1})\, \mathcal{F}_{N-2n}^\beta}{N-1-2n} \left(\frac{4\pi}{\varepsilon^2}\right)^{\frac{N-1-2n}{2}} - \frac{\pi}{12\beta^2}\zeta'(0|L_{N-1}), \quad (12)$$

where $\mathcal{F}_{N-2n}^\beta$ indicates the free energy density for a scalar field with (negative) square mass $-\varrho^2_{N-2n}$ on an ultrastatic manifold with hyperbolic H^{N-2n} spatial section.

It has to be noted that the parameter β is the inverse of the physical temperature only if $b = 1$. More generally, before to interpret β^{-1} as the temperature in Eqs. (11) and (12), one has to make the substitution $\beta \to b\beta$. The reason is due to the fact that, in order to write the metric (1) in the form (7), we have changed the time coordinate according to $\tau = bx_0$. Independently on the manifold $\mathcal{M}^{N-1}$, we see that the (non renormalized) free energy has a leading divergence of the kind $\varepsilon^{-(D-2)}$ proportional to the transverse area V_{D-2}, since K_0 and $\mathcal{F}_N^\beta$ are always non vanishing. More generally, one has $\left[\frac{D-1}{2}\right]$ divergences of the kind $\varepsilon^{-(D-2-2n)}$ (depending on the manifold and the operator L_{N-1}) and, for even D, also a possible logarithmic divergence. All these divergences are also present in the expressions of internal energy and entropy and their expressions can be obtained by means of Eqs. (6).

With regard to the internal energy, we have at disposal a renormalization procedure, which is well understood for $D = 4$. In fact, in Rindler and black hole space-times it is known that the renormalized stress-energy tensor is finite at the horizon in the Hartle-Hawking state [9], corresponding to the temperature $\beta = \beta_H$. This is equivalent to write

$$U(\beta)^{ren} = U_\varepsilon(\beta) - U_\varepsilon(\beta_H) + \text{finite part}\,,$$

where $U_\varepsilon(\beta)$ is the divergent part of the internal energy. Thus, the divergences are present in the expression of the renormalized internal energy, but only for some particular value of β, say β_H (β_U). For example, in the case of Rindler space-time, such a value is $\beta_U = 2\pi a^{-1}$, the Unruh temperature (here a is the acceleration), while in the 4-dimensional black hole background one has $\beta_H = 8\pi MG$, the Hawking temperature. In the general case, we may use the same renormalization procedure. Note, however, that the corresponding renormalized partition function, free energy and entropy remain divergent also at the distinguished temperature $\beta = \beta_H$.

As simple physical applications of the general formulae derived above, first we consider the case in which $\mathcal{M}^{N-1} = \mathbb{R}^{N-1}$, namely the Rindler case. One has

$$K_{2n}(L_{N-1}) = \frac{(-m^2)^n}{n!}\frac{V_{N-1}}{(4\pi)^{\frac{N-1}{2}}}\,, \qquad \zeta(z|L_{N-1}) = \frac{V_{N-1}\Gamma\left(z - \frac{N-1}{2}\right)}{(4\pi)^{\frac{N-1}{2}}\Gamma(z)} m^{N-1-2z}\,,$$

where $C = m^2$ has been put since Rindler is a flat manifold. Now, using Eqs. (11) and (12) together with the two equations above, we could write down the general expressions (see Ref. [6]).

Here we shall report the $D = 4$ case, for which we have

$$F_\beta^{Rind} = -\frac{A\pi^2}{180\beta^4\varepsilon^2} + \frac{Am^2(1 - \ln\frac{m^2\varepsilon^2}{4})}{48\beta^2}$$
$$+ \frac{Am^2}{4\pi}\left[\frac{\ln\beta}{4\beta} - \frac{1}{2\pi\beta}\int_0^\infty [\psi(ir) + \psi(-ir)][1 - e^{-\beta r}]\,dr\right]\,,$$

where the transverse area $A = V_2$ has been introduced to compare the latter formula with well known results (see for example Refs. [10, 11]). There is agreement in the massless case, but not in the massive case, where we also obtain a finite contribution. We conclude this section with some remarks on renormalization. As we have seen above, in our formalism, massive scalar fields in Rindler space-time can be easily treated, because the optical spatial section turns out to be the hyperbolic space H^3 and the harmonic analysis on such a manifold is well known. The formulae are particularly simple in the massless case. For example, in 4-dimensions, the total free energy may be chosen in the form

$$F^{ren}(\beta) = -\frac{A}{45(8\pi)^2\varepsilon^2}\left[\left(\frac{\beta_U}{\beta}\right)^4 + 3\right],$$

where $\beta_U = 2\pi$ is the Unruh temperature ($a = 1$). As a consequence, the entropy turns out to be

$$S_\beta = \frac{8\pi^2 A}{45\varepsilon^2\beta^3}$$

and it diverges for every finite β, but is zero at zero temperature (the Fulling-Rindler state), which is correct, since we are dealing with a pure state. At $\beta = \beta_U$, corresponding to the Minkowski vacuum, we have a divergent entropy proportional to the area, regardless of the fact that the Minkowski vacuum is a pure state. This is also to be expected, since an uniformly accelerated observer cannot observe the whole Minkowski space-time. With this renormalization prescription, the internal energy should read

$$U^{ren}(\beta) = \frac{A}{15(8\pi)^2\varepsilon^2}\left[\left(\frac{\beta_U}{\beta}\right)^4 - 1\right]$$

and this is vanishing and a fortiori finite at $\beta = \beta_U$, as it should be. Furthermore, at $\beta = \infty$, namely in the Fulling-Rindler vacuum, it is in agreement with the result obtained in Ref. [12].

As a second example we consider the case in which $\mathcal{M}^{N-1} = S^{N-1}$. This is a toy model for a D-dimensional black hole. In fact it is described by the static metric (we assume $D > 3$ and vanishing cosmological constant)

$$ds^2 = -\left[1 - \left(\frac{r_H}{r}\right)^{D-3}\right]dx_0^2 + \left[1 - \left(\frac{r_H}{r}\right)^{D-3}\right]^{-1}dr^2 + r^2\,d\Omega_{D-2}\,,$$

where we are using polar coordinates, r being the radial one and $d\Omega_{D-2}$ the $D-2$-dimensional spherical unit metric. The horizon radius is given by

$$r_H = \hat{G}_D M^{\frac{1}{D-3}}\,, \qquad \hat{G}_D = \left[\frac{2\pi^{\frac{D-3}{2}}G_D}{(D-2)\Gamma(\frac{D-1}{2})}\right]^{\frac{1}{D-3}}\,,$$

M being the mass of the black hole and G_D the generalized Newton constant. The associated Hawking temperature reads $\beta_H = 4\pi r_H/(D-3)$. The corresponding Bekenstein-Hawking entropy may be computed by making use of $\beta_H = \frac{\partial S_H}{\partial M}$. Thus we have

$$S_H = \frac{4\pi\hat{G}_D}{D-2}M^{\frac{D-2}{D-3}}$$

From now on, we put $r_H = 1$. It is possible to show that, redefining the radial Schwarzschild coordinate $r = r(\rho)$ and time $x_0 = x_0'/b$, $b = (D-3)/2$, one has $g_{00} = \rho^2 + O(\rho^4)$, and the optical metric related to this new set of coordinates reads

$$d\bar{s}^2 = -dx_0'^2 + \frac{1}{\rho^2}\left[d\rho^2 + G(\rho)\,d\Omega_{D-2}\right], \qquad G(\rho) = 1 + O(\rho^2).$$

From the latter equation we see that, near the horizon $\rho = 0$, we can set $G(\rho) = 1$ and so the optical metric assumes the form considered in previous Sections. In this approximation the manifold $\mathcal{M}^{N-1}$ becomes the unit sphere S^{N-1} and we have

$$d\bar{s}^2 \simeq -dx_0'^2 + \frac{1}{\rho^2}\left[d\rho^2 + d\Omega_{D-2}\right]. \tag{13}$$

The metric (13) is related to a manifold with curvature $R^{\bar{g}} = -(D-1)(D-2) + O(\rho^2)$, the relevant operator becomes

$$\bar{L}_N = -\Delta_N^{\bar{g}} - \varrho_N^2 + C\rho^2 + O(\rho^4), \tag{14}$$

where now C is a positive constant, which takes into account the mass and curvature contributions to this order. Note that since for the original manifold $R^g = 0$, ξ does not appear in the formulae.

As an explicit example, we consider the case $D = 4$ [13]. We have $r_H = 2MG$, $b = 1/2$,

$$\rho = 2(r-1)^{\frac{1}{2}}e^{(r-1)/2}, \qquad r = 1 + \frac{\rho^2}{4} - \frac{\rho^4}{16} + O(\rho^6).$$

Then, according to Eq. (14), the relevant operator becomes $\bar{L}_3 = -\bar{\Delta}_3 - 1 + C\rho^2$, where $C = m^2 + 1/3$ takes into account the curvature $R^{\bar{g}} = -6 + 2\rho^2$ of the optical manifold.

Now, directly using Eqs. (4), (5) and (11), after the replacement $\beta \to \beta/2$ due to the redefinition of the Schwarzschild time (remember that $b = 1/2$), for the total free energy we obtain

$$\begin{aligned}
F^{bh}(\beta) &= -Aj_\varepsilon + \frac{1}{4}\zeta(-\frac{1}{2}|\bar{L}_3) - \frac{2\pi^2 A}{45\varepsilon^2\beta^4} - \frac{A}{12\beta^2}\left[\frac{\zeta'(0|L_2)}{2} - \zeta(0|L_2)\ln\frac{\varepsilon}{2}\right] \\
&\quad -\frac{A\zeta(0|L_2)}{16\pi\beta}\ln\frac{\beta}{2} + \frac{A\zeta(0|L_2)}{8\pi^2\beta}\int_0^\infty \ln\left(1 - e^{-\beta r/2}\right)\left[\psi(ir) + \psi(-ir)\right]dr,
\end{aligned}$$

where we have written the Jacobian contribution to the partition function due to the conformal transformation in the form $A\beta j_\varepsilon$, and now $A = 4\pi r_H^2$ is the transverse area of the black hole.

The leading divergence, due to the optical volume, is proportional to the horizon area [4], but in contrast with the Rindler case, a logarithmic divergence is also present, similar to the one found in Refs. [14]. We have seen that this is a feature of even dimensions.

Let us briefly discuss the renormalization of the internal energy in this particular case. We recall that one needs a renormalization in order to remove the vacuum divergences. These divergences, as well as the Jacobian conformal factor, do not contribute to the entropy. However the situation presented here is complicated by the presence of horizon divergences, controlled by the cutoff parameter ε. In the 4-dimensional Schwarzschild space-time, it is known that the renormalized stress-energy tensor is well defined at the horizon in the Hartle-Hawking state, which in our formalism corresponds to the Hawking temperature $\beta = \beta_H$. The renormalized internal energy reads (the dots stay for finite contributions at the horizon, which we do not write down because their value depend on the approximation made)

$$U^{ren}(\beta) \;=\; \frac{A}{30(8\pi)^2\varepsilon^2}\left[\left(\frac{\beta_H}{\beta}\right)^4 - 1\right] - \frac{A}{3(8\pi)^2}\ln\varepsilon\left[\left(\frac{\beta_H}{\beta}\right)^2 - 1\right] + \ldots ,$$

which has no divergences for $\beta = \beta_H$, the Hawking temperature, while the entropy

$$S_\beta = \frac{8\pi^2 A}{45\varepsilon^2\beta^3} - \frac{A\ln\varepsilon}{6\beta} + \ldots ,$$

also for $\beta = \beta_H$ contains the well known divergent term proportional to the horizon area [4]. In the general case, the discussion is quite similar and it can be performed by using Eqs. (11) or (12) with the replacement $\beta \to (D-3)\beta/2$.

5   Conclusions

In this contribution, the first quantum corrections to the thermodynamic quantities of fields in a D-dimensional Rindler-like space have been investigated making use of conformal transformation techniques and ζ-function regularization.

The general form of the horizon divergences of the free energy has been obtained as a function of free energy densities of fields having negative square masses (absence of the gap in the Laplace operator spectrum) on ultrastatic manifolds with hyperbolic spatial section H^{N-2n} and of the Seeley-DeWitt coefficients $K_{2n}(L_{N-1})$ of the Laplace operator on $\mathcal{M}^{N-1}$. Since there exist recurrence relations for free energy densities, it is sufficient to study the cases $D = 3$ and $D = 4$ ($D = 4$ and $D = 5$ for applications to black holes). The leading divergence can be seen to be given by the volume of the spatial section of the optical manifold. For $D = 4$, our results are consistent with the ones obtained with brick wall, path-integral and canonical methods.

With regard to physical applications, we have used the general results on finite temperature field theory in order to investigate the quantum corrections to the Bekenstein-Hawking entropy for massive fields in a large mass black hole background. This approach gives rise to a leading divergence for the entropy similar to the one obtained for the Rindler case background, but in this case other divergent contributions are present and their structure depend on the dimension of the space-time considered. Here we have shown how it is possible to get the general form valid for an arbitrary dimension and we have explicitly considered the case $D = 4$.

As far as the horizon divergences are concerned we recall that they may be interpreted physically in terms of the infinite gravitational redshift existing between the spatial infinity, where one measures the generic equilibrium temperature and the horizon, which is classically unaccessible for the Schwarzschild external observer. Furthermore, we have argued that they are absent in the internal energy at the Unruh-Hawking temperature. However, they remain in the entropy and in the other thermodynamical quantities, as soon as one assumes the validity of the usual thermodynamical relations. For $D = 4$, a possible way to deal with such divergences has been suggested in Refs. [4, 3], where it has been argued that the quantum fluctuations at the horizon might provide a natural cutoff. In particular, choosing the horizon cutoff parameter of the order of the Planck length $(\varepsilon^2 \sim G)$, the leading "divergence", evaluated at the Hawking temperature, turns out to be of the form of the "classical" Bekenstein-Hawking entropy. This seems a reasonable assumption, because we have worked within the fixed background approximation. However one should remark that other terms are present, giving contributions which violate the area law. A more elaborate discussion can be found in the Frolov's contribution.

Finally, we mention that there has also been the proposal to absorb the horizon divergences, at least for $D = 2, 3, 4$, by making use of the standard ultraviolet gravitational constant renormalization [10]. This proposal, is essentially based on the use of Euclidean section with a conical singularity and associated heat kernel expansion.

References

[1] J.D. Bekenstein. *Do we understand black hole entropy?* In Proceedings Seventh Marcell Grossmann Meeting, Stanford, Califo+ EDITOR=, *General Relativity.*

[2] A. O. Barvinsky,V.P. Frolov and A.I. Zelnikov. Phys. Rev. **D 51**, 1741 (1995).

[3] V.P. Frolov. Phys. Rev. Lett. **74**, 3319 (1995).

[4] G.'t Hooft. Nucl. Phys. **B 256**, 727 (1985).

[5] M. Banados, C. Teitelboim and J. Zanelli. Phys. Rev. Lett. **72**, 957 (1995); J. S. Dowker. Class. Quantum Grav. **11**, L55 (1994).

[6] A.A. Bytsenko, G. Cognola and S. Zerbini, *Finite temperature effects for massive fields in D-dimensional Rindler-like spaces.* Preprint Trento University UTF–357 (1995), to be published in Nucl. Phys. B.

[7] J.S. Dowker and G. Kennedy. J. Phys. **A 11**, 895 (1978); J.S. Dowker and J.P. Schofield. Phys. Rev. **D 38**, 3327 (1988).

[8] A.A. Bytsenko, G. Cognola, L. Vanzo and S. Zerbini, *Quantum fields and extended objects in space-times with constant curvature spatial section.* Phys. Rep. (1995), (to be published).

[9] D. W. Sciama, P. Candelas and D. Deutsch. Adv. in Physics **30**, 327 (1981); M.R. Brown and A.C. Ottewill. Phys. Rev. **D 31**, 2514 (1985).

[10] L. Susskind and J. Uglum. Phys. Rev. **D 50**, 2700 (1994).

[11] D. Kabat and M. J. Strassler. Phys. Lett. **B 329**, 46 (1994).

[12] M.R. Brown, A.C. Ottewill and D. Page. Phys. Rev. **D 33**, 2840 (1986).

[13] G. Cognola, L. Vanzo and S. Zerbini, *One-Loop quantum corrections to the entropy for a 4-dimensional eternal black hole.* Class. Quantum Grav. **12**, 1927 (1995).

[14] S. Solodukhin, *The Conical Singularity and Quantum Corrections to BH Entropy.* Dubna JINR E2–94–246 (1994); S.P. De Alwis and N. Ohta, *On the Entropy of Quantum Field in Black Hole Backgrounds.* Colorado Univ. COLO–HEP–347 (1994).

On the Quantum Stability of the Time Machine

Sergey V. Krasnikov

1 Introduction

One of the most serious attempts to find a mechanism "protecting causality" was made in a number of papers beginning from [1, 2] where it is argued that vacuum fluctuations prevent the creation of a time machine by making the expectation value of the stress-energy tensor $\langle \mathbf{T} \rangle^{\text{ren}}$ diverge at the Cauchy horizon. In the present talk I show that

1. There *are* spacetimes with causality violated and with $\langle \mathbf{T} \rangle^{\text{ren}}$, nevertheless, bounded in causal regions.

2. There is actually *no* reason to expect that $\langle \mathbf{T} \rangle^{\text{ren}}$ diverges in the general case.

2 Why one could think that the instability exists

The basic idea of [1, 2] is the transition from a multiply connected space M containing a time machine to its universal covering $\widetilde{M}$ by the use of the formula

$$G^{(1)}(X, X') = G^{\Sigma} \equiv \sum_n G^{(1)}_{\widetilde{M}}(X, \gamma^n X'). \tag{1}$$

where $G^{(1)}_{M}$ and $G^{(1)}_{\widetilde{M}}$ are the Hadamard functions on M and $\widetilde{M}$ respectively. $\gamma^n X'$ is the nth inverse image of X'. Applying then

$$\langle T_{\mu\nu} \rangle = D_{\mu\nu} G^{(1)}_{M}$$

(here $D_{\mu\nu}$ is some known differential operator [3]) to (1) and neglecting all terms but the leading in the expansion of $G^{(1)}_{\widetilde{M}}(X, \gamma^n X')$:

$$G^{(1)}_{\widetilde{M}}(X, \gamma^n X') = \tilde{u}\sigma_n^{-1} + \tilde{v} \ln |\sigma_n| + \text{nonsingular terms}, \tag{2}$$

where σ_n is the geodetic distance between X and X', one we might conclude following [1, 2] that

$$\begin{aligned}
\langle T_{\mu\nu} \rangle^{\text{ren}}_{M} &= \langle T_{\mu\nu} \rangle^{\text{ren}}_{\widetilde{M}} + \lim_{X' \to X} D_{\mu\nu} \sum_{n \neq 0} \widetilde{G}^{(1)}(X, \gamma^n X') \\
&\to \langle T_{\mu\nu} \rangle^{\text{ren}}_{\widetilde{M}} + \sum_{\substack{n \neq 0}} \lim_{\substack{X' \to X \\ X \to \text{horizon}}} D_{\mu\nu}(\tilde{u}\sigma_n^{-1} + \tilde{v} \ln |\sigma_n|).
\end{aligned} \tag{3}$$

The Cauchy horizon is a null geodesic. So, the last series diverges and the conclusion is made that the time machine in any quantum state must be quantum unstable.

3   Counterexample

Consider the two-dimensional cylinder M obtained from the plane $\widetilde{M}$ with the metric

$$ds^2 = C(-d\tau^2 + d\chi^2) = C\,du\,dv. \tag{4}$$

by identifying $\chi \to \chi + H$. When $C = e^{2W\tau}$, M is the Misner spacetime and, when $C = \cosh^{-2}W\tau$, it is the "standard model" [1]. Determine the Hadamard function $G_M^{(1)}$ on M by choosing the following set of orthonormal modes (with arbitrary real f_0):

$$u_n^{(M)}\Big|_{n\neq 0} \equiv \frac{1}{2\sqrt{\pi|n|}}\exp\{2\pi i H^{-1}(n\chi - |n|\tau)\}, \quad u_0^{(M)} \equiv (2H)^{-1/2}(f_0\tau + i/f_0)$$

Now define on $\widetilde{M}$ the Hadamard function $G_{\widetilde{M}}^{(1)}$ by choosing the modes

$$u_p^{\widetilde{M}} \equiv \begin{cases} \dfrac{1}{2\sqrt{\pi\omega}}e^{ip\chi - i\omega\tau}, & \omega \geq 1/2 \ (\text{"usual" modes}) \\[2mm] \dfrac{1}{2\sqrt{\pi}}e^{ip\chi}(f^{-1}\cos\omega\tau - i\omega^{-1}f\sin\omega\tau), & \omega < 1/2. \end{cases}$$

$\forall f(\omega):\ f$ — smooth, positive; $f(\omega \geq 1/2 \leftleftarrows \sqrt{\omega})$, $f(0) = f_0$.

Considering the so defined quantum states one can readily find that

1. *The renormalized energy density is bounded on M for some f_0.*

2. Eqs. (1,2) hold (though their validity in the general case is disputable). However, eq. (3) *does not hold.* The reason is that the series (1) converges nonuniformly (due to (2) this must be true for the general case as well). Thus summation, differentiation and taking limits do not commute and the second line in (3) does not follow, in fact, from the first.

4   Conclusion

We see that depending on the quantum state the energy density can be bounded or unbounded near the Cauchy horizon, just as it does in the case of the Minkowski plane. Thus it is most likely that there is no specific quantum instability inherent to the time machine.

References

[1] V.P. Frolov, 1991 *Phys. Rev* **43D** 3878

[2] S.W. Kim and K.S. Thorne, 1991 *Phys. Rev* **43D** 3929

[3] N.D. Birrel and P.C.V. Davies, 1982 *Quantum Fields in Curved Space* (Cambridge: Cambridge University Press)

Field Interaction Effects of a Charged String in a Magnetic Background

Efrain J. Ferrer and <u>Vivian de la Incera</u>

Exactly solvable models are models of strings in non-trivial background fields constructed under the requirement of the conformal invariance of the theory. Although these models are obtained by using an expansion in powers of the derivatives of the fields, their solutions are exact in the sense that they contain all orders in α'. In this way they provide nonperturbative information about the system and one expects they might be a good guide to discover the hidden symmetries of the theory.

In this work we consider an open charged string propagating in a constant magnetic background. We obtain the Casimir operators of the algebra of the global symmetries and show that two independent linear combinations of these invariants give rise to the squared string mass and the squared rest energy. Using them it is found that the quantized string energy can be written as

$$\alpha' : E^2 := \alpha' \sum_{i=1}^{\frac{D-2}{2}} \left[(2\overline{a}_0^i a_0^i + 1)qH^i - 2qH^i \sum_{n=1}^{\infty} \left(\overline{b}_n^i b_n^i - \overline{a}_n^i a_n^i \right) \right] + \alpha' : \mathcal{M}^2 :,$$

and, as a consequence, at any arbitrary strength of the magnetic field components H^i, it has the same form as the Schwinger energy of a charged particle interacting with a constant magnetic field. The decomposition of the energy in terms of the invariants provides a method to identify and interpret each energy term, as well as a consistent criterion for normal-ordering it. Then, it can be shown that the Regge trajectories of the system are affected by the spin-field interaction in such a way that zero Landau level states lying in the first Regge trajectory satisfy

$$S = \alpha' m^2 + 1 - \frac{\gamma}{2} + \gamma S_H$$

and

$$S = \alpha' E^2 + 1 - \frac{\epsilon}{2} + \epsilon S_H$$

where S, S_H, m, and E are the spin, spin projection, mass and rest energy eigenvalues respectively, and ϵ and γ are functions of the background field. We conclude that the first Regge trajectory has been transformed into a family of infinitely many straight lines, each one associated with a different spin projection S_H along the magnetic field.

A Selfconsistent Semiclassical Solution with a Wormhole in the Theory of Gravity

Arkadiy Popov and Sergey Sushkov

We have considered the selfconsistent theory of gravity with quantum fields in the framework of the Killing ansatz, giving the approximate vacuum expectation value of the renormalized stress-energy tensor $T_{\mu\nu}$ in static spacetimes which was obtained by Frolov and Zelnikov (Phys.Rev. D **35**, 3031 (1987)). We have found an explicit example of the Lorentz wormhole with the metric in the form

$$ds^2 = -e^{2\nu}dt^2 + e^{2\mu}[dr^2 + (r^2 + a^2)(d\theta^2 + \sin^2\theta \, d\varphi^2)].$$

The asymptotic of the metric coefficients is

$$e^{2\nu} \simeq \exp\left(\frac{C_1^2}{3D_1^2} - 2\frac{q_2}{\alpha} - 2\frac{q_1}{\alpha} - \frac{7}{3}\right) +$$

$$+\frac{C_2}{r} + \frac{2}{r}\left[\lambda_1^2\left(C_1 e^{ir/\lambda_1} + \overline{C_1}e^{-ir/\lambda_1}\right) - 2\lambda_2^2\left(D_1 e^{ir/\lambda_2} + \overline{D_1}e^{-ir/\lambda_2}\right)\right],$$

$$e^{2\mu} \simeq 24D_1^2 \ln\frac{r}{\sqrt{8\pi\alpha}} + D_3 +$$

$$+\frac{D_2}{r} + \frac{2}{r}\left[\lambda_1^2\left(C_1 e^{ir/\lambda_1} + \overline{C_1}e^{-ir/\lambda_1}\right) + \lambda_2^2\left(D_1 e^{ir/\lambda_2} + \overline{D_1}e^{-ir/\lambda_2}\right)\right],$$

$$\lambda_1 = \left[\frac{9}{2}\pi^{-1}\alpha^{-1}D_1^2\left(\ln\frac{r}{\sqrt{8\pi\alpha}} + D_3\right)\right]^{-1/2}, \quad \lambda_2 = 6\left(\frac{C_1^2}{6D_1^2} - \frac{1}{2}\right)\lambda_1,$$

D_i and C_j are the complex constants of integration, q_1 and q_2 are the arbitrary constants of ansatz,

$$\alpha = \frac{1}{2^9 \cdot 45\pi^2}\left[12h\left(0\right) + 18h\left(\frac{1}{2}\right) + 72h\left(1\right)\right],$$

$h\left(s\right)$ is the number of helisity states for the conformal massless field of a spin s.

We have shown that the obtained solution describes the Minkowski spacetime in the limit $|r| \gg a$, where a is a throat's radius, whose value has an order of the Planck's length.

Part V

Topics in (Quantum) Gravity and in Quantum Optics

General Quantization Anomaly in Bosonic String Theory Interacting with Background Gravitational Field

Ioseph L. Buchbinder, B.R. Mistchuk[1] and V.D. Pershin[2]

Abstract

The problem of anomaly at the generalized canonical quantization (BFV - quantization) of bosonic string coupled to background fields is considered. The equation for symbol of anomaly operator is obtained. The general solution of this equation is found and the arbitrariness in general form of anomaly is investigated.

1 Introduction

It is well known that all conventional string models are the anomalous theories at the quantum level and only under certain conditions on the parameters of the theory the anomalies can be canceled. In the string theory coupled to background fields the roles of the parameters are played by these fields and hence conditions of anomaly cancelation can be treated as the equations of motion for background fields.

The string models interacting with massless background gravitational fields have been introduced in refs. [1, 2, 3] and investigated by a numbers of authors. This approach is based on the principle of quantum Weyl invariance according to which the renormalized operator corresponding to trace of energy-momentum tensor should vanishes[3]. The general form of renormalized composite operator of trace of energy-momentum tensor for bosonic string has been found in refs. [4, 5] (see also the reviews [6, 7]). The conditions under of which the renormalized trace is equal to zero are expressed in terms of so called modified beta-functions leading to strongly modified gravitational field equations.

Another approach to string theory coupled to background fields is associated with canonical quantization and it is mainly based on BFV - procedure [8, 9, 10, 11]. BFV - quantization procedure provides a powerful and universal approach to formulation of quantum gauge theories. In framework of this procedure the anomalies are connected with violation of the nilpotency conditions for the canonical BRST - charge.

BFV - quantization of string models has been initiated by the papers [12, 13, 14] and was discussed by a number of authors. However practically all results have been obtained for free string or in linear approximation. Application of the BFV - method to

---

[1] Department of Theoretical Physics, Tomsk State Pedagogical University, Russia.

[2] Department of Theoretical Physics, Tomsk State University, Russia

[3] The attempts to generalize this approach to take into account the strings interacting with massive backgrounds fields have been undertaken in refs. [19, 20].

                                    I.L. Buchbinder, B.R. Mistchuk, V.D. Pershin

string models interacting with background fields faces the problems connected with the very complicated structure of these theories. The model of above type are essentially non-linear and therefore they require a suitable perturbation scheme. An attempt to construct such a scheme within canonical approach has been undertaken in ref.[15] where the one-loop approximation was considered and correspondence with covariant approach was discussed.

In our paper [18] the general structure of the anomaly within canonical approach has been investigated using only the general algebraic properties of canonical BRST - charge. The algebraic analysis of anomaly for free string models in BFV - method has been studied ref. [16, 18]. In case of interacting string models the situation is much more complicated. The models under consideration are the theories with non-polynomial interaction. Therefore they lead to a number of new aspects and details in comparison with free model. Nevertheless we will see that the general form of anomaly can be established and its general structure can be investigated within the BFV - approach[4].

2   Canonical BRST - charge

The closed bosonic string interacting with background gravitational field is described by the action

$$ S = -\frac{1}{2} \int d^2\sigma \, \sqrt{-g} \, g^{ab} \partial_a x^\mu \partial_b x^\nu G_{\mu\nu}(x). \tag{1} $$

Here $G_{\mu\nu}(x)$ is the metric of $d-$dimensional space-time with coordinates x^μ; $\mu, \nu = 0, 1, ..., d - 1$. g_{ab} is the metric of the two-dimensional world sheet of the string, $\sigma^a = (\tau, \sigma)$ are the coordinates on the world sheet; $a, b = 0, 1$.

The theory possesses two first class constraints [15]

$$ \begin{aligned}
L(\sigma) &= \frac{1}{4} G^{\mu\nu} (p_\mu - G_{\mu\alpha} x'^\alpha)(p_\nu - G_{\nu\beta} x'^\beta), \\
\bar{L}(\sigma) &= \frac{1}{4} G^{\mu\nu} (p_\mu + G_{\mu\alpha} x'^\alpha)(p_\nu + G_{\nu\beta} x'^\beta),
\end{aligned} \tag{2} $$

where the p_μ are the momenta canonically conjugated to x^μ and $x'^\mu = \partial_\sigma x^\mu$. These constraints satisfy the Virasoro algebra in terms of Poisson brackets

$$ \begin{aligned}
\{L(\sigma)\, L(\sigma')\} &= -\left(L(\sigma) + L(\sigma')\right) \delta'(\sigma - \sigma'), \\
\{\bar{L}(\sigma)\, \bar{L}(\sigma')\} &= \left(\bar{L}(\sigma) + \bar{L}(\sigma')\right) \delta'(\sigma - \sigma'), \\
\{L(\sigma)\, \bar{L}(\sigma')\} &= 0.
\end{aligned} \tag{3} $$

---

[4]In recent paper [21] the analogous problem has been considered within BV - method.

The canonical BRST charge is constructed on the base of the algebra (3) and looks like this

$$\Omega = \int_{0}^{2\pi} d\sigma \left(L(\sigma)\eta(\sigma) + \bar{L}(\sigma)\bar{\eta}(\sigma) - \mathcal{P}(\sigma)\eta(\sigma)\eta'(\sigma) + \bar{\mathcal{P}}(\sigma)\bar{\eta}(\sigma)\bar{\eta}'(\sigma) \right). \qquad (4)$$

Here $\eta, \mathcal{P}$ and $\bar{\eta}, \bar{\mathcal{P}}$ are the canonically conjugated ghost fermion pairs corresponding to constraints L and $\bar{L}$ respectively.

The charge Ω is a generator of BRST transformations

$$\delta B = \{B, \Omega\}\epsilon \qquad (5)$$

where ϵ is a fermionic parameter and B is an arbitrary functional on extended phase space. In the theory under consideration the eq.(5) leads to the following transformation properties for the fundamental variables:

$$
\begin{aligned}
\delta\eta &= -\eta\,\eta'\,\epsilon, \quad \delta\bar{\eta} = \bar{\eta}\,\bar{\eta}'\,\epsilon, \\
\delta x^\mu &= \frac{1}{2}\left(\eta V^\mu + \bar{\eta}\bar{V}^\mu\right)\epsilon, \\
\delta p_\mu &= \left(\frac{1}{2}\,\partial_\sigma(\bar{\eta}\,\bar{V}_\mu - \eta\,V_\mu) - \frac{1}{4}\,\partial_\mu G_{\alpha\beta}V^\alpha\bar{V}^\beta(\eta+\bar{\eta})\right)\epsilon, \\
\delta V^\mu &= \left(-(\eta V^\mu)' - \frac{1}{2}\,\Gamma^\mu_{\alpha\beta}V^\alpha\bar{V}^\beta(\eta+\bar{\eta})\right)\epsilon, \\
\delta\bar{V}^\mu &= \left((\bar{\eta}\bar{V}^\mu)' - \frac{1}{2}\,\Gamma^\mu_{\alpha\beta}V^\alpha\bar{V}^\beta(\eta+\bar{\eta})\right)\epsilon.
\end{aligned}
\qquad (6)
$$

Here $\Gamma^\mu_{\alpha\beta}$ are Christoffel symbols corresponding to the metric $G_{\mu\nu}$ and we have introduced new fields

$$V^\mu = G^{\mu\nu}p_\nu - x'^\mu, \qquad \bar{V}^\mu = G^{\mu\nu}p_\nu + x'^\mu, \qquad (7)$$

which are the generalization of well known Fubini fields. In terms of these fields the constraints have the simple form

$$L = \frac{1}{4}\,G_{\mu\nu}V^\mu V^\nu, \qquad \bar{L} = \frac{1}{4}\,G_{\mu\nu}\bar{V}^\mu\bar{V}^\nu. \qquad (8)$$

3 Anomaly Operator

The quantum aspect of the theory under consideration have been discussed in ref. [15]. The operators of the constraints L and $\bar{L}$ are represented as the complicated functions of creation and annihilation operators for oscillating modes of the string and coordinate and momentum operators for zero mode. We will assume that the operators L and $\bar{L}$ are expressed in some normal form (for example, Wick ordering for oscillating modes and Weyl ordering for zero mode [15]).

The theory is called anomalous if the following relations are fulfilled

$$[\,\Omega,\,\Omega\,] \;\equiv\; A \neq 0.$$
$$[\,\Omega, H\,] \;\equiv\; A_H \neq 0 \tag{9}$$

where H is so called unitarizing Hamiltonian [8, 9, 10, 11] and $[\,,\,]$ is a super-commutator. The operators A and A_H are called the anomaly operators. One can show that in string theory due to reparametrization invariance of action the form of operator A_H is defined by the operator A. It means, the only anomaly operator will be in string theory.

Since the operator Ω must obey the (super)Jacobi identity:

$$[\,\Omega,\,[\,\Omega,\,\Omega\,]] = 0 \tag{10}$$

we obtain the equation for the anomaly operator

$$[\,\Omega,\,A\,] = 0. \tag{11}$$

This equation is a very powerful restriction on the possible form of anomaly and allows to write the operator A explicitly up to an arbitrary function depending only on x^μ.

For further analysis we will use a technique of symbols of operators (see f.e. [22]). Let the anomaly operator is written in normal form, the corresponding symbol is denoted as $\mathcal{A}$. We also take into account the dimensions of the basic objects of the theory

$$\dim x^\mu = 0,\ \dim p_\mu = 1,\ \dim \eta = \dim \bar\eta = c,\ \dim \mathcal{P} = \dim \bar{\mathcal{P}} = 1 - c,$$
$$\dim L = \dim \bar L = 2,\quad \dim \Omega = 1 + c,\quad \dim A = 2 + 2c. \tag{12}$$

where c is an arbitrary parameter (ghost number) and the symmetry properties of the commutators of the constraints. As a result we obtain the symbol $\mathcal{A}$ of anomaly operator as follows

$$
\begin{aligned}
\mathcal{A} \;&=\; \int d\sigma d\sigma' \left\{ i\hbar^2 \frac{13}{12\pi}(\eta\eta''' + \bar\eta\bar\eta''') + a_0 + a_1 + a_2 + a_3 \right\}, \\
a_0 \;&=\; 2\eta\bar\eta\, h_0, \\
a_1 \;&=\; \eta\eta'\, f_1 + \bar\eta\bar\eta'\, g_1 + 2(\eta\bar\eta' - \bar\eta\eta')\, h_1, \\
a_2 \;&=\; 2(\eta\bar\eta'' - \bar\eta\eta'')\, h_2, \\
a_3 \;&=\; \eta\,\eta'''f_3 + \bar\eta\,\bar\eta'''g_3 + 2(\eta\bar\eta''' - \bar\eta\eta''')\, h_3.
\end{aligned}
\tag{13}
$$

where h_i $(i = 1, ..., 3)$, f_k, g_k $(k = 1, ..., 3)$ are the arbitrary functions depending on x^μ, p_μ with define dimensions: $\dim h_i = 3 - i, \dim f_k = \dim g_k = 3 - k$. The term $i\hbar^2 \frac{13}{12\pi}(\eta\eta''' + \bar\eta\bar\eta''')$ is a ghost contribution.

Let us consider now the eq.(11) for anomaly operator. Being rewritten in terms of symbols of operators this equations leads to equation of the following type

$$\hat{\delta}\mathcal{A} = 0, \tag{14}$$

where the operator $\hat{\delta}$ has the structure

$$\hat{\delta} = \sum_{n=0}^{\infty} \hbar^n \delta_n.$$
$$\delta_0 = \delta \tag{15}$$

Here δ is the operator of classical BRST-transformations (5,6).

The symbol $\mathcal{A}$ can be represented as follows

$$\mathcal{A} = \sum_{n=1}^{\infty} \hbar^{n+1} \mathcal{A}_n, \tag{16}$$

where $\mathcal{A}_n$ can be treated as the loop quantum corrections to the anomaly. After some transformation we obtain

$$\delta \mathcal{A}_1 = 0 \tag{17}$$
$$\delta \mathcal{A}_n = G_n, \quad n \geq 2 \tag{18}$$
$$G_n = -\sum_{m=1}^{n} \delta_m \mathcal{A}_{n-m}.$$

These equations, in principle, allow to find all coefficients $\mathcal{A}_n$ and to define the symbol $\mathcal{A}$.

4 General Form of Anomaly

Let us start with the one-loop contribution $\mathcal{A}_1$. The eq.(17) means

$$\sum_k \delta a_k = \partial_\sigma \chi \tag{19}$$

with some function χ.

We note that the functions f_k, g_k, h_i in eq.(13) can depend only on x^μ, p_μ, x'^μ, p'_μ, x''^μ, p''_μ or equivalently depend on $V_\mu, \bar{V}_\mu, V'_\mu, \bar{V}'_\mu, V''_\mu, \bar{V}''_\mu$. Taking into account the dimensions we see that these functions are the polinoms in $V_\mu, \bar{V}_\mu, V'_\mu, \bar{V}'_\mu$. For example

$$f_1 = \beta_{\mu\nu} V^\mu \bar{V}^\nu + \alpha_{\mu\nu} V^\mu V^\nu + \bar{\alpha}_{\mu\nu} \bar{V}^\mu \bar{V}^\nu + \gamma_\mu V'^\mu + \bar{\gamma}_\mu \bar{V}'^\mu, \tag{20}$$

where $\beta_{\mu\nu}, \alpha_{\mu\nu}, \bar{\alpha}_{\mu\nu}, \gamma_\mu, \bar{\gamma}_\mu$ are arbitrary functions of x.

Taking into account the ghost structure of a_k and δa_k we one find

$$\mathcal{A}_1 = i\hbar^2 \int d\sigma \left\{ (\eta + \bar{\eta})(\eta - \bar{\eta})' \, \beta^{(1)}_{\mu\nu}(x) V^\mu \bar{V}^\nu + \eta\eta'''\beta^{(1)} + \bar{\eta}\bar{\eta}'''\bar{\beta}^{(1)} \right\}. \tag{21}$$

where $\beta^{(1)}$ and $\bar{\beta}^{(1)}$ are the arbitrary constants, $\beta^{(1)}_{\mu\nu}(x)$ are the arbitrary functions of x^μ. The index $^{(1)}$ indicates the one-loop contribution.

We want to pay attention that the function $\mathcal{A}_1$ (21) is a general solution of the equation $\delta \mathcal{A} = 0$ where $\dim \mathcal{A} = 2 + 2c$.

Let us consider the equation (18) defining the higher quantum corrections. If we write

$$\mathcal{A}_n = \int d\sigma a^{(n)}, \quad G_n = \int d\sigma g^{(n)}, \tag{22}$$

then one gets

$$\delta a^{(n)} = g^{(n)} + \partial_\sigma \chi^{(n)} \tag{23}$$

with some functions $\chi^{(n)}$. Since $\delta^2 = 0$, $\delta^2 a^{(n)} = 0$ one obtains

$$\delta g^{(n)} = -\partial_\sigma \delta \chi^{(n)}.$$

It leads to

$$g^{(n)} = g_0^{(n)} + \partial_\sigma \, \kappa^{(n)}, \tag{24}$$

where $\kappa^{(n)}$ are some functions and $g_0^{(n)}$ is a general solution of homogeneous equation $\delta g_0^{(n)} = 0$. Taking into account the ghost structure of $g_0^{(n)}$ and $\dim g_0^{(n)} = 4 + 3c$ we find $g_0^{(n)} = 0$. It means $G_n = 0$. Thus we see in order to find high-loop corrections $\mathcal{A}$ we should again to solve the equation $\delta \mathcal{A} = 0$. But the general solution of this equation is already known. Therefore we can write the general form of symbol of anomaly operator

$$\mathcal{A} = i\hbar^2 \int d\sigma \left\{ (\eta + \bar{\eta})(\eta - \bar{\eta})' \, \beta_{\mu\nu}(x) V^\mu \bar{V}^\nu + \eta\eta'''\beta + \bar{\eta}\bar{\eta}'''\bar{\beta} \right\}, \tag{25}$$

Here $\beta_{\mu\nu}(x)$ are the arbitrary functions on x^μ and $\beta, \bar{\beta}$ are the arbitrary constants.

As a result we see that the most general form of anomaly admissible by quantization procedure can be established in explicit form up to the functions $\beta_{\mu\nu}(x)$ and the constants β and $\bar{\beta}$.

5 Arbitrariness in the Definition of Anomaly

We have obtained the anomaly $\mathcal{A}$ as the local solution of the equation

$$\delta \mathcal{A} = 0$$
$$\mathcal{A} = i\hbar^2 \int d\sigma a, \quad \dim \mathcal{A} = 2 + 2c \tag{26}$$

It is evident that if $\mathcal{A}$ is a solution of the eq.(26) then $\tilde{\mathcal{A}} = \mathcal{A} + \delta\Theta$ is also a solution where Θ is an arbitrary functional. Writing

$$\tilde{\mathcal{A}} = i\hbar^2 \int d\sigma \tilde{a}, \quad \Theta = i\hbar^2 \int d\sigma \theta,$$

one gets

$$\tilde{a} = a + \delta\theta + \partial_\sigma \omega,$$
$$(\eta + \bar\eta)(\eta - \bar\eta)'\, (\tilde{\beta}_{\mu\nu}(x) - \beta_{\mu\nu}(x))V^\mu \bar{V}^\nu + \eta\eta'''(\tilde{\beta} - \beta) + \bar\eta\bar\eta'''(\tilde{\bar\beta} - \bar\beta) = \delta\theta + \partial_\sigma \omega \tag{27}$$

with some function ω, $\dim\omega = 2 + 2c$, $\dim\theta = 2 + c$.

Let us consider the term $\partial_\sigma \omega$. Its ghost structure is defined by the ghost structure of a. If we write a most general form ω taking into account $\dim\omega = 2 + 2c$, calculate $\partial_\sigma \omega$ and compare the ghost structure with the ghost structure of a we find $\omega = 0$.

Let us consider $\delta\theta$. Its ghost structure is defined by the ghost structure of a. If we write a most general form of θ taking into account $\dim\theta = 2 + c$, calculate $\delta\theta$ and compare the obtained ghost structure with the ghost structure of a ones get

$$\delta\theta = (\eta + \bar\eta)(\eta - \bar\eta)'\nabla_\mu \xi_\nu V^\mu \bar{V}^\nu \tag{28}$$

where $\xi_\mu(x)$ is an arbitrary vector field and ∇_μ is a standard covariant derivative.

As a result we obtain

$$\tilde{\beta}_{\mu\nu}(x) = \beta_{\mu\nu}(x) + \nabla_\mu \xi_\nu(x),$$
$$\tilde{\beta} = \beta, \quad \tilde{\bar\beta} = \bar\beta. \tag{29}$$

The eqs.(29) define the arbitrariness in solution of the equation $\delta\mathcal{A} = 0$, $\dim\mathcal{A} = 2 + 2c$.

6 Summary

We have considered the problem of anomaly in closed bosonic string theory coupled to background gravitational field. The general form of symbol of anomaly operator has found and it is given by the eq.(25). The anomaly depends on the arbitrary functions $\beta_{\mu\nu}(x)$ and the arbitrary constants β and $\bar\beta$. The arbitrariness in definition of anomaly operators has investigated and it shown that the above arbitrariness is given by eq.(29).

The problem of anomaly has formulated as a problem of solution of the equation

$$\delta \mathcal{A} = 0, \quad \dim \mathcal{A} = 2 + 2c$$
$$\delta^2 = 0 \tag{30}$$

where δ is the classical canonical BRST - transformation (5) and $\mathcal{A}(x, p, \eta, \bar{\eta})$ is a function on extended phase space undepending on ghost momenta $\mathcal{P}$ and $\bar{\mathcal{P}}$. It means that the above problem can be reformulated as a cohomology problem for the operator δ and apparently the solution of this problem can be described in suitable geometrical terms.

It is interesting to compare the found anomaly with result based on quantum Weyl invariance principle. The general structure of renormalized trace of energy-momentum tensor for bosonic string in field of graviton and dilaton has been given in refs. [4, 5] where it has been shown

$$[T] \sim [\sqrt{g}\, g^{ab}\, \partial_a x^\mu \partial_b x^\nu\, \bar{\beta}^G_{\mu\nu}(x)] + [\sqrt{g} R^{(2)} \bar{\beta}^\Phi(x)] \tag{31}$$

here $R^{(2)}$ is a scalar curvature of world sheet, the brackets denote the renormalized composite operators. The $\bar{\beta}^G_{\mu\nu}$ and $\bar{\beta}^\Phi$ are the modified beta-functions satisfying Curci-Paffuti relations [23] leading to $\bar{\beta}^\Phi = const$ at $\bar{\beta}^G_{\mu\nu} = 0$.

Let us compare the $[T]$ (31) and $\mathcal{A}$ (25). We see that the term $(\eta + \bar{\eta})(\eta - \bar{\eta})'\, V^\mu \bar{V}^\nu$ is analogous to $[\sqrt{g}\, g^{ab}\, \partial_a x^\mu \partial_b x^\nu \bar{\beta}^G_{\mu\nu}]$ and the term $\beta \eta \eta''' + \bar{\beta} \bar{\eta} \bar{\eta}'''$ is analogous to $[\sqrt{g} R^{(2)} \bar{\beta}^\Phi]$. Although we had not considered the interaction with dilaton field we can conclude that the arbitrary functions $\beta_{\mu\nu}$ and the constants β, $\bar{\beta}$ are the canonical analogs of the modified beta-functions. Thus, we see the room for modified beta-functions in canonical formulation. We pay an attention that we considered not Weyl anomaly but rather Virasoro one and could not expect from the very beginning that there can be so closed analogy between form of trace of energy-momentum tensor in covariant approach and form of Virasoro anomaly in canonical approach.

The approach under consideration is quite general, it is based only on algebraic properties of Ω operator and it can be applied to other string models coupled to background fields (strings coupled to dilatonic field, massive background fields, fermionic strings).

Acknowledgements

I.L.B is grateful to DFG for support of his visit to Institute of Physics at Humboldt Berlin University. The research was supported in part by ISF (grant No RI1300) and RFFR (project No 94-02-03234).

References

[1] E.S. Fradkin, A.A. Tseytlin, Phys. Lett. 158B (1985) 316; Nucl. Phys. B261 (1985) 1.

[2] C. Callan, D. Friedan, E. Martinec, M. Perry, Nucl. Phys. B 262 (1985) 593.

[3] A. Sen, Phys. Rev. Lett. 55 (1985) 1846; Phys. Rev. D32 (1985) 210.

[4] A.A. Tseytlin, Nucl. Phys. B294 (1987) 383.

[5] H. Osborn, Nucl. Phys. B294 (1987) 595.

[6] A.A. Tseytlin, Int. J. Mod. Phys. A 4 (1989) 4249.

[7] H. Osborn, Ann. Phys. 200 (1990) 1.

[8] E.S. Fradkin, G.A. Vilkovisky, Phys. Lett. 55B (1975) 224.

[9] I.A. Batalin, G.A.Vilkovisky, Phys. Lett. 69B (1977) 409.

[10] E.S. Fradkin, T.A. Fradkina, Phys. Lett. 72B (1978) 343.

[11] I.A. Batalin, E.S. Fradkin, Phys. Lett. 128B (1983) 303; 180B (1986) 157; J. Math. Phys. 25 (1984) 2426; Nucl. Phys. B279 (1989) 514; Riv. Nuovo Cim. 9 (1986) 1; Ann. Inst. Henri Poincare 49 (1988) 145.

[12] M. Kato, K. Ogawa, Nucl. Phys. B212 (1983) 443.

[13] S. Hwang, Phys. Rev. D28 (1983) 2614.

[14] R. Marnelius, Nucl. Phys. B211 (1983) 14; B221 (1983) 409.

[15] I.L. Buchbinder, E.S. Fradkin, S.L. Lyakhovich, V.D. Pershin, Int. J. Mod. Phys. A6 (1991) 1211.

[16] T. Fujiwara, Y. Igarashi, J. Kubo, K. Maeda, Nucl. Phys. B391 (1993) 211.

[17] T. Fujiwara, Y. Igarashi, M. Koseki, R. Kuriki, T. Tabei, Nucl. Phys. B425 (1994) 289.

[18] I.L. Buchbinder, B.R. Mistchuk, V.D. Pershin, Phys. Lett. 353B (1995) 457.

[19] I.L. Buchbinder, E.S. Fradkin, S.L. Lyakhovich, V.D. Pershin, Phys. Lett. 304B (1993) 239.

[20] I.L. Buchbinder, V.A. Krykhtin, V.D. Pershin, Phys. Lett. 348B (1995) 63.

[21] F. Brandt, W. Troost, A. van Proeyen, The BRST-antibracket cohomology of 2d gravity conformally coupled to scalar matter. Preprint KUL-TF-95117; hep-th/9509035

[22] F.A. Berezin, M.A. Shubin, Schrödinger equation, Moscow Univ. Press., 1983 (in Russian)

[23] C. Curci, G. Paffuti, Nucl.Phys. B286 (1987) 399.

Non-Local Properties in Euclidean Quantum Gravity

Giampiero Esposito

Abstract

In the one-loop approximation for Euclidean quantum gravity, the boundary conditions which are completely invariant under gauge transformations of metric perturbations involve both normal and tangential derivatives of the metric perturbations h_{00} and h_{0i}, while the h_{ij} perturbations and the whole ghost one-form are set to zero at the boundary. The corresponding one-loop divergency for pure gravity has been recently evaluated by means of analytic techniques. It now remains to compute the contribution of all perturbative modes of gauge fields and gravitation to the one-loop effective action for problems with boundaries. The functional determinant has a non-local nature, independently of boundary conditions. Moreover, the analysis of one-loop divergences for supergravity with non-local boundary conditions has not yet been completed and is still under active investigation.

1 Introduction

This paper is devoted to local and non-local properties which are relevant for the analysis of Euclidean quantum gravity in the presence of boundaries. Before presenting the technical details, it is necessary to describe why boundaries are so important in quantum gravity. As far as we can see, there are at least two main motivations:

(i) The propagator of quantum gravity may be expressed formally as a path integral over all Riemannian four-geometries matching the boundary data on two (compact) Riemannian three-geometries $\left(\Sigma_1, h_1\right)$ and $\left(\Sigma_2, h_2\right)$, where h_i is the metric induced on the surface Σ_i, with $i = 1, 2$. It is then necessary to understand how to fix the boundary data on $\left(\Sigma_1, h_1\right)$ and $\left(\Sigma_2, h_2\right)$. In quantum cosmology, this analysis leads to a prescription for the quantum state of the universe, i.e. a functional of the three-geometry which solves the Wheeler-DeWitt equation and represents the probability amplitude of having data on a compact Riemannian three-geometry (Σ, h) [1,2]. These data consist of the metric configuration on (Σ, h), and of matter field configurations (e.g. fermionic fields or bosonic gauge fields).

(ii) The effective action remains the main tool of perturbative quantum field theory [3-5]. Its one-loop approximation contains relevant information about trace anomalies and one-loop divergences [6,7], and is at the heart of symmetry-breaking phenomena [8-13]. The general form of volume terms in the corresponding asymptotic heat kernel has been obtained, after many years of dedicated work, by DeWitt, Gilkey, Avramidi [14-16]. In the presence of boundaries, surface terms occur which have a rich geometric structure and are necessary to obtain the correct values of the trace anomalies and to investigate the non-local nature of the one-loop effective action. For spinor fields, gauge fields and gravitation, the correct values of these one-loop divergences have been obtained for the first time only very recently, by using analytic or geometric techniques [6,17-25]. There is agreement, by now, between the analytic (mode-by-mode) and geometric (space-time covariant) calculations of the trace anomalies for gauge fields and one-loop divergences for gravitation subject to local boundary conditions, when the Faddeev-Popov formalism is used with manifestly covariant gauges [22-25]. However, the presence of boundaries leads to severe technical complications, and the geometric form of such divergences with non-covariant gauges and other families of boundary conditions is not yet completely understood.

In section 2 we derive in detail a set of mixed boundary conditions in Euclidean quantum gravity. In section 3 we discuss the open problems in this branch of perturbative quantum gravity.

2   Mixed Boundary Conditions for Euclidean Quantum Gravity

For gauge fields and gravitation, the boundary conditions are mixed, in that some components of the field (more precisely, a one-form or a two-form) obey a set of boundary conditions, and the remaining part of the field obeys another set of boundary conditions. Moreover, the boundary conditions are invariant under local gauge transformations providing suitable boundary conditions are imposed on the corresponding ghost zero-form or one-form.

We are here interested in the derivation of mixed boundary conditions for Euclidean quantum gravity. The knowledge of the classical variational problem, and the principle of gauge invariance, are enough to lead to a highly non-trivial quantum boundary-value problem. Indeed, it is by now well known that, if one fixes the three-metric at the boundary in general relativity, the corresponding variational problem is well posed and leads to the Einstein equations, providing the Einstein-Hilbert action is supplemented by a boundary term whose integrand is proportional to the trace of the second fundamental form [6,26-27]. In the corresponding quantum boundary-value problem, which is relevant for the one-loop approximation in quantum gravity [6], the perturbations h_{ij} of the induced three-metric are set to zero at the boundary. Moreover, the whole set of metric perturbations $h_{\mu\nu}$ are subject to the so-called *gauge transformations* [25]

$$\widehat{h}_{\mu\nu} \equiv h_{\mu\nu} + \nabla_{(\mu}\,\varphi_{\nu)} \quad , \tag{2.1}$$

where ∇ is the Levi-Civita connection of the background four-geometry with metric g, and $\varphi_\nu dx^\nu$ is the ghost one-form [25]. In geometric language, the difference between $\widehat{h}_{\mu\nu}$ and $h_{\mu\nu}$ is given by the Lie derivative along φ of the four-metric g.

For problems with boundaries, Eq. (2.1) implies that

$$\widehat{h}_{ij} = h_{ij} + \varphi_{(i|j)} + K_{ij}\varphi_0 \quad , \tag{2.2}$$

where the stroke denotes, as usual, three-dimensional covariant differentiation tangentially with respect to the intrinsic Levi-Civita connection of the boundary, while K_{ij} is the extrinsic-curvature tensor of the boundary. Of course, φ_0 and φ_i are the normal and tangential components of the ghost one-form, respectively. Note that boundaries make it necessary to perform a 3+1 split of space-time geometry and physical fields. As such, they introduce non-covariant elements in the analysis of problems relevant for quantum gravity. This seems to be an unavoidable feature, although the boundary conditions may be written in a covariant way [28].

In the light of (2.2), the boundary conditions

$$\left[h_{ij}\right]_{\partial M} = 0 \qquad\qquad\qquad (2.3a)$$

are gauge invariant, i.e.

$$\left[\widehat{h}_{ij}\right]_{\partial M} = 0 \quad , \qquad\qquad\qquad (2.3b)$$

if and only if the whole ghost one-form obeys homogeneous Dirichlet conditions, so that

$$\left[\varphi_0\right]_{\partial M} = 0 \quad , \qquad\qquad\qquad (2.4)$$

$$\left[\varphi_i\right]_{\partial M} = 0 \quad . \qquad\qquad\qquad (2.5)$$

To prove necessity and sufficiency of the conditions (2.4)-(2.5), one has to bear in mind the independent expansions in harmonics of φ_0 and φ_i. These obey a factorization property, and hence the three-dimensional covariant derivatives only act on the spatial harmonics, so that $\varphi_{(i|j)}$ vanishes at the boundary if and only if (2.5) holds [25].

The problem now arises to impose boundary conditions on the remaining set of metric perturbations. The key point is to make sure that the invariance of such boundary conditions under the transformations (2.1) is again guaranteed by (2.4)-(2.5), since otherwise one would obtain incompatible sets of boundary conditions on the ghost one-form. Indeed, on using the Faddeev-Popov formalism for the amplitudes of quantum gravity, it is necessary to use a gauge-averaging term in the Euclidean action, of the form [24]

$$I_{\text{g.a.}} \equiv \frac{1}{32\pi G\alpha} \int_M \Phi_\nu \Phi^\nu \sqrt{\det g}\; d^4x \quad , \qquad\qquad (2.6)$$

where Φ_ν is any relativistic gauge-averaging functional which leads to self-adjoint elliptic operators on metric and ghost perturbations. In particular, if the de Donder gauge is chosen [24],

$$\Phi_\nu^{dD}(h) \equiv \nabla^\mu\left(h_{\mu\nu} - \frac{1}{2}g_{\mu\nu}g^{\rho\sigma}h_{\rho\sigma}\right) \quad , \qquad\qquad (2.7)$$

one finds that [25]

$$\Phi_\nu^{dD}(h) - \Phi_\nu^{dD}(\widehat{h}) = -\frac{1}{2}\left(g_{\mu\nu}\Box + R_{\mu\nu}\right)\varphi^\mu = \frac{\lambda}{2}\varphi_\nu \quad , \qquad\qquad (2.8)$$

where $\Box \equiv g^{\mu\nu}\nabla_\mu\nabla_\nu$, $R_{\mu\nu}$ is the Ricci tensor of the background, and λ denotes the eigenvalues of the elliptic operator $-\left(g_{\mu\nu}\Box + R_{\mu\nu}\right)$. Indeed, our notation in the second equality of (2.8) is loose, and it is enough to emphasize the elliptic nature of the operator acting on the ghost one-form. Thus, if one imposes the boundary conditions

$$\left[\Phi_0^{dD}(h)\right]_{\partial M} = 0 \quad , \tag{2.9a}$$

$$\left[\Phi_i^{dD}(h)\right]_{\partial M} = 0 \quad , \tag{2.10a}$$

their invariance under (2.1) is guaranteed when (2.4)-(2.5) hold, by virtue of (2.8). Hence one also has

$$\left[\Phi_0^{dD}(\widehat{h})\right]_{\partial M} = 0 \quad , \tag{2.9b}$$

$$\left[\Phi_i^{dD}(\widehat{h})\right]_{\partial M} = 0 \quad . \tag{2.10b}$$

Note that the boundary conditions on the ghost one-form become redundant if one also imposes the conditions (2.3b), (2.9b) and (2.10b). Nevertheless, we shall always write them explicitly, since the ghost one-form plays a key role in quantum gravity.

The most general scheme does *not* depend on the choice of the de Donder term, so that it relies on (2.3a)-(2.3b), (2.4)-(2.5), jointly with [7,25]

$$\left[\Phi_0(h)\right]_{\partial M} = 0 \quad , \tag{2.11a}$$

$$\left[\Phi_0(\widehat{h})\right]_{\partial M} = 0 \quad , \tag{2.11b}$$

$$\left[\Phi_i(h)\right]_{\partial M} = 0 \quad , \tag{2.12a}$$

$$\left[\Phi_i(\widehat{h})\right]_{\partial M} = 0 \quad . \tag{2.12b}$$

Again, it is enough to write (2.3a), (2.11a), (2.12a), (2.4)-(2.5), or (2.3a)-(2.3b) jointly with (2.11a)-(2.11b) and (2.12a)-(2.12b).

In the particular and relevant case of flat Euclidean four-space bounded by a three-sphere [6,29], the de Donder gauge and the boundary conditions (2.3a), (2.9a), (2.10a) lead to [25]

$$\left[\frac{\partial h_{00}}{\partial\tau} + \frac{6}{\tau}h_{00} - \frac{\partial}{\partial\tau}\left(g^{ij}h_{ij}\right) + \frac{2}{\tau^2}h_{0i}^{|i}\right]_{\partial M} = 0 \tag{2.13}$$

$$\left[\frac{\partial h_{0i}}{\partial\tau} + \frac{3}{\tau}h_{0i} - \frac{1}{2}\frac{\partial h_{00}}{\partial x^i}\right]_{\partial M} = 0 \quad , \tag{2.14}$$

where τ is the Euclidean-time coordinate, which here represents the radius of three-spheres centered at the origin. These boundary conditions have three basic properties:

(i) They involve both normal and tangential derivatives of the h_{00} and h_{0i} metric perturbations.

(ii) The ghost boundary conditions cannot be expressed in terms of complementary projection operators on φ_0 and φ_i, and are instead Dirichlet on both φ_0 and φ_i.

(iii) They can be combined with non-local boundary conditions in $N = 1$ supergravity, when half of the tangential part of the gravitino potential is set to zero at the boundary. This is a non-local operation, since it makes it necessary to pick out the modes which multiply harmonics having positive eigenvalues for the intrinsic three-dimensional Dirac operator at the boundary. What is non-local is the separation of the spectrum of an elliptic operator into its positive and negative parts, and this leads to one of the two possible sets of mixed boundary conditions for spin-3/2 potentials [6,18].

When a three-sphere bounds flat Euclidean four-space, the symmetries of the problem and the use of zeta-function regularization [30,31] make it possible to evaluate the corresponding one-loop divergency for pure gravity, which is found to be [25]

$$\zeta(0) = -\frac{241}{90} \quad . \tag{2.15}$$

3 Achievements and Unsolved Problems

Over the past six years, impressive progress has been made in our understanding of boundary terms in the asymptotic heat kernel. Thus, the trace anomalies for massless scalar fields subject to Dirichlet, Neumann or Robin boundary conditions, or massless spin-1/2 fields with local or spectral boundary conditions, or Euclidean Maxwell theory in vacuum with magnetic or electric boundary conditions, or one-loop divergences for linearized gravity with three different sets of mixed boundary conditions, are by now well known [17-25,32,33]. Moreover, when boundaries are present, the $\zeta'(0)$ values for scalar and spin-1/2 fields, and the contributions of physical degrees of freedom to $\zeta'(0)$ for spin-1, spin-3/2 and spin-2 fields have also been obtained [34-37]. Despite this very encouraging progress, where one should also mention the contribution of physical degrees of freedom to the one-loop divergency for gravitino potentials [6,38], many important problems remain unsolved. They are as follows.

(i) What is the *geometric* form of the one-loop divergency for linearized gravity subject to the boundary conditions of section 2, and of the trace anomaly for massless spin-1/2 fields subject to non-local boundary conditions ? Can one find a suitable generalization of the Schwinger-DeWitt ansatz (cf. [39,40]) ?

(ii) What is the contribution of gauge modes and ghost modes to the one-loop divergency for gravitino potentials with non-local boundary conditions ?

(iii) What is the correct form of $\zeta'(0)$ for gauge fields and gravitation, when *all* perturbative modes are taken into account in the presence of boundaries ? [The work in [37] is restricted to the so-called physical degrees of freedom, e.g. the transverse part of the electromagnetic potential, or transverse-traceless perturbations for pure gravity.]

(iv) Can one evaluate the one-loop divergences with *arbitrary* relativistic gauges ? [These gauges [41,42] lead to non-minimal operators, and the presence of boundaries makes it very difficult to perform a mode-by-mode analysis of the quantized field in such a case.]

(v) Can one prove essential self-adjointness of the various elliptic operators occurring in the semi-classical amplitudes ?

(vi) Can one pick out a preferred choice of mixed boundary conditions for Euclidean quantum gravity, among the four different sets studied so far in the literature [24,25,33,43-46] ?

To make sense of the quantum state of the universe [2], of the path-integral approach to quantum gravity [47] and of the corresponding semi-classical approximation [6], it is essential to understand the various aspects of the problem of boundary conditions, i.e. the form of the boundary, the four-geometries summed over in the path integral and the boundary data chosen in the one-loop calculation. For this purpose, the old geometrodynamical framework [1], jointly with the key principles of modern quantum field theory, e.g. gauge invariance and BRST symmetry, is still the most appropriate for the active investigation of these issues. Hence we hope that, in the near future, the scientific community will come to appreciate how deep are the issues raised, and possibly solved, by a rigorous analysis of quantum field theories in the presence of boundaries [6,28].

Acknowledgements

I would like to thank Michael Bordag and Klaus Kirsten for their warm hospitality during the Leipzig Workshop. I am indebted to Peter D'Eath, Alexander Kamenshchik, Igor Mishakov and Giuseppe Pollifrone for scientific collaboration on the evaluation of one-loop divergences, and to Andrei Barvinsky, Stuart Dowker, Hugh Luckock, Ian Moss, Stephen Poletti and Dmitri Vassilevich for correspondence. I am also indebted to Ivan Avramidi for comments and criticisms which led to an improvement of the original manuscript. The work of Ian Moss [48,49] has been of invaluable importance to obtain a correct approach to one-loop quantum cosmology. Last, but not least, I am grateful to the European Union for partial support under the Human Capital and Mobility Programme.

References

[1] J.A. Wheeler, (1962) *Geometrodynamics* (New York: Academic Press).

[2] S.W. Hawking, (1984) *Nucl. Phys.* **B 239**, 257.

[3] G.A. Vilkovisky, (1984) *Nucl. Phys.* **B 234**, 125.

[4] B.S. DeWitt, (1987) *The Effective Action*, in *Architecture of Fundamental Interactions at Short Distances*, Les Houches Session XLIV, eds. P. Ramond and R. Stora (Amsterdam: North-Holland) p. 1023.

[5] I.L. Buchbinder, S.D. Odintsov and I.L. Shapiro, (1992) *Effective Action in Quantum Gravity* (Bristol: IOP).

[6] G. Esposito, (1994) *Quantum Gravity, Quantum Cosmology and Lorentzian Geometries*, Lecture Notes in Physics, New Series m: Monographs, Vol. m12, second corrected and enlarged edition (Berlin: Springer-Verlag).

[7] G. Esposito, *The Impact of Quantum Cosmology on Quantum Field Theory*, to appear in Proceedings of the 1995 Moscow Quantum Gravity Seminar (DSF preprint 95/27, GR-QC 9506046).

[8] S. Coleman and E. Weinberg, (1973) *Phys. Rev.* **D 7**, 1888.

[9] B. Allen, (1983) *Nucl. Phys.* **B 226**, 228.

[10] F. Buccella, G. Esposito and G. Miele, (1992) *Class. Quantum Grav.* **9**, 1499.

[11] G. Esposito, G. Miele and L. Rosa, (1993) *Class. Quantum Grav.* **10**, 1285.

[12] G. Esposito, G. Miele and L. Rosa, (1994) *Class. Quantum Grav.* **11**, 2031.

[13] G.Esposito, G. Miele, L. Rosa and P. Santorelli, *Quantum Effects in FRW Cosmologies* (DSF preprint 95/6, GR-QC 9508010).

[14] B.S. DeWitt, (1965) *Dynamical Theory of Groups and Fields* (New York: Gordon and Breach).

[15] P.B. Gilkey, (1975) *J. Diff. Geom.* **10**, 601.

[16] I. Avramidi, (1991) *Nucl. Phys.* **B 355**, 712.

[17] P.D. D'Eath and G. Esposito, (1991) *Phys. Rev.* **D 43**, 3234.

[18] P.D. D'Eath and G. Esposito, (1991) *Phys. Rev.* **D 44**, 1713.

[19] A.Yu. Kamenshchik and I.V. Mishakov, (1992) *Int. J. Mod. Phys.* **A 7**, 3713.

[20] A.Yu. Kamenshchik and I.V. Mishakov, (1993) *Phys. Rev.* **D 47**, 1380.

[21] A.Yu. Kamenshchik and I.V. Mishakov, (1994) *Phys. Rev.* **D 49**, 816.

[22] I.G. Moss and S.J. Poletti, (1994) *Phys. Lett.* **B 333**, 326.

[23] G. Esposito, A.Yu. Kamenshchik, I.V. Mishakov and G. Pollifrone, (1994) *Class. Quantum Grav.* **11**, 2939.

[24] G. Esposito, A.Yu. Kamenshchik, I.V. Mishakov and G. Pollifrone, (1994) *Phys. Rev.* **D 50**, 6329.

[25] G. Esposito, A.Yu. Kamenshchik, I.V. Mishakov and G. Pollifrone, (1995) *Phys. Rev.* **D 52**, 3457.

[26] J.W. York, (1972) *Phys. Rev. Lett.* **28**, 1072.

[27] J.W. York, (1986) *Found. Phys.* **16**, 249.

[28] G.Esposito, A.Yu. Kamenshchik and G. Pollifrone, *Quantum Field Theories in the Presence of Boundaries* (book in preparation).

[29] K. Schleich, (1985) *Phys. Rev.* **D 32**, 1889.

[30] S.W.Hawking, (1977) *Commun. Math. Phys.* **55**, 133.

[31] A.O. Barvinsky, A.Yu. Kamenshchik and I.P. Karmazin, (1992) *Ann. Phys.* **219**, 201.

[32] M. Bordag, E. Elizalde and K. Kirsten, *Heat-Kernel Coefficients of the Laplace Operator on the D-Dimensional Ball* (UB-ECM-PF preprint 95/3, HEP-TH 9503023).

[33] G. Esposito and A.Yu. Kamenshchik, *Mixed Boundary Conditions in Euclidean Quantum Gravity* (DSF preprint 95/23, GR-QC 9506092).

[34] J.S. Dowker and J.S. Apps, (1995) *Class. Quantum Grav.* **12**, 1363.

[35] M. Bordag, B. Geyer, K. Kirsten and E. Elizalde, *Zeta-Function Determinant of the Laplace Operator on the D-Dimensional Ball* (UB-ECM-PF preprint 95/10, HEP-TH 9505157).

[36] J.S. Dowker, *Spin on the 4-Ball* (MUTP preprint 95/13, HEP-TH 9508082).

[37] K. Kirsten and G. Cognola, *Heat-Kernel Coefficients and Functional Determinants for Higher-Spin Fields on the Ball* (UTF preprint 354, HEP-TH 9508088).

[38] G. Esposito, (1994) *Int. J. Mod. Phys.* **D 3**, 593.

[39] D.M.McAvity and H. Osborne, (1991) *Class. Quantum Grav.* **8**, 603.

[40] D.M. McAvity and H. Osborne, (1991) *Class. Quantum Grav.* **8**, 1445.

[41] G. Esposito, (1994) *Class. Quantum Grav.* **11**, 905.

[42] G. Esposito, A.Yu. Kamenshchik, I.V. Mishakov and G. Pollifrone, (1995) *Phys. Rev.* **D 52**, 2183.

[43] A.O. Barvinsky, (1987) *Phys. Lett.* **B 195**, 344.

[44] H.C. Luckock, (1991) *J. Math. Phys.* **32**, 1755.

[45] I.G. Moss and S.J. Poletti, (1990) *Nucl. Phys.* **B 341**, 155.

[46] V.N. Marachevsky and D.V. Vassilevich, (1995) *Diffeomorphism Invariant Eigenvalue Problem for Metric Perturbations in a Bounded Region* (paper in preparation).

[47] S.W. Hawking, (1979) *The Path-Integral Approach to Quantum Gravity*, in *General Relativity, an Einstein Centenary Survey*, eds. W. Israel and S.W. Hawking (Cambridge: Cambridge University Press) p. 746.

[48] I.G. Moss, (1989) *Class. Quantum Grav.* **6**, 759.

[49] I.G. Moss, (1995) *One-Loop Quantum Cosmology and the Vanishing Ghost*, talk given at the 1995 Moscow Quantum Gravity Seminar.

Geroch–Kinnersley–Chitre Group for Dilaton–Axion Gravity

Dmitri V. Gal'tsov

Abstract

A Kinnersley–type representation is constructed for the four–dimensional Einstein–Maxwell–dilaton–axion system restricted to space–times possessing two non–null commuting Killing symmetries. The new representation essentially uses the matrix–valued $SL(2,R)$ formulation and effectively reduces the construction of the Geroch group to the corresponding problem for the vacuum Einstein equations. An infinite hierarchy of potentials is introduced in terms of 2×2 real symmetric matrices directly generalizing the scalar hierarchy of Kinnersley–Chitre known for the vacuum Einstein equations.

"Geroch–Kinnersley–Chitre group" is a folk name for the infinite symmetry arising in the Einstein and Einstein–matter theories enjoying a complete integrability property after reduction to two dimensions via imposition of a rank two Abelian isometry group on the four–dimensional space-time. An original observation by Geroch [1], essentially based on the earlier works of Ehlers [2] and Harrison [3], was then developed by Kinnersley [4] and Kinnersley and Chitre [5]. Important insights in establishing the integrability of the two–dimensional Einstein and Einstein–Maxwell (EM) theories were due to the works of Neugebauer and Kramer [6], [7] in which the relevance of the three–dimensional σ–models on symmetric spaces has been discovered. A concise complex potential representation given by Ernst [8] for both Einstein and Einstein–Maxwell theories provided new structures particularly useful in this research. Different proofs of the complete integrability of two–dimensional reductions of the Einstein and EM systems were given in the late 70–ths by Harrison [9], Maison [10], Belinskii and Zakharov [11], and Hauser and Ernst [12]. These alternative formulations as well as some other approaches [13], [14] were compared and extended by Cosgrove [15]. Extensive application of related methods in General Relativity was very fruitful in the search of exact classical solutions [16] (for more recent development and review see [17], [18], [19], [20], [21], [22], [23], [24], [25], [26]).

In more general and modern language, Geroch group may be seen as an infinite-dimensional symmetry acting on harmonic maps of Riemann surfaces into symmetric spaces, the associated methods being related to the study of deformation of harmonic maps. In the gravity context such maps are connected to the three–dimensional σ–models to which the Einstein and some Einstein–matter systems are reduced for space–times admitting a non–null Killing vector field. If the corresponding target space is a symmetric Riemannian space, its finite isometries undergo infinite affine extension when one goes to two dimensions. Some other physically interesting systems possess such a property, the most notable examples being the Kaluza–Klein [6], [27], [28] and certain supergravity models [29], [30], [31].

It is worth noting that similar infinite symmetries are encountered in a different class of theories, two–dimensional conformal field models, in particular, in the theory of superstrings. In the low energy limit such theories lead to the gravity–matter systems which may have in their turn the associated infinite symmetries. The investigation of links between two theories seems to be a promising field of research. Here we discuss Geroch–type symmetries found recently in the heterotic string low energy effective theory [32], [33], [34]. In the bosonic sector of this theory one finds the Einstein gravity coupled to massless vector and scalar fields. The simplest model of this kind incorporating basic features of the full effective action — "dilaton–axion gravity", or Einstein–Maxwell–dilaton–axion (EMDA) system — can be formulated directly in four dimensions where it includes one $U(1)$ vector and two scalar fields coupled in such a way that the theory possesses non–abelian $SL(2, R)$ duality symmetry.

To begin, let us review briefly how Geroch symmetry emerges in the vacuum Einstein theory. Consider a four–dimensional manifold admitting a two–parameter group of motions generated by the Abelian algebra of Killing vectors K_A, $A = 1, 2$. In the adapted coordinates $K_A = \partial_A$ the interval can be written as

$$ ds^2 = f_{AB} dx^A dx^B - h_{MN} dx^{M+2} dx^{N+2}, \tag{1} $$

where $A, B, M, N = 1, 2$ and f_{AB}, h_{MN} are real symmetric 2×2 matrices depending only on x^3, x^4. In the case of stationary axisymmetric metrics which we choose here for definiteness a particular gauge (Lewis–Papapetrou) is appropriate:

$$ f_{11} = f, \quad f_{12} = -f\omega, \quad f_{22} = f\omega^2 - \rho^2/f, \tag{2} $$

and

$$ h_{11} = h_{22} = e^{2\gamma} f^{-1}, \tag{3} $$

where f, ω, γ are functions of $x^3 = \rho$, $x^4 = z$. The resulting two–dimensional equations can be conveniently written down using the Hodge–conjugated operators

$$ \nabla = \partial_{N+2} = (\partial_\rho, \ \partial_z), \quad \tilde{\nabla} = \tilde{\partial}_{N+2} = (-\partial_\rho, \ \partial_z). \tag{4} $$

The vacuum Einstein equations lead to the following system of equations for f_{AB}

$$ \nabla \left(\rho^{-1} f^{AB} \nabla f_{BC} \right) = 0, \tag{5} $$

                                                     D.V. Gal'tsov

where rising and lowering of indices is effected using the alternating symbol

$$\epsilon_{AB} = -\epsilon^{AB} = \begin{pmatrix} 0 & -1 \\ 1 & 0 \end{pmatrix}.\tag{6}$$

Once f_{AB} is found, the function γ can be constructed by solving a simple system of partial differential equations of the first order (what remains true for the EMDA system too, so we will not concentrate on this problem here).

These equations can be shown to possess two finite symmetry groups. The first is the so–called Matzner-Misner $SL(2,R)$ group G

$$x^A \rightarrow G^A{}_B\, x^B, \quad \det G = 1,\tag{7}$$

under which f_{AB} transforms as the $SL(2,R)$ tensor

$$f_{AB} \rightarrow G_A{}^C G_B{}^D f_{CD}.\tag{8}$$

The second, known as the Ehlers group H (also $SL(2,R)$), acts on the initial metric variables in a non–local way. To make this explicit one introduces a twist potential χ, the existence of which is implied by the 11–component of the Eq. (5), via duality relation

$$\tilde{\nabla}\chi = \rho^{-1} f^2 \nabla\omega.\tag{9}$$

The set of variables f, χ gives rise to another real symmetric matrix m_{AB},

$$m_{11} = f^{-1}, \quad m_{12} = \chi f^{-1}, \quad m_{22} = \chi^2 f^{-1} + f,\tag{10}$$

in terms of which the equivalent to (5) set of equations reads

$$\nabla\left(\rho\, m^{AB}\, \nabla m_{BC}\right) = 0.\tag{11}$$

This tensor transforms nonlinearly under the Ehlers group consisting of a gauge ($\chi \rightarrow \chi + \lambda_g$), scale ($f, \chi \rightarrow e^{2\lambda_s} f,\ e^{2\lambda_s}\chi$) and a proper Ehlers transformation, which can conveniently be expressed in terms of the Ernst potential $E = if - \chi$ as

$$E \rightarrow \frac{E}{1 + \lambda_E E}.\tag{12}$$

Acting by these transformations on some solution to the Eqs. (5) one obtains new solutions. Now, if G–covariance is taken into account, one can construct new symmetries by conjugation of H with $G:\ GHG^{-1}$. Repeating this operation one is led to an infinite–dimensional group

$$K = ...H \times G \times H \times G \times H,\tag{13}$$

which is the Geroch–Kinnersley–Chitre group. It can be realized on an infinite hierarchy of potentials which may be introduced via the G–covariant dualization procedure [4].

The same reasoning was shown to hold for the EM system (Kinnersley and Chitre [5]) where the hierarchy is more complicated and involves some additional structures. Remarkably, the EMDA system turns out to be in this respect simpler than the EM one, and involves only the structures typical for the vacuum Einstein equations.

Consider the EMDA action containing a metric $g_{\mu\nu}$, a $U(1)$ vector field A_μ, a Kalb-Ramond antisymmetric tensor field $B_{\mu\nu}$, and a dilaton ϕ in four dimensions

$$S = \frac{1}{16\pi} \int \left\{ -R + \frac{1}{3}e^{-4\phi}H_{\mu\nu\lambda}H^{\mu\nu\lambda} + 2\partial_\mu\phi\partial^\mu\phi - e^{-2\phi}F_{\mu\nu}F^{\mu\nu} \right\} \sqrt{-g}d^4x, \quad (14)$$

where

$$H_{\mu\nu\lambda} = \partial_\mu B_{\nu\lambda} - A_\mu F_{\nu\lambda} \quad + \quad \text{cyclic}, \quad (15)$$

and $F_{\mu\nu} = \nabla_\mu A_v - \nabla_\nu A_\mu$. After reduction to two dimensions the set of dynamical quantities, in addition to metric variables described above, contains two components of the 4–potential

$$A_B = \frac{1}{\sqrt{2}}(v,\, a), \quad (16)$$

one non–trivial component of the Kalb–Ramond field

$$B_{12} = b, \quad (17)$$

and the dilaton ϕ. The corresponding system in presence of one Killing vector field was studied in [32] and in the two–dimensional case in [33] in G–noncovariant way. Here we give G–covariant formulation which opens a way to construct the Geroch–Kinnersley group explicitly. From the results of [35] it is clear that the underlying H group is a symplectic group $Sp(4, R)$, which fits well with the $SL(2, R)$ structure of the group G. One can find G–covariant description of the equations of motion simply by introducing an additional 2×2 matrix structure into the Kinnersley formalism described above.

First we introduce the following two 2×2 real symmetric matrices instead of the one–component quantities f and ω above:

$$P = \begin{pmatrix} f - e^{-2\phi}v^2 & -e^{-2\phi}v \\ -e^{-2\phi}v & -e^{-2\phi} \end{pmatrix}, \quad \Omega = \begin{pmatrix} \omega & -q \\ -q & qv - b \end{pmatrix} \quad (18)$$

where $q = a + \omega v$. Using them as building blocks, one can now construct a G–tensor

$$F_{AB} = \begin{pmatrix} P & -P\Omega \\ -\Omega P & \Omega P\Omega - \rho^2 P^{-1} \end{pmatrix}. \quad (19)$$

Raising and lowering the indices now has to be performed with the matrix–valued alternating tensor

$$\epsilon_{AB} = -\epsilon^{AB} = \begin{pmatrix} 0 & -I \\ I & 0 \end{pmatrix}, \quad I = \begin{pmatrix} 1 & 0 \\ 0 & 1 \end{pmatrix}. \tag{20}$$

Note that the inverse tensor

$$F^{-1}{}_{AB} = \rho^{-2} F^{AB} = \rho^{-2} \begin{pmatrix} \rho^2\, P^{-1} - \Omega\, P\, \Omega & -\Omega\, P \\ -P\, \Omega & -P \end{pmatrix}. \tag{21}$$

In terms of this matrix the dynamical equations of motion assume essentially the same form as (5):

$$\nabla \left(\rho^{-1} F^{AB} \nabla F_{BC} \right) = 0. \tag{22}$$

Four 2×2 equations here are not algebraically independent. As two independent equations the 11 and 22 ones can be conveniently chosen

$$\nabla \left(\rho^{-1} P \nabla \Omega P \right) = 0, \tag{23}$$

$$\nabla \left(\rho \nabla P P^{-1} + \rho^{-1} P \nabla \Omega P \Omega \right) = 0, \tag{24}$$

The 22 component is related to (24) by transposition, while the 21 one is satisfied automatically due to both (23) and (24).

Ehlers group for the EMDA system was first described in [32] as some ten–parameter semisimple Lie group and then identified with $Sp(4, R)$ in [33]. It acts non–linearly on the set of six variables consisting of three initial variables f, v, ϕ and three dualized variables χ, u, κ. Here κ is the pseudoscalar Pecci–Quinn counterpart to the Kalb–Ramond field

$$H^{\mu\nu\lambda} = \frac{1}{2} e^{4\phi} E^{\mu\nu\lambda\tau} \frac{\partial \kappa}{\partial x^\tau}, \tag{25}$$

u is the magnetic potential related to the Maxwell tensor as

$$e^{-2\phi} F^{N2} + \kappa \tilde{F}^{N2} = \frac{f^2 e^{-2\gamma}}{\sqrt{2}\rho} \tilde{\partial}_{N+2} u, \tag{26}$$

while χ is the twist potential which now is the solution of the equation

$$\nabla \chi = u \nabla v - v \nabla u - \rho^{-1} f^2 \tilde{\nabla} \omega. \tag{27}$$

From six real variables entering the problem as a set of scalar fields one can build three complex potentials of the Ernst type

$$\Phi = u - zv, \quad E = if - \chi + v\Phi, \quad z = \kappa + i e^{-2\phi}. \tag{28}$$

It is worth noting that neither Φ, nor E reduce to the original Ernst potential for Einstein–Maxwell system. However their role in formulation of integrable system of equations is very similar to that of the original Ernst potentials (more detailed discussion can be found in [35]). Now we have one more complex variable (complex dilaton–axion field z) to describe the system. However the theory has additional symmetries which effectively reduce its complexity. One can show, in particular, that the whole theory (not only with two, but also with one Killing symmetry imposed on a four–dimensional manifold) is invariant under a discrete transformation

$$z^{'} = E, \quad E^{'} = z, \quad \Phi^{'} = \Phi, \tag{29}$$

i.e. an interchange of z and E. Within the EMDA system the complex dilaton–axion field and the Ernst variable form quite similar algebraic structures.

New dualized variables can be combined within the 2×2 matrix Q related to Ω by matrix dualization [36]

$$\nabla Q = -\rho^{-1} P \nabla \Omega P. \tag{30}$$

Explicitly it reads

$$Q = \begin{pmatrix} wv - \kappa & w \\ w & -\kappa \end{pmatrix}, \tag{31}$$

where $w = u - \kappa v$. Using P, Q pair one can now build 4×4 matrix transforming as an $Sp(2, R)$ matrix–valued object which is a direct analog of the matrix m_{AB} (10)

$$M_{AB} = \begin{pmatrix} P^{-1} & P^{-1}Q \\ QP^{-1} & P + QP^{-1}Q \end{pmatrix}. \tag{32}$$

In new terms the equations of motion read

$$\nabla \left(\rho \, M^{AB} \, \nabla M_{BC} \right) = 0. \tag{33}$$

The H–transformations are most conveniently described in terms of the complex combination

$$Z = Q + iP = \begin{pmatrix} E & \Phi \\ \Phi & -z \end{pmatrix}. \tag{34}$$

They consist of the shift on the constant real symmetric matrix

$$Z \to Z + R, \qquad R = \begin{pmatrix} \lambda_g & \lambda_m \\ \lambda_m & -\lambda_{d_1} \end{pmatrix}, \tag{35}$$

matrix "scale" transformation

$$Z \to S^T Z S, \qquad S = \begin{pmatrix} e^{\lambda_s} & \lambda_{H_1} \\ -\lambda_e & e^{\lambda_{d_3}} \end{pmatrix}, \tag{36}$$

and the shift of the inverted matrix

$$Z^{-1} \to Z^{-1} + L, \qquad L = \begin{pmatrix} \lambda_E & \lambda_{H_2} \\ \lambda_{H_2} & -\lambda_{d_2} \end{pmatrix}. \tag{37}$$

The parameters λ introduced here correspond to ten H–transformations (forming $Sp(4, R)$) generalizing the Ehlers group to the EMDA system. It is worthwhile to list them separately. Parameters λ_g, λ_e, λ_m correspond to the gravitational, electric and magnetic gauge transformations:

$$E = E_0 + \lambda_g, \quad \Phi = \Phi_0, \quad z = z_0, \tag{38}$$

$$E = E_0 - 2\lambda_e \Phi_0 - \lambda_e^2 z_0, \quad \Phi = \Phi_0 + \lambda_e z_0, \quad z = z_0, \tag{39}$$

$$E = E_0, \quad \Phi = \Phi_0 + \lambda_m, \quad z = z_0. \tag{40}$$

The scale transformation now reads

$$E = e^{2\lambda} E_0, \quad \Phi = e^{\lambda} \Phi_0, \quad z = z_0. \tag{41}$$

The $SL(2, R)$ S–duality subgroup is

$$\begin{aligned} E &= E_0, \quad \Phi = \Phi_0, \quad z = z_0 + \lambda_{d_1}, \\ E &= E_0 + \lambda_{d_2}\frac{\Phi_0^2}{1 + \lambda_{d_2} z_0}, \quad \Phi = \frac{\Phi_0}{1 + \lambda_{d_2} z_0}, \quad z = \frac{z_0}{1 + \lambda_{d_2} z_0}, \\ E &= E_0, \quad \Phi = e^{\lambda_{d_3}} \Phi_0, \quad z = e^{2\lambda_{d_3}} z_0. \end{aligned} \tag{42}$$

The most nontrivial part of the group in the physical terms consists of the pair of the electric and magnetic Harrison transformations

$$E = E_0, \quad \Phi = \Phi_0 + \lambda_{H_1} E_0, \quad z = z_0 - 2\lambda_{H_1}\Phi_0 - \lambda_{H_1}^2 E_0, \tag{43}$$

$$\begin{aligned} E &= \frac{E_0}{(1 + \lambda_{H_2}\Phi_0)^2 + \lambda_{H_2}^2 E_0 z_0}, \\ \Phi &= \frac{\Phi_0(1 + \lambda_{H_2}\Phi_0) + \lambda_{H_2} E_0 z_0}{(1 + \lambda_{H_2}\Phi_0)^2 + \lambda_{H_2}^2 E_0 z_0}, \\ z &= \frac{z_0}{(1 + \lambda_{H_2}\Phi_0)^2 + \lambda_{H_2}^2 E_0 z_0}, \end{aligned} \tag{44}$$

and the Ehlers transformation

$$E = \frac{E_0}{1 + \lambda_E E_0}, \quad \Phi = \frac{\Phi_0}{1 + \lambda_E E_0}, \quad z = z_0 + \frac{\lambda_E \Phi_0^2}{1 + \lambda_E E_0}. \tag{45}$$

Now let us turn to construction of the infinite hierarchy of potentials. The advantage of the present formulation consists in the close similarity to the Kinnersley–Chitre description of the vacuum Einstein equations (rather than electrovacuum). So we can directly repeat the procedure in terms of matrix–valued quantities. The first step consists in writing down the whole system of equations as a manifestly G–covariant condition of self–duality. To this end one introduces the matrix-valued twist tensor

$$\tilde{\nabla}\Psi_B^A = \rho^{-1}\, F^{AC}\, \nabla F_{CB}, \tag{46}$$

whose existence is implied by the equation (22). Taking complex combination

$$H_{AB} = F_{AB} + i\Psi_{AB}, \tag{47}$$

one can directly check that it obeys the self–duality condition

$$\nabla H_{AB} = -i\rho^{-1}F_A^C\,\tilde{\nabla}H_{CB}. \tag{48}$$

Infinite hierarchy of potentials $\overset{n}{H}_{AB}$ now can be generated via recursive equations

$$\overset{n+1}{H}_{AB} = i\left(\overset{1n}{N}_{AB} + H_A^C\,\overset{n}{H}_{CB}\right), \tag{49}$$

where the quadratic potentials $\overset{mn}{N}_{AB}$ are introduced through the relation

$$\nabla\overset{mn}{N}_{AB} = \overset{m}{H}{}^*_{CA}\nabla\overset{n}{H}{}^C{}_B. \tag{50}$$

These relations are valid for all $n,\, m = 0, 1, 2, \ldots$ and it is understood that

$$\overset{0}{H}_{AB} = i\epsilon_{AB}, \qquad \overset{0n}{N}_{AB} = -i\overset{n}{H}_{AB}. \tag{51}$$

Although formally these relations coincide with those for the vacuum Einstein equations, an essential complication comes from the fact that now the potentials are non–commuting matrices. Hence the problem of explicit construction of finite trans- formations acting on the hierarchy is much more tedious (details will be given else- where).

Alternatively, various linear deformation problems were suggested to deal with the Einstein equations, which can be probed in the present case too. We briefly describe here the formulation of the Belinskii–Zakharov inverse scattering transform method appropriate to the EMDA system. It can be derived starting with the null–curvature representation of the equations of motion following from the symmetric space nature of three–dimensional σ–model target space. Such representation was derived in [33] in terms of the $P,\, Q$ variables. Now we are in a position to give similar formulation directly in terms of the initial physical quantities entering the Eqs. (22).

Rewriting the system (22) as the modified chiral equation for the symmetric 4×4 matrix $F = F_{AB}$

$$(\rho F_{,\rho}F^{-1})_{,\rho} + (\rho F_{,z}F^{-1})_{,z} = 0, \tag{52}$$

one can obtain the corresponding Lax pair with a complex spectral parameter λ in the original Belinskii–Zakharov form:

$$D_1 \Psi = \frac{\rho U - \lambda V}{\rho^2 + \lambda^2} \Psi, \quad D_2 \Psi = \frac{\rho V + \lambda U}{\rho^2 + \lambda^2} \Psi. \tag{53}$$

Here $V = \rho F_{,\rho} F^{-1}$, $U = \rho F_{,z} F^{-1}$, Ψ is a matrix "wave function", and

$$D_1 = \partial_z - \frac{2\lambda^2}{\rho^2 + \lambda^2} \partial_\lambda, \quad D_2 = \partial_\rho + \frac{2\lambda\rho}{\rho^2 + \lambda^2} \partial_\lambda \tag{54}$$

are commuting operators; then the nonlinear system (52) can be regarded as the compatibility condition of the linear system $[D_1, D_2]\Psi = 0$. Similar linear system may be written in terms of the dualized variables, i.e. the matrix M satisfying Eq. (33).

An infinite algebra of the Geroch–Kinnersley–Chitre group can be obtained from here via an expansion of the linear system in power series in terms of the complex spectral parameter λ. Practical use of the 4×4 matrix Lax pair requires further study. We note another 4×4 linear problem (with different group structure) discussed recently [37].

Acknowledgements

The author thanks the organizers of this Conference for kind hospitality and support. Useful discussions with L. Bordag are gratefully acknowledged. At different stages of this research the author profited much from conversations with G.A. Alekseev, G. Neugebauer, D. Kramer, and J.B. Griffiths. The research was supported in part by the Russian Foundation for Fundamental Research Grant 93–02–16977, the International Science Foundation and Russian Gov. Grant M79300 and the INTAS Grant 93-3262.

References

[1] R. Geroch, Journ. Math. Phys. **13**, 394 (1972).

[2] J. Ehlers, in *Les Theories Relativestes de la Gravitation*, CNRS, Paris, p. 275, 1959.

[3] B.K. Harrison, J. Math. Phys. **9**, 1774 (1968).

[4] W. Kinnersley, Journ. Math. Phys. **14**, 651 (1973); **18**, 1529 (1977).

[5] W. Kinnersley and D. Chitre, Journ. Math. Phys. **18**, 1538 (1977); **19**, 1926, 2037 (1978).

[6] G. Neugebauer, Habilitationschrift, Jena, 1969, (unpublished).

[7] G. Neugebauer and D. Kramer, Ann. der Physik (Leipzig) **24**, 62 (1969).

[8] F.J. Ernst, Phys. Rev. **167** (1968) 1175; **168** (1968) 1415.

[9] B.K. Harrison, Phys. Rev. Lett., **41**, 1197, (1978).

[10] D. Maison, Journ. Math. Phys. **21**, 871 (1979).

[11] V.A. Belinskii and V.E. Zakharov. Sov. Phys. JETP, **48**, 985 (1978); **50**, 1 (1979).

[12] I. Hauser and F.J. Ernst, Phys. Rev. **D 20**, 362, 1783 (1979).

[13] G. Neugebauer, Journ. Phys. A: **12**, L67; **1**, L19 (1979).

[14] G.A. Alekseev, Pis'ma Zh. Eksp. Teor. Fiz. **32**, 301 (1980).

[15] C. Cosgrove, Journ. Math. Phys., **21**, 2417 (1981); **22**, 2624 (1981); **23**, 615 (1982).

[16] D. Kramer, H. Stephani, M. MacCallum, and E. Herlt, Exact Solutions of the Einstein's field equations, CUP, 1980.

[17] A. Eris, M. Gürses, and A. Karasu, Journ. Math. Phys. **25**, 1489 (1984); M. Gürses, in [18], p. 199.

[18] *"Exact solutions of Einstein's equations: Techniques and results."* Ed.W. Dietz and C. Hoenselaers. Lecture Notes in Physics, **205**, 1985.

[19] D. Kramer in *Unified field theories of more than four dimensions*, Eds. V. De Sabbata and E. Schmutzer, WS, 1983, p. 248; G. Neugebauer and D. Kramer, in *Galaxies, Axisymmetric systems and Relativity*, ed. by M. MacCallum, CUP, 1986, p.149; D. Kramer and G. Neugebauer in [18], p. 1.

[20] P. Breitenlohner and D. Maison, in [18], p. 276; Ann. IHP **46**, p. 215, (1987).

[21] P. S. Letelier, Journ. Math. Phys., **23**, 467. (1985).

[22] G.A. Alekseev, in Trudi MIAN , **176**, (1987), p. 211.

[23] N.R. Sibgatullin, Oscillations and waves in strong gravitational fields, Moscow, Nauka, 1984.

[24] R.K. Dodd, in *Soliton Theory, a survey of results*, Eds. A.P. Fordy, Manchester UP, 1990. p.174.

[25] E. Verdaguer, Phys. Repts, **229**, 2, (1993)

[26] J.B. Griffiths, Colliding Plane Waves in General Relativity, Clarendon Press 1991.

[27] D. Maison, Gen. Rel. and Grav. **10**, 717 (1979); V. Belinskii and R. Ruffini, Phys. Lett. **B89**, 195 (1980).

[28] B. Clement, Gen. Rel. and Grav., **18**, 137, 861, (1986); Phys. Lett. **A 118**, 11, (1986).

[29] B. Julia, in:*"Proceedings of the John Hopkins Workshop on Particle Theory"*, Baltimore (1981); in *"Superspace and Supergravity"*, Eds. S. Hawking and M. Rocek, Cambridge, 1981; in *"Unified Theories of More than Four Dimensions"* Eds. V. De Sabbata and E. Shmutzer, WS, Singapore 1983.

[30] P. Breitenlohner, D. Maison, and G. Gibbons, Comm. Math. Phys. **120**, 253 (1988).

[31] H. Nicolai, Phys. Letts. **B194** (1987) 402.

[32] D.V. Gal'tsov and O.V. Kechkin, Phys. Rev. **D50** (1994) 7394; (hep–th/9407155).

[33] D.V. Gal'tsov, Phys. Rev. Lett. **74** (1995) 2863, (hep-th/9410217).

[34] I. Bakas, *Nucl. Phys.* **B428** (1994) 374; *Phys. Lett.* **B343** (1995) 103; J. Maharana, *Hidden Symmetries of Two Dimensional String Effective Action*, hep-th/9502001; *Symmetries of the Dimensionally Reduced String Effective Action*, hep-th/9502002; A. Sen, *Duality Symmetry Group of Two Dimensional Heterotic String Theory*, preprint TIFR-TH-95-10, hep-th/9503057. J.H. Schwarz, *Classical Symmetries of Some Two-Dimensional Models*, CALT-68-1978, hep-th/9503078.

[35] D.V. Gal'tsov and O.V. Kechkin, *U–Duality and Symplectic Formulation of Dilaton–Axion gravity*, Preprint IC/95/155, Hep-th/9507005; D.V. Gal'tsov, A.A. Garcia, O.V. Kechkin, J. Math. Phys., 1995, v. 36, n. 9, p. 1 – 19, D.V. Gal'tsov, O.V. Kechkin, Matrix Dilaton–Axion for Heterotic String in Three Dimensions, preprint DTP–MSU 95/24, hep–th/9507164.

[36] D.V. Gal'tsov, O.V. Kechkin, *Hidden symmetries in dilaton–axion gravity*, in press.

[37] L.A. Bordag, A.B. Yanovski, J. Phys. **A 28**, 4007, (1995).

Geometric Quantization Of Kähler Spaces Admitting H-Projective Mappings

Asya V. Aminova and **Dmitri A. Kalinin**

1 Introduction

The n-dimensional oscillator algebra $\mathbf{osc}(n)$ is the Lie algebra with $(2n+1)$ basic generators $T^\alpha, T^{\tilde\alpha}$ and T which obey the following commutation relations

$$[T^\alpha, T^{\bar\beta}] = [T^\alpha, T^\beta] = [T^{\bar\alpha}, T^{\bar\beta}] = 0,$$

$$[T^\alpha, T] = -iT^\alpha, \qquad [T^{\bar\alpha}, T] = iT^{\bar\alpha}, \qquad \alpha = 1, ..., n. \tag{1}$$

Let $\mathcal{A}_m$ be the manifold of Lie algebra structures in an m-dimensional vector space V_m. Since each Lie algebra is completely defined by its structure constants C^i_{jk}, the manifold $\mathcal{A}_m$ can be referred to coordinates C^i_{jk}. Thus every point in $\mathcal{A}_m$ determines some Lie algebra. The curve $\ell(t)$ in $\mathcal{A}_m$ passing through the algebra $\ell = \ell(0)$ is called the deformation of the Lie algebra ℓ.

Deformations of the one-dimensional oscillator algebra are closely related with different physical and mathematical objects: anti-de Sitter quantum mechanics [4, 6], symplectic geometry of 1-dimensional complex disk [5, 7, 14], quantization of spinning particle [12] etc.

In this talk the quantization of a mechanical system for which the Hamiltonian vector fields of observables form the deformations of n-dimensional oscillator algebra where the role of deformation parameter plays curvature are considered. Because of this fact these systems can be considered as "deformations" of the harmonic oscillator. The talk consists of four parts. In the second part a short review of the geometric quantization procedure is given. In the third part the set of above mentioned mechanical systems is constructed which is realized at the classical level in the form of Kähler symplectic manifolds of constant holomorphic curvature [9]. Such mechanical systems are quantized later with the help of the geometric quantization approach [12]. As it is known the Kähler manifold of constant holomorphic curvature is H-projectively flat, i.e. it admits H-projective mappings on the flat space. In the fourth part of the talk the quantization of more general Kähler manifolds (of non-constant holomorphic curvature) admitting H-projective holomorphic mappings are discussed.

2 Review of geometric quantization procedure

To start with, we recall some relevant facts about the geometric quantization procedure [8, 12]. Let (M, ω) be a symplectic manifold. According to Dirac *quantization* is the linear map $Q : f \to \hat{f}$ of Poisson (sub)algebra $C^\infty(M)$ into the set of operators in some (pre)Hilbert space $\mathcal{H}$ poses sing the properties:

1. $\hat{1} = 1$;

2. $\widehat{\{f,g\}}_h = \frac{i}{h}(\hat{f}\hat{g} - \hat{g}\hat{f})$;

3. $\hat{\bar{f}} = (\hat{f})^*$;

4. for some complete set of functions $f_1, ..., f_n$ the operators $\hat{f}_1, ..., \hat{f}_n$ also form a complete set,

where bar is the complex conjugation, star denotes the conjugation of the operator and $h = 2\pi\hbar$ is the Planck constant. The linear map $\mathcal{P} : f \to \check{f}$ possessing the first three properties is called *prequantization*. For the case $M = T^*M$, $\omega = d\alpha$ prequantization was constructed by Koopman, Van Hove and Segal. It has the form

$$\mathcal{P}f = \check{f} = f - i\hbar V(f) - \alpha(V(f)), \tag{2}$$

where vector field $V(f)$ is the Hamiltonian vector field of the function $f \in C^\infty(M)$ defined by the condition

$$V(f) \rfloor \omega = -df,$$

where $\rfloor$ is the internal product. In local coordinates x^i, $i = 1, ..., 2n$ from here we have

$$V(f) = \omega^{ij}\partial_j f \partial_i. \tag{3}$$

Let $\mathcal{L}$ be the Hermitian line bundle with connection D and D-invariant Hermitian structure $<, >$. Recall that D-invariance means that for each pair of sections λ and μ of $\mathcal{L}$ and each real vector field X on M holds

$$X < \lambda, \mu > = < D_X\lambda, \mu > + < \lambda, D_X\mu > . \tag{4}$$

Let (x, U) be the local coordinate system on M. If μ_0 is a nonvanishing section of $\mathcal{L}$ over U, then we can identify the space $\Gamma(\mathcal{L}, U)$ of sections with $C^\infty(M)$ by the formula $C^\infty(U) \ni \varphi \leftrightarrow \varphi\mu_0 \in \Gamma(\mathcal{L}, U)$. The operator D_X in this case takes the next form

$$D_X\varphi = X\varphi - i\hbar^{-1}\alpha(X)\varphi, \tag{5}$$

where the 1-form α is given by the relation

$$D_X\mu_0 = -i\hbar^{-1}\alpha(X)\mu_0. \tag{6}$$

By comparing (2) and (6) we have the prequantization formula of Souriau-Kostant

$$\check{f} = f - i\hbar D_{V(f)}. \tag{7}$$

The curvature form Ω of D is defined by the identity

$$\Omega(X,Y) = \frac{1}{2\pi i}([D_X, D_Y] - D_{[X,Y]}),$$

and locally we have

$$\Omega = h^{-1}d\alpha. \tag{8}$$

Theorem 1 [8]. *The Souriau-Kostant formula (6) defines the prequantization if and only if the curvature form Ω coincides with $h^{-1}\omega$.*

The construction of a Hilbert space $\mathcal{H}$ in the geometric quantization procedure essentially involves the choice of the *polarization* that is the involutive Lagrange distribution F in $TM \otimes_{\mathbf{R}} \mathbf{C}$. The polarization F is called *Kähler* if $F \cap \overline{F} = \emptyset$, and the Hermitian form $b(X,Y) = i\omega(X, \overline{Y})$ is positively defined for $X \in F$.

If the polarization F on $TM \otimes \mathbf{C}$ is chosen then the Hilbert space $\mathcal{H}$ consists of the sections λ of $\mathcal{L}$ which are covariantly constant along F

$$D_X\lambda = 0, \quad \lambda \in \mathcal{L}, \quad X \in F. \tag{9}$$

It is said that the function $f \in C^\infty(M)$ preserves the polarization F if the flow $V(f)$ of f obeys the condition

$$[V(f), X^\alpha] = a[f]^\alpha_\beta X^\beta, \tag{10}$$

where $a[f]^\alpha_\beta$ are some smooth functions on M and the vector fields X^α, $\alpha = 1, ..., n$ span F. Let us consider the particular case when the polarization F is spanned by the complex Hamiltonian vector fields. In this case the functions preserving polarization can be quantized with the help of the next formula [12]:

$$Qf = -i\hbar D_{V(f)} + f - \frac{i\hbar}{2}a[f], \tag{11}$$

where $a[f] = \Sigma^n_{\alpha=1}a[f]^\alpha_\alpha$.

In the general situation f does not preserve the polarization and for quantizing f one have to use the *Blattner-Kostant-Sternberg (BKS) kernel* which connects representations for different polarizations [8, 12]. It is easy to see that in the case of an n-dimensional oscillator the flow of the Hamiltonian

$$H = \frac{1}{2}\sum_{\alpha=1}^{n}((p^\alpha)^2 + (q^\alpha)^2) \tag{12}$$

does not preserve the polarization spanned by the Hamiltonian vector fields of both position q^α and momentum p^α variables. Therefore we must use BKS kernel to quantize H. However, when we introduce the complex coordinates

$$z^\alpha = \frac{1}{\sqrt{2}}(p^\alpha + iq^\alpha), \qquad z^{\overline{\alpha}} = \frac{1}{\sqrt{2}}(p^\alpha - iq^\alpha)$$

and Kähler polarization spanned by the vector fields $V(z^\alpha)$ we can see that $H = \Sigma z^\alpha z^{\overline{\alpha}}$ preserves the polarization and one can use (11) to quantize H. As a result they obtain the differential operator $\hat{H}$

$$\hat{H} = Q_{\mathcal{FB}}\, H = \hbar(z^\alpha \frac{\partial}{\partial z^\alpha} + \frac{n}{2}) \tag{13}$$

on the space $\mathcal{O}(U)$ of holomorphic functions on $U \in \mathbf{C}^n$ which we denote $Q_{\mathcal{FB}}$ because the corresponding representation is called Fock-Bargmann representation. In the considered case the representation space $\mathcal{H}$ consists of the sections of $\mathcal{L}$ which have the form $\psi(z)\mu_0$, where μ_0 is a nonvanishing section of $\mathcal{L}$ and $\psi(z) \in \mathcal{O}(U)$.

3   Generalized Fock-Bargmann representation

Consider the mechanical system whose quantization is connected with a generalization of the Fock-Bargmann representation. Let (M, ω) be a $2n$-dimensional Kähler manifold with fundamental form $-\omega$ and positively definite Kähler metric g to be given in local complex coordinates $(z^\alpha, z^{\overline{\alpha}})$ by the formula

$$- \omega = -\omega^{\alpha\overline{\beta}} dz^\alpha \wedge dz^{\overline{\beta}} = -i\partial_{\alpha\overline{\beta}}\Phi\, dz^\alpha \wedge dz^{\overline{\beta}}, \tag{14}$$

$$g = g_{\alpha\overline{\beta}} dz^\alpha dz^{\overline{\beta}} = \partial_{\alpha\overline{\beta}}\Phi\, dz^\alpha dz^{\overline{\beta}}, \tag{15}$$

where Φ is the Kähler potential. As the 2-form ω is closed and nondegenerate, it defines the symplectic structure on M and we can consider (M, ω) as a symplectic manifold and the phase space of some mechanical system. The classical observables [8] of such system form a Lie algebra $C^\infty(M)$ with respect to Poisson brackets.

Let as define the Kähler polarization F on $TM \otimes \mathbf{C}$ in the form

$$F = \{X \in TM \otimes \mathbf{C}\,|\,X = \xi_\alpha V(z^\alpha), \xi_\alpha \in C^\infty(M)\}, \tag{16}$$

where $V(z^\alpha) = \omega^{\overline{\sigma}\alpha}\partial_{\overline{\sigma}}$ is given by (3). By (10) the function f preserves the polarization F if and only if

$$[V(f), V(z^\alpha)] = a[f]^\alpha_\mu V(z^\mu) \tag{17}$$

where according to (3) and (14) $V(f) = \omega^{\mu\overline{\nu}}(\partial_{\overline{\nu}} f \partial_\mu - \partial_\mu f \partial_{\overline{\nu}})$, whence it follows

$$\partial_{\overline{\sigma}}\omega^{\mu\overline{\nu}}\partial_{\overline{\nu}} f + \omega^{\mu\overline{\nu}}\partial_{\overline{\nu}}\partial_{\overline{\sigma}} f \partial_\nu = 0. \tag{18}$$

or

$$\nabla_X \nabla_Y f = 0, \qquad X, Y \in F \tag{19}$$

where ∇ denotes the covariant derivation with respect to Kähler metric g. By Theorem 1 we find from (8) $\omega = d\alpha$ and from (14)

$$\alpha = -i\partial_\alpha \Phi \, dz^\alpha \tag{20}$$

modulo to the exact 1-form $d\beta$.

If we choose both μ and λ in (4) equal to the nonvanishing section μ_0, then using (6) we obtain

$$X < \mu_0, \mu_0 > = i\hbar^{-1}(\alpha(X) - \overline{\alpha(X)}) < \mu_0, \mu_0 > .$$

Evaluating this formula on the vector fields ∂_α, $\partial_{\overline{\alpha}}$, $\alpha = 1, ..., n$ we find with the help of (20)

$$< \mu_0, \mu_0 > = exp(-\hbar^{-1}\Phi) \tag{21}$$

up to a constant multiplier which we omitted.

Now we determine the sections of Hermitian line bundle $\mathcal{L}$ which form the representation space $\mathcal{H}$. Being covariantly constant with respect to $D_{X \in F}$ these sections must obey the equation (9):

$$D_{V(z^\alpha)}\mu = 0, \quad \mu \in \Gamma(\mathcal{L}).$$

From here we find with the help of (5), (6), (16) and (20) that $\mu = \psi(z)\mu_0$, where $\psi(z)$ is holomorphic function on $U \subset \mathbf{C}^n$.

If M is contractible then using the Hermitian structure $<,>$ in $\mathcal{L}$ we can define the scalar product in $\mathcal{H}$ by the formula [8, 12]

$$(\mu_1, \mu_2) = \int \psi_1(z)\overline{\psi_2(z)} < \mu_0, \mu_0 > \omega^n, \tag{22}$$

where $\mu_1 = \psi_1(z)\mu_0$, $\mu_2 = \psi_2(z)\mu_0$ and ω^n is n-th external degree of ω. Using (21) we find

$$(\mu_1, \mu_2) = \int \psi_1(z)\overline{\psi_2(z)}exp(-\hbar^{-1}\Phi)\,\omega^n. \tag{23}$$

In this case the representation the Hilbert space associated with polarization F given by (16) can be identified with the Fock space $L_2^{hol}(U, dm)$ of holomorphic functions on $U \subset \mathbf{C}^n$ quadratically integrable with the measure $dm = \exp(-\hbar^{-1}\Phi)\omega^n$.

Let us consider the Kähler space $\mathcal{K}_{2n}$ of constant holomorphic curvature k (see for example [9]). As it is known the space $\mathcal{K}_{2n}$ is isometric to the projective space $\mathbf{CP}^n$ for $k > 0$, to the disk $D_n^R = \{z \in \mathbf{C}^n | z\overline{z} < R\}$ for $k < 0$ and to $\mathbf{C}^n$ for $k = 0$. The metric of the space $\mathcal{K}_{2n}$ in the local complex coordinates is

$$ds^2 = 2g_{\alpha\overline{\beta}}dz^\alpha dz^{\overline{\beta}}, \qquad g_{\alpha\overline{\beta}} = \partial_{\alpha\overline{\beta}} = (A\delta_{\alpha\beta} - \frac{k}{4}z^{\overline{\alpha}}z^\beta)A^{-2}, \tag{24}$$

$$\Phi = \frac{4}{k} ln\, A, \quad A = 1 + \frac{k}{4}\Sigma z^\nu z^{\overline\nu}.$$

The curve $x(t)$ on the Kähler manifold M is called *H-planar* (or *holomorphical planar*) [11] if it obeys the following equation

$$\nabla_\chi \chi = a(t)\chi + b(t)J(\chi), \qquad \chi \equiv \dot{x}$$

where $a(t)$, $b(t)$ are some real-valued functions and J is the complex structure operator in TM.

Let M and M' be two Kähler manifolds. The mapping $f : M \to M'$ is called *H*-projective (see for example [13]) if it transforms *H*-planar curves of M into *H*-planar curves of M'.

The contravariant components and nonvanishing Christoffel symbols of the metric (24) in local complex coordinates are given by the formula

$$g^{\alpha\overline\beta} = (\delta^{\alpha\beta} + \frac{k}{4}z^\alpha z^{\overline\beta})A^2 \equiv -i\omega^{\alpha\overline\beta}, \qquad \Gamma^\alpha_{\beta\gamma} = -\frac{k}{4}A^{-1}(z^{\overline\alpha}\delta^\beta_\gamma + z^{\overline\beta}\delta^\alpha_\gamma) = \overline{\Gamma^{\overline\alpha}_{\overline\beta\overline\gamma}}.$$

From here it is follows that the considered metrics are *H*-projectively flat. Now we find from (19)

$$\partial_{\mu\overline\nu}f + 2A^{-1}\partial_{(\overline\mu}A\partial_{\overline\nu)}f = 0. \tag{25}$$

After substitution $f = WA^{-1}$ in this equation it takes the form

$$\partial_{\overline\mu\overline\nu}W = 0$$

whence

$$W = u_{\overline\alpha}(z)z^{\overline\alpha} + v(z),$$

where $u_{\overline\alpha}$ and v are arbitrary holomorphic functions.

In [7] the system of observables

$$\tilde{H} = \frac{1 + z\overline z}{1 - z\overline z}, \quad N = \frac{z}{1 - z\overline z}, \qquad \overline N = \frac{\overline z}{1 - z\overline z} \tag{26}$$

was considered when quantizing 1-dimensional harmonic oscillator. The Hamiltonian vector fields $V(\tilde H)$, $V(N)$ and $V(\overline N)$ form the basis of holomorphic isometries Lie algebra in the space $\mathcal{K}_2$ of holomorphic curvature $k = -4$ (see for example [14]). We use the next system of observables

$$H = \frac{\Sigma z^\nu z^{\overline\nu}}{A}, \quad (u_{\overline\alpha} = z^\alpha, \quad v = 0), \tag{27}$$

$$N^\beta = \frac{z^\beta}{A}, \quad (u_{\overline\alpha} = 0, \quad v = z^\beta), \tag{28}$$

$$N^{\overline\beta} = \frac{z^{\overline\beta}}{A}, \quad (u_{\overline\alpha} = \delta^\beta_\alpha, \quad v = 0). \tag{29}$$

One can easily check that H, N^α and $N^{\overline\alpha}$ are the solutions of equation (25). The Hamiltonian vector fields of this functions define infinitesimal isometries in $\mathcal{K}_{2n}$ and H-projective transformations in the flat Kähler manifold $\mathbf{C}^n$. Note that these isometries do not form a Lie algebra. In the 1-dimensional case N^1, $N^{\overline 1}$ coincide with $N, \overline N$ from (24) and H can be obtained from $\tilde H$ by the linear substitution. The use of H is more preferable from the point of view of the limit transition to the flat space ($k = 0$). In the limit $k \to 0$ we obtain $H = \Sigma z^\nu z^{\overline\nu}$, i.e. the Hamiltonian of the harmonic oscillator (12) written in complex coordinates.

The Hamiltonian vector fields of the functions H, N^α and $N^{\overline\alpha}$ have the form

$$T \equiv V(H) = \omega^{\mu\overline\nu}(\partial_{\overline\nu}H\partial_\mu - \partial_\nu H\partial_{\overline\mu}) = i(z^\nu\partial_\nu - z^{\overline\nu}\partial_{\overline\nu}),$$

$$T^\alpha \equiv V(N^\alpha) = -i\left(\frac{k}{4}z^\alpha z\nu\partial_\nu + \partial_{\overline\alpha}\right),\tag{30}$$

$$T^{\overline\alpha} \equiv V(N^{\overline\alpha}) = i\left(\frac{k}{4}z^{\overline\alpha}z\overline\nu\partial_{\overline\nu} + \partial_\alpha\right).$$

Using this formulae we can calculate the commutators of the vector fields T, T^α and $T^{\overline\alpha}$

$$[T^\alpha . T^\beta] = 0, \qquad [T^\alpha, T^{\overline\beta}] = i\frac{k}{4}(\delta^\alpha_\beta T + T^{\alpha\overline\beta}),\tag{31}$$

$$[T^\alpha, T] = -iT^\alpha, \qquad [T^{\overline\alpha}, T] = -iT^{\overline\alpha},$$

where $T^{\alpha\overline\beta} = V(z^\alpha z^{\overline\beta}A^{-1}) = i(z^\alpha\partial_\beta - z^{\overline\beta}\partial_{\overline\alpha})$.

The generators T, T^α and $T^{\overline\alpha}$ do not form a Lie algebra but if we join to them the generator $T^{\alpha\overline\beta}$ then we obtain

$$[T^\alpha, T^{\beta\overline\gamma}] = i\delta^\alpha_\gamma T^\beta, \qquad [T^{\overline\alpha}, T^{\beta\overline\gamma}] = -i\delta^\alpha_\beta T^{\overline\gamma},\tag{32}$$

$$[T^{\alpha\overline\beta}, T^{\gamma\overline\nu}] = i(\delta^\gamma_\beta T^{\alpha\overline\nu} - \delta^\nu_\alpha T^{\gamma\overline\beta}).$$

Because $T^\alpha, T^{\overline\alpha}, T^{\alpha\overline\beta}$ are linearly independent and $T = \Sigma T^{\alpha\overline\alpha}$ from (29), (30) it follows that T^α, $T^{\overline\alpha}$ and $T^{\alpha\overline\beta}$ form a basis of a $n(n+4)$-dimensional (over $\mathbf{R}$) Lie algebra $\ell(k)$ which is the Lie algebra of infinitesimal isometries of the space $\mathcal{K}_{2n}$ preserving the complex structure.

The Poisson brackets $\{f, g\} = \omega^{\alpha\overline\beta}(\partial_\alpha f\partial_{\overline\beta}g - \partial_{\overline\beta}f\partial_\alpha g)$ of the functions $H^{\alpha\overline\beta} = z^\alpha z^{\overline\beta}A^{-1}$, $N^\alpha = z^\alpha A^{-1}$ and $N^{\overline\beta} = z^{\overline\beta}A^{-1}$ are

$$\{N^\alpha, N^\beta\} = 0, \qquad \{N^\alpha, N^{\overline\beta}\} = i\frac{k}{4}(\delta^\alpha_\beta H + N^{\alpha\overline\beta}) - i\delta^\alpha_\beta,$$

$$\{N^\alpha, N^{\beta\overline\gamma}\} = i\delta^\alpha_\gamma N^\beta, \qquad \{N^{\overline\alpha}, N^{\beta\overline\gamma}\} = -i\delta^\alpha_\beta N^{\overline\gamma},$$

$$\{N^{\alpha\overline\beta}, N^{\gamma\overline\nu}\} = i(\delta^\alpha_\nu N^{\gamma\overline\beta} - \delta^\gamma_\beta N^{\alpha\overline\nu}).$$

Note that if we take the limit $k \to 0$ (31), (32) turns to

$$[T^\alpha, T^\beta] = 0, \qquad [T^\alpha, T^{\bar\beta}] = 0,$$

$$[T^\alpha, T^{\beta\bar\gamma}] = i\delta_\gamma^\alpha T^\beta, \qquad [T^{\bar\alpha}, T^{\beta\bar\gamma}] = -i\delta_\beta^{\bar\alpha} T^{\bar\gamma}, \tag{33}$$

$$[T^{\alpha\bar\beta}, T^{\gamma\bar\nu}] = i(\delta_\nu^\alpha T^{\gamma\bar\beta} - \delta_\beta^\gamma T^{\alpha\bar\nu}).$$

and define the $n(n+4)$-dimensional Lie algebra $\ell(0)$. The curve $\ell(k)$ in the manifold of the $n(n+4)$-dimensional Lie algebra structures is the deformation of the algebra $\ell(0)$ defined by the commutation relations (33) and containing the n-dimensional harmonic oscillator algebra $\mathbf{osc}(n)$ as the Lie subalgebra. That is why we can consider the mechanical system with the phase space $\mathcal{K}_{2n}$, symplectic form ω and the observables H, N^α and $N^{\bar\alpha}$ as the "deformation" of classical n-dimensional harmonic oscillator.

Now we quantize the classical mechanical systems obtained in the preceding sections using the polarization F defined by (16). We calculate now $a[f] = \Sigma a[f]_\nu^\nu$ (see (11)) for $f = H$, N^α and $N^{\bar\alpha}$. Substituting in (10) $V(z^\alpha) = \omega^{\bar\nu\alpha}\partial_{\bar\nu}$ instead of X^α and T, T^α, $T^{\bar\alpha}$ instead of $V(f)$ we find using formulae (23) and (30)

$$[T, V(z^\beta)] = iV(z^\alpha),$$

$$[T^\alpha, V(z^\beta)] = -i\frac{k}{4}(z^\alpha\delta_\nu^\beta + z^\beta\delta_\nu^\alpha)V(z^\nu),$$

$$[T^{\bar\alpha}, V(z^\beta)] = 0$$

whence

$$a[N^\alpha] = -i\frac{k}{4}z^\alpha(n+1), \qquad a[N] = 0, \qquad a[H] = in.$$

After this from (27)-(29) using (5), (20) and (22) we obtain the following expressions for differential operators in $L_2^{hol}(U, dm)$ which are the quantizations of the observables H, N^α and $N^{\bar\alpha}$

$$QH \equiv \hat{H} = \hbar(z^\nu\partial_\nu\psi + \frac{n}{2}\psi), \tag{34}$$

$$QN^\alpha \equiv \hat{N}^\alpha = -\hbar\frac{k}{4}z^\alpha z^\nu\partial_\nu\psi + z^\alpha(1 - \frac{\hbar k}{8}(n+1))\psi,$$

$$QN^{\bar\alpha} \equiv \hat{N}^{\bar\alpha} = \hbar\partial_\alpha\psi.$$

Let $B : \mathcal{H} \to \mathcal{H}$ be the selfadjoint operator in the Hilbert space $\mathcal{H}$. The set $\sigma(B) = \{\rho \in \mathbf{R} | \exists \mu_\rho \in \mathcal{H} : B\mu_\rho = \rho\mu_\rho\}$ is called the *spectrum* of the operator B. The number $\rho \in \sigma(B) \subset \mathbf{R}$ and section $\mu_\rho \in \mathcal{H}$ are called the *eigenvalue* of B and *eigenstate* with eigenvalue ρ.

Let us consider the eigenstate ψ_E of the operator QH with the eigenvalue E. Equation (34) yields

$$\hbar(z^\nu\partial_\nu\psi_E + \frac{n}{2}\psi_E) = E\psi_E$$

which is equivalent to

$$z^\nu \partial_\nu \psi_E = (E\hbar^{-1} - \frac{n}{2})\psi_E$$

whence ψ_E is a homogeneous function of z of degree $l = (E\hbar^{-1} - \frac{n}{2})$. Since ψ_E is a holomorphic it follows that l is a non-negative integer, so that the spectrum of QH is given by

$$E_l = (l + \frac{n}{2})\hbar, \quad l \in \{0\} \cup \mathbf{N}$$

and coincides with the spectrum of the n-dimensional harmonic oscillator Hamiltonian (12) (see for example [12]).

4 Quantization of Kähler manifolds admitting H-projective mappings

In this section we consider the quantization of Kähler spaces admitting H-projective mappings onto other Kähler spaces. Let (M, ω) and (M', ω') be two Kähler manifolds with fundamental forms $-\omega$ and $-\omega'$. Let $\varrho : M \to M'$ be H-projective mapping, it is well known that a H-projective mapping preserves the complex structure. Therefore we can choose the local complex chart $(z^\alpha, z^{\overline\alpha}, U)$ in M such that for each point $p \in U$ with coordinates $(z^\alpha, z^{\overline\alpha})$ its image $\varrho(p) \in \varrho(U)$ has the same coordinates. The necessary and sufficient condition for the mapping $\varrho : M \to M'$ to be H-projective is expressed with the following equation [13]

$$b_{\alpha\overline\beta;\gamma} = 2\phi'_\alpha g_{\overline\beta\gamma}, \tag{35}$$

where

$$b_{\alpha\overline\beta} = e^{2\phi} g'^{\overline\mu\nu} g_{\overline\mu\alpha} g_{\nu\overline\beta}, \qquad b_{\alpha\beta} = b_{\overline\alpha\overline\beta} = 0,$$

$$\phi'_\alpha = \partial_\alpha \phi' = \partial_\mu \phi e^{2\phi} g'^{\mu\overline\nu} g_{\alpha\overline\nu},$$

$$J^i_k = J'^i_k, \qquad J^\mu_\nu = -J^{\overline\mu}_{\overline\nu} = i\delta^\mu_\nu, \qquad J^\mu_{\overline\nu} = J^{\overline\mu}_\nu = 0,$$

ϕ is some function on U and semicolon denotes the covariant derivation with respect to g.

Because of the positive definiteness of g we can define the complex frame $\{Z_A, Z_{\overline A}\}$ which is adapted for the Hermitian structure of M [10]. Then for the frame components of g we have

$$g_{A\overline B} = \delta^A_B.$$

Transformations, preserving this form of g, belong to the unitary group $U(n)$ for each point $p \in M$. With the help of such transformations we can choose the frame $\{Y_A, Y_{\overline B}\}$, so that

$$g_{A\overline B} = \delta^A_B, \qquad b_{A\overline B} = \lambda_A \delta^A_B, \tag{36}$$

where $\lambda_A = \overline{\lambda}_A$ are the roots of the λ-matrix $(b - \lambda g)$. Written in the frame (see for example [1, 2]) (19) and (35) have the form

$$Y_{\overline{A}} Y_{\overline{B}} f - \sum_S \gamma_{\overline{BS}\overline{A}} Y_S f = 0, \tag{37}$$

$$\delta_{AB} Y_A \lambda_A + \sum_S (\gamma_{\overline{S}AC} \lambda_S \delta_{SB} + \gamma_{S\overline{BC}} \lambda_S \delta_{SA}) = 2\delta_{CB} Y_A \phi', \tag{38}$$

where $\gamma_{\overline{S}AC}$ $(\gamma_{SAC} = \gamma_{\overline{SAC}}) = 0$ are Ricci rotation coefficients of the frame.

Let $\{\theta^A, \theta^{\overline{A}}\}$, $\theta^A, \theta^{\overline{A}} \in T^*M$ be the coframe dual to the frame $\{Y^A, Y^{\overline{A}}\}$. Then the connection form α in the Hermitian line bundle $\mathcal{L}$ (see §1) can be written in the following form

$$\alpha = -iY_A \Phi \theta^A$$

as in §2. From (14) and (36) it follows $\omega_{A\overline{B}} = \omega^{A\overline{B}} = i\delta^A_B$. Then for $Y_A = \xi^\mu_A \partial_\mu$ we obtain

$$V(z^\alpha) = -i\Sigma \xi^\alpha_B Y_{\overline{B}}.$$

If the function $F \in C^\infty(U)$ preserves polarization F it obeys the condition (17) which we can write in the form

$$[V(f), V(z^\alpha)] = a[f]^\alpha_\mu V(z^\mu) = \sum_{A,B} (Y_{\overline{A}} f Y_A \xi^\alpha_B - Y_A f Y_{\overline{A}} \xi^\alpha_B + \xi^\alpha_A (Y_{\overline{A}} Y_B f)) Y_{\overline{B}},$$

and here we find

$$a[f] = i\zeta^A_\alpha \sum_B (Y_{\overline{B}} f Y_B \xi^\alpha_A - Y_B f Y_{\overline{B}} \xi^\alpha_A + \xi^\alpha_B Y_{\overline{B}} Y_A f),$$

where ζ^B_α are the components of the inverse (ξ^α_B) matrix: $\zeta^A_\mu \xi^\nu_A = \delta^\nu_\mu$.

At last we can evaluate the differential operator $\mathcal{Q}f$ in $L^{hol}_2(U, dm)$ for the function $f \in C^\infty(U)$ obeying (37)

$$(\mathcal{Q})f\psi \equiv \hat{f}\psi = \hbar \sum_A Y_{\overline{A}} f Y_A \psi - Y_A \Phi Y_{\overline{A}} f \psi + f\psi - \frac{i\hbar}{2} a[f]\psi.$$

To obtain concrete results we have to use specific expressions for Ricci rotation coefficients and frame vector fields. In particular, for 4-dimensional Kähler manifold admitting H-projective mappings there are two possibilities

$$1) \quad \lambda_1 = \lambda_2, \qquad 2) \quad \lambda_1 \neq \lambda_2.$$

In the first case we have $b = \lambda g$ and, hence, $g' = \mu g$ where $\mu = (\lambda e^{2\phi})^{-1}$. Then (38) implies $\mu = const$, and H-projective mappings are only rescalings of the metric. The second case is more interesting. If $\lambda_1 \neq \lambda_2$ then from (38) it follows [3]

$$Y_1 \lambda_1 = Y_{\overline{1}} \lambda_2 = Y_2 \lambda_1 = Y_{\overline{2}} \lambda_1 = 0, \qquad \gamma_{1\overline{2}2} = Y_1 \ln |\lambda_1 - \lambda_2|,$$

$$\gamma_{\bar{1}21} = -Y_2 \ln |\lambda_1 - \lambda_2|, \qquad \gamma_{\bar{2}11} = \gamma_{\bar{1}22}.$$

This work was partially supported by grant 1749 of International Science Foundation and grant RFFI-94-01-01118-a of Russian Foundation for Fundamental Investigati ons.

References

[1] A.V. Aminova: *Tensor, N.S.*, **45** (1987), 1-13.

[2] A.V. Aminova: *Uspekhi matem. nauk*, **48** (1993), 107-159.

[3] A.V. Aminova and D.A. Kalinin: *IZV. VUZ. Matem.*, No.8 (1994), 11-21.

[4] R. Balbinot, A.El Gradechi, J.-P. Gazeau and B. Giorgini: *J. Phys. A*, **25** (1992), 1185-1210.

[5] F.A. Berezin: *Comm. Math. Phys.*, **40** (1975), 153-174.

[6] J.-P. Gazeau and V. Hussin: *J.Phys. A*, **25** (1992), 1549-1573.

[7] J.-P. Gazeau and J. Renaud: *Preprint Université Paris VII, PAR-LPTM-92*.

[8] A.A. Kirillov:"Modern problems of mathematics", **4**, VINITI SSSR, Moscow, 1985, 141-204

[9] S. Kobayashi and K. Nomizu: Foundation of differential geometry, V. II, Intersci. Publ., N.Y., 1969.

[10] A. Lichnerowicz: Theorie globale des connexiones et des groupes d'holonomie, Cremonese, Roma. 1955.

[11] T. Otsuki and Y. Tashiro: Math. J. Okayama Univ., 4(1954), 57-78.

[12] J. Sniatycki: Geometric quntization and quantum mechanics, Springer, Berlin etc, 1980.

[13] N.S. Sinyukov: Geodesic mappings of Riemannian spaces, *Moscow, Nauka*, 1979.

[14] J.M. Tuynman: Quantization. Towards a comparison between methods, *J. Math. Phys.*, **28**(1987). 2829-2840.

Renormalization Group Flow of the Chern-Simons Parameter

Martin Reuter

Abstract

The effective average actions for gauge theories and their associated exact renormalization group equations are briefly reviewed and an application to Chern-Simons theory is described.

1 Introduction

The purpose of these notes is twofold. First we shall give a brief introduction to the method of the effective average actions and their associated exact renormalization group equations [1, 2, 16] and then we shall apply these ideas to a system which is also quite interesting in its own right, namely pure Chern-Simons theory in three dimensions [4].

The effective average action Γ_k can be thought of as a continuum version of the block spin action for spin systems [1]. The functional Γ_k is the action relevant to the physics at (mass) scale k. It has the quantum fluctuations with momenta larger than k integrated out already, but those with momenta smaller than k are not yet included. Γ_k interpolates between the classical action S for large values of k, and the conventional effective action for k approaching zero: $\Gamma_{k\to\infty} = S, \Gamma_{k\to 0} = \Gamma$. In many important cases where perturbation theory is inapplicable due to infrared divergences the limit $k \to 0$ exists and can be computed by various methods. This includes for instance massless theories in low dimensions or the high temperature limit of 4 dimensional theories. The functional Γ_k can be obtained by solving an exact renormalization group equation which describes its evolution while k is lowered from infinity to zero. In the approach of ref. [2], and for models with a scalar field ϕ only, this evolution equation reads

$$\frac{\partial}{\partial t}\Gamma_k[\phi] = \frac{1}{2}\mathrm{Tr}\left[\frac{\partial}{\partial t}R_k\left(\Gamma_k^{(2)}[\phi] + R_k\right)^{-1}\right]. \tag{1}$$

Here $t \equiv \ln k$ is the "renormalization group time" and $\Gamma_k^{(2)}$ denotes the matrix of the second functional derivatives of Γ_k. The operator $R_k \equiv R_k(-\partial^2)$ or, in momentum space, $R_k \equiv R_k(q^2)$ describes the details of how the small momentum modes are cut off and it is to some extent arbitrary. It has to vanish for $q^2 \gg k^2$ and to become a mass-like term proportional to k^2 for small momenta $q^2 \ll k^2$. The derivation of (1) proceeds as follows. In the euclidean functional integral for the generating functional of the connected Green functions one adds a momentum-dependent mass term (playing the role of a smooth IR cutoff) $\frac{1}{2}\int \phi R_k(-\partial^2)\phi$ to the classical action S. Then, up to

where $\lambda, \omega, \gamma, \Lambda$ are the gravitational coupling constants. In particulary, γ^2, Λ are the Newton and cosmological constants, $\mathcal{L}_m$ is the Lagrangian for matter including some scalar multiplet φ, ξ is scalar-gravitational coupling constant. $\mathcal{L}_{m(2)}$ is the conformally invariant version of (1), where now $\mathcal{L}_{m(2)}$ describes massless matter.

Our purpose here will be to find the effective potential for the scalar field with account for QG corrections. The general structure of such a one-loop effective potential in the linear curvature approximation ($\varphi^2 \gg R$) may be presented in the following form

$$
\begin{aligned}
V^{(1)} &= \frac{1}{4!} f \varphi^4 - \frac{1}{2} \xi R \varphi^2 \\
&+ \frac{1}{2 \cdot 4!} (\beta_f - 4 f \gamma_\varphi) \varphi^4 \left(\log \frac{\varphi^2}{\mu^2} - \frac{25}{6} \right) \\
&- \frac{1}{4} (\beta_\xi - 2 \xi \gamma_\varphi) R \varphi^2 \left(\log \frac{\varphi^2}{\mu^2} - 3 \right).
\end{aligned}
\tag{3}
$$

Here $\beta_f, \beta_\xi, \gamma_\varphi$ are the RG beta functions for the effective couplings f, ξ, φ, respectively. Coleman-Weinberg renormalization conditions [3] have been used in the explicit solution of the RG equations for the effective potential, μ^2 is a mass parameter. We also suppose that the theory under discussion is massless or, if massive, the background field satisfies $\varphi^2 \gg m^2$, where m^2 is the biggest effective mass of the model.

The general form of the beta functions in the presence of QG effects is following

$$
\begin{aligned}
(4\pi)^2 \beta_f &= (4\pi)^2 \beta_f^{(0)} + \lambda^2 \xi^2 \left(-15 + \frac{3}{4\omega^2} - \frac{9\xi}{\omega^2} + \frac{27\xi^2}{\omega^2} \right) \\
&\qquad\qquad - \lambda f \left(5 + 3\xi^2 + \frac{33}{2\omega}\xi^2 - \frac{6}{\omega}\xi + \frac{1}{2\omega} \right) \\
&\equiv (4\pi)^2 (\beta_f^{(0)} + \Delta\beta_f), \\
(4\pi)^2 \beta_\xi &= (4\pi)^2 \beta_\xi^{(0)} + \lambda\xi \left(-\frac{3}{2}\xi^2 + 4\xi + 3 + \frac{10}{3}\omega - \frac{9}{4\omega}\xi^2 + \frac{5}{\omega}\xi + \frac{1}{\omega} \right) \\
&\equiv (4\pi)^2 (\beta_\xi^{(0)} + \Delta\beta_\xi), \\
(4\pi)^2 \gamma_\varphi &= (4\pi)^2 \gamma_\varphi^{(0)} + \frac{\lambda}{4} \left(\frac{13}{3} - 8\xi - 3\xi^2 - \frac{1}{6\omega} - \frac{2\xi}{\omega} + \frac{3\xi^2}{2\omega} \right) \\
&\equiv (4\pi)^2 (\gamma_\varphi^{(0)} + \Delta\gamma_\varphi),
\end{aligned}
\tag{4}
$$

where $\beta_f^{(0)}, \beta_\xi^{(0)}, \gamma_\varphi^{(0)}$ are the corresponding beta functions in the absence of QG. Note that QG corrections in (4) are universal for every matter theory. However, we will restrict ourselves to the case where the theory does not contain Yukawa interactions, i.e. we will consider at most a free fermionic sector (or no fermions at all) in $\mathcal{L}_m$, what gives us less cumbersome expressions . Notice also that the above expressions (4) are given only for model (1), and the harmonic-type gauge has been used for the calculation of γ_φ.

For its conformally invariant version (supposing special conformal regularization) we have

$$(4\pi)^2\beta_f = (4\pi)^2\beta_f^{(0)} + \frac{5}{12}\lambda^2 - \frac{41}{8}\lambda f,$$
$$(4\pi)^2\gamma_\varphi = (4\pi)^2\gamma_\varphi^{(0)} + \frac{27}{32}\lambda. \tag{5}$$

Thus, we have obtained the one-loop effective potential for a renormalizable matter theory coupled with quantum R^2- gravity.

Following the seminal paper by Coleman and Weinberg [3] we may extend the range of validity for the scalar field in the effective potential using the technique of RG improvement. Because this technique can be straightforwardly generalized to the case when the spacetime is curved and no big complications appear, we will give here the final result only, omitting the details of its derivation (see also [4]):

$$V = \frac{f(t)}{4!}\varphi^4(t) - \frac{1}{2}\xi(t)R\varphi^2(t), \tag{6}$$

where

$$\dot{f}_i(t) = \beta_{f_i}(t), \quad \dot{\xi}(t) = \beta_\xi(t), \quad \frac{d\varphi(t)}{dt} = -\gamma_\varphi(t)\varphi(t),$$
$$f_i(0) = f_i, \quad \xi(0) = \xi, \quad \varphi(0) = \varphi, \tag{7}$$

and the f_i's stand for all the couplings of the theory. Note that we again consider the situation where the effective masses are much smaller than φ^2, hence $t = \frac{1}{2}\log\frac{\varphi^2}{\mu^2}$ and one can effectively neglect the terms of the form $m^2(t)\varphi^2(t)$, $\Lambda(t)$, $\gamma^2(t)R$, which contain the effective masses. Having at hand the general expressions (3), (6), one can apply them to the analysis of the effective potential in specific models.

3 $f\varphi^4$-theory with R^2-gravity

Let us start from the simplest conformally invariant theory with matter as the second example

$$\mathcal{L}_{m(2)} = \frac{1}{2}g^{\mu\nu}\partial_\mu\varphi\partial_\nu\varphi - \frac{1}{4!}\,f\varphi^4.$$

As a first step, we will choose also the flat background. Then, using (3) and (5) we get

$$V^{(1)} = \frac{1}{4!}f\varphi^4 + \frac{1}{48(4\pi)^2}\left[3f^2 + \frac{5}{12}\lambda^2 - \frac{17}{2}\lambda f\right]\varphi^4\left(\log\frac{\varphi^2}{\mu^2} - \frac{25}{6}\right). \tag{8}$$

As one can see, there exists a region of the parameter space for this model where quantum corrections become negative and the potential is unbounded from below. But the λ^2 QG correction is positive like in scalar QED. The form of (8) is similar to that of scalar QED, and similar effects —dimensional transmutation— may take place. Indeed, choosing the renormalization mass μ to be equal to the minimum φ_m of $V^{(1)}$, from the equation $\dfrac{\partial V^{(1)}}{\partial \varphi} = 0$ one has

$$2(4\pi)^2 f = \frac{11}{3}\left(\frac{5}{12}\lambda^2 - \frac{17}{2}\lambda f + 3f^2\right).\tag{9}$$

There are two ways from this point. First, we may consider $f \sim \lambda^2$, then $2(4\pi)^2 f = \dfrac{55}{36}\lambda^2$, and all the remaining terms in (9) become higher order corrections. Then,

$$V^{(1)} = \frac{1}{48(4\pi)^2}\left[\frac{5}{12}\lambda^2\right]\varphi^4\left(\log\frac{\varphi^2}{\varphi_m^2} - \frac{1}{2}\right).\tag{10}$$

Alternatively, one can also solve eq. (9) for f in terms of λ:

$$\begin{aligned}
V^{(1)} &= \frac{1}{88}\left[\frac{1}{2}\left(\frac{17}{6}\lambda + \frac{2(4\pi)^2}{11}\right) \pm \frac{1}{2}\sqrt{\left(\frac{17}{6}\lambda + \frac{2(4\pi)^2}{11}\right)^2 - \frac{5}{9}\lambda^2}\right]\\
&\quad \times \varphi^4\left(\log\frac{\varphi^2}{\varphi_m^2} - \frac{1}{2}\right).
\end{aligned}\tag{11}$$

Hence, the dimensional transmutation (10) naturally takes place in QG on flat background.

Let us consider now the simplest version of a scalar-tensor theory (or dilaton gravity) where the scalar part is given by the $f\varphi^4 - theory$ interacting with higher-derivative gravity (1). Similarly, one can repeat the above analysis for such a theory. Using expression (3) one obtains

$$\begin{aligned}
V^{(1)} = {}& \frac{1}{4!}f\varphi^4 - \frac{1}{2}\xi R\varphi^2\\
& + \frac{1}{48(4\pi)^2}\left[3f^2 + \lambda^2\xi^2\left(-15 + \frac{3}{4\omega^2} - \frac{9\xi}{\omega^2} + \frac{27\xi^2}{\omega^2}\right)\right.\\
& \left.\quad - \lambda f\left(\frac{28}{3} + 18\frac{\xi^2}{\omega} - \frac{8\xi}{\omega} - 8\xi + \frac{1}{3\omega}\right)\right]\varphi^4\left(\log\frac{\varphi^2}{\mu^2} - \frac{25}{6}\right)\\
& - \frac{1}{4(4\pi)^2}\left\{f\left(\xi - \frac{1}{6}\right) + \lambda\xi\left[8\xi + \frac{5}{6} + \frac{10}{3}\omega + \frac{1}{\omega}\left(-3\xi^2 + 6\xi + \frac{13}{12}\right)\right]\right\}\\
& \quad R\varphi^2\left(\log\frac{\varphi^2}{\mu^2} - 3\right).
\end{aligned}\tag{12}$$

Starting from flat background and supposing that $f \sim \lambda^2$ (we remember that $|\xi| < 1$) one has

$$V^{(1)} = \frac{1}{48(4\pi)^2}\left[\lambda^2\xi^2\left(-15 + \frac{3}{4\omega^2} - \frac{9\xi}{\omega^2} + \frac{27\xi^2}{\omega^2}\right)\right]\varphi^4\left(\log\frac{\varphi^2}{\varphi_m^2} - \frac{1}{2}\right),\tag{13}$$

and the $f\lambda$, f^2 terms are of higher order, like before. It is interesting to note that, if we work in linear curvature approximation, *i.e.* with (12), then dimensional transmutation may fix other parameters of the theory. Indeed, aplying the same technique, *i.e.* choosing $\mu^2 = \varphi_m^2$, applying $\dfrac{\partial V^{(1)}}{\partial \varphi} = 0$ to expression (12) $(R \neq 0)$ and bearing in mind that we have already fixed f in (13), one may get a similar condition to determine ξ.

Now let us study a more reliable RG improved effective potential for the same theory. Note that we improve tree-level potential only, however there are no problems to start from the one-loop effective potential and apply the improving (Wilsonian) procedure here. Then, one can get next-to- leading approach to RG improved effective potential.

a) Conformal version. In this case the RG improved potential is

$$
V = \frac{\lambda(1 - Cx^B)x^{-1}}{Cx^B(9.79 - \sqrt{9.79^2 - 7.2}) - (9.79 + \sqrt{9.79^2 - 7.2})} \frac{\varphi^4(t)}{24}
- \frac{1}{6}R\varphi^2(t),
\tag{14}
$$

where $x = 1 + \dfrac{\alpha^2\lambda t}{(4\pi)^2}$, $\lambda(t) = \dfrac{\lambda}{1 + \dfrac{\alpha^2\lambda t}{(4\pi)^2}}$, $\alpha^2 = \dfrac{797}{60}$; $B = \dfrac{5\sqrt{9.79^2 - 7.2}}{6\alpha^2} < 1$,

$t = \dfrac{1}{2}\log\dfrac{\varphi^2}{\mu^2}$ and the integration constant C is defined by

$$
C = \frac{\lambda + f(9.79 + \sqrt{9.79^2 - 7.2})}{f(9.79 - \sqrt{9.79^2 - 7.2}) + \lambda},
$$

$\varphi(t) = \varphi x^{-\frac{27}{32\alpha^2}}$. As one can see from (14), at extremely large t's the first term becomes the leading one. But the effective scalar coupling in (14) becomes negative for large t, and the potential (14) turns out to be unstable. This explicit example shows that QG corrections may drive the scalar coupling constant to become negative, making the potential unstable, as happens with Yukawa theories. Stability ,in principle, may put restrictions to initial values of QG couplings. However, in difference with Yukawa theory this QG driven instability is very weak.It corresponds to extremely large (rather non-realistic) t.

It is very interesting to note that stability of this potential maybe studied also in the region between Planck mass scale and GUT scale (not very large negative t).Then one can see that RG improved potential is stable one and no restrictions to QG couplings appear.

b) General case. In this case one cannot solve the RG equations for the coupling constants analytically. The non-explicit expression looks as follows

$$
V = \frac{1}{4!}f(t)\varphi^4(t) - \frac{1}{2}\xi(t)R\varphi^2(t),
\tag{15}
$$

where the RG equations are [1]:

$$\lambda^2(t) = \frac{\lambda}{1 + \dfrac{799\lambda t}{60(4\pi)^2}},$$

$$(4\pi)^2 \frac{d\omega}{dt} = -\lambda \left[\frac{10}{3}\omega^2 + \left(18 + \frac{19}{60}\right)\omega + \frac{5}{12} + \frac{3}{2}\left(\xi - \frac{1}{6}\right)^2 \right], \tag{16}$$

$$\frac{d\xi}{dt} = \beta_\xi(t), \qquad \frac{df}{dt} = \beta_f(t), \qquad \frac{d\varphi}{dt} = -\gamma_\varphi(t)\varphi,$$

where $\beta_\xi, \beta_f, \gamma_\varphi$ are given by (4) with $\beta_\xi^{(0)} = \dfrac{f\left(\xi - \frac{1}{6}\right)}{(4\pi)^2}$, $\beta_f^{(0)} = \dfrac{3f^2}{(4\pi)^2}$, $\gamma_\varphi^{(0)} = 0$. Qualitative analysis of the above equations shows that the scalar coupling constant may be driven to become negative due to QG corrections at extremely large t. One can solve (16) numerically for different initial values of the coupling constants and to show that for large but reasonable t the potential is stable one.

Using the RG-improved effective potential, one can study the possibility of curvature-induced phase transitions. For the one-loop effective potential in a matter theory interacting with quantum R^2-gravity, such an analysis has been given for example in [1]. The result was that a first kind curvature transition is possible with inducing of Einstein gravity at the minimum $\varphi = \varphi_m$. Moreover, for some choices of higher-derivative gravity parameters, the cosmological constant may be equal to zero.

4 Gauge theories with quantum R^2-gravity

The formalism above developed is very useful and may be straightforwardly applied to gauge theories. As an example, let us consider massless scalar electrodynamics. For simplicity we will give only the one-loop effective potential. The Lagrangian is given by

$$\mathcal{L}_m = \frac{1}{2}\left(\partial_\mu\varphi_1 - eA_\mu\varphi_2\right)^2 + \frac{1}{2}\left(\partial_\mu\varphi_2 - eA_\mu\varphi_1\right)^2 - \frac{1}{4!}f\varphi^4 - \frac{1}{4}F_{\mu\nu}^2, \tag{17}$$

where $\varphi^2 = \varphi_i\varphi_i$. For interaction of $\mathcal{L}_m$ with the conformal version we will get (in Landau gauge for vectors)

$$\begin{aligned}
V^{(1)} = {} & \frac{1}{4!}f\varphi^4 + \frac{1}{48(4\pi)^2}\left[\frac{10}{3}f^2 + 36e^4 + \frac{5}{12}\lambda^2 - \frac{17}{2}\lambda f\right]\varphi^4\left(\log\frac{\varphi^2}{\mu^2} - \frac{25}{6}\right) \\
& - \frac{1}{12}R\varphi^2 + \frac{1}{12(4\pi)^2}\left[-3e^2 + \frac{27}{32}\lambda\right]R\varphi^2\left(\log\frac{\varphi^2}{\mu^2} - 3\right).
\end{aligned} \tag{18}$$

Working on flat background, choosing $\mu^2 = \varphi_m^2$ and using $\dfrac{\partial V^{(1)}}{\partial\varphi} = 0$ one gets

$$2f(4\pi)^2 = \frac{11}{3}\left(\frac{5}{12}\lambda^2 - \frac{17}{2}\lambda f + 36e^4 + \frac{10}{3}f^2\right). \tag{19}$$

 S. Odintsov

Now the natural question appears: which from the gauge or gravitational couplings is leading?. If $e^4 > \lambda^2$ we are in the situation described by Coleman and Weinberg [3], and QG corrections are negligible. However, if λ^2 is of the same order as e^4 or bigger, we get:

$$2f(4\pi)^2 = \frac{11}{3}\left(\frac{5}{12}\lambda^2 + 36e^4\right), \tag{20}$$

and the rest of the terms are of higher order. Then, after dimensional transmutation

$$V^{(1)} = \frac{1}{48(4\pi)^2}\left[\frac{5}{12}\lambda^2 + 36e^4\right]\varphi^4\left(\log\frac{\varphi^2}{\varphi_m^2} - \frac{1}{2}\right). \tag{21}$$

From this expression one can get the QG corrections for the scalar-to-vector mass ratio. Indeed, after shifting the field, the mass of the scalar meson is given by

$$m^2(S) = V''(\varphi_m) = \frac{1}{6(4\pi)^2}\left[\frac{5}{12}\lambda^2 + 36e^4\right]\varphi_m^2. \tag{22}$$

After spontaneous symmetry breaking, the scalar makes the photon a massive vector meson; its mass is the same as in the absence of QG corrections: $m^2(V) = e^2\varphi_m^2$. Then,

$$\frac{m^2(S)}{m^2(V)} = \frac{1}{6(4\pi)^2}\left[\frac{5}{12}\frac{\lambda^2}{e^2} + 36e^2\right]. \tag{23}$$

Its is interesting to note that, if $\lambda^2 \gg e^4$, then the ratio (23) is defined by QG effects. One can also estimate the small-curvature corrections to this relation.

It is not difficult to find the analog of (23) in the general model of R^2-gravity (1) with scalar QED

$$\frac{m^2(S)}{m^2(V)} = \frac{1}{6(4\pi)^2}\left[36e^2 + \frac{\lambda^2\xi^2}{e^2}\left(-15 + \frac{3}{4\omega^2} - \frac{9\xi}{\omega^2} + \frac{27\xi^2}{\omega^2}\right)\right]. \tag{24}$$

Again, one can see that QG effects may become dominant in this relation.

The formalism developed here may be easily applied to any GUT. Of course, many questions are still left for further study, among of them numerical study of stability of the scalar coupling in different theories, study of curvature-induced phase transitions using RG-improved potentials and inducing Einstein gravity, clear understanding of how the above formalism may influence GUT phenomenology, and so on.

In summary, let us mention few more situations where above formalism may be applied.

a).In presented discussion we limited ourselves to the case of the effective potential. However, it is straitforward to repeat above analysis for study of RG improved effective Lagrangian in QG with matter [5]. Moreover, such technique maybe used to find RG improved effective Lagrangian in non-local form [6].

b).RG technique maybe applied to four-fermion models in curved spacetime [7] where Schwinger-Dyson equations are getting extremely difficult unlike the situation in flat space.

c).Study of effective equations of motion obtained from RG improved effective Lagrangian in quantum cosmology.

d). Investigations of the RG improved effective action in curved spacetime with non-trivial boundaries where surface running couplings are getting essential [8].

Acknowledgements

I would like to thank I. Antoniadis, I.Buchbinder, M. Einhorn, E.Elizalde, E. Mottola, A. Romeo and A.Wipf for discussions on related topics and M.Bordag and B.Geyer for kind invitation to give a talk on results of our research on Third Workshop on Quantum Field Theory under the influence of the External Conditions. This work has been supported by DGICYT (Spanish Government) and CIRIT (Generalitat de Catalunya), and partly by RFFR project 94 02-03234 and ISF project RI1000(Russia).

References

[1] I.L. Buchbinder, S.D. Odintsov and I.L. Shapiro, *Effective Action in Quantum Gravity*, IOP Publishing, Bristol and Philadephia, 1992).

[2] K.S. Stelle, *Phys. Rev.* **D 16** (1977) 953;

[3] S. Coleman and E. Weinberg, *Phys. Rev.* **D7** (1973) 1888.

[4] E. Elizalde and S.D. Odintsov, *Phys. Lett.* **B 303** (1993) 240; *Z. Phys.* **C**: Particles and Fields, **64** (1994) 699.

[5] E. Elizalde, S.D. Odintsov and A. Romeo, *Phys. Rev.* **D51** (1995) 1680.

[6] E. Elizalde and S.D. Odintsov, *Mod. Phys. Lett.* **A** to appear;

[7] E. Elizalde and S.D. Odintsov, *Phys. Rev.* **D51** (1995) 5950.

[8] S.D. Odintsov and A. Wipf, *Phys. Lett.* **B 356** (1995) 26.

Phase Structure of $2D$ Gross-Neveu Model in Spacetimes of Constant Curvature

Yurii I. Shil'nov

It is known that simple $2D$ four-fermion models reproduce some important features of modern quantum field theory. More than that, this kind of models describes sometimes a real particle physics. From the other hand, most of them are solvable analitically.

All of above-mentioned facts let us believe that the four-fermion models in curved spacetimes have the similar attractive properties and could be fruitful to understand the evolution of early Universe in more detail.

One of the most important phenomena of quantum field theory are the dynamical symmetry breaking and dynamical generation of fermionic mass. However, the curvature influence has been found to cause sometimes the symmetry restoration. Therefore, it should be paid a particular attention to the calculation of effective potential of composite fermionic fields on curved manifolds. It provides us the most reliable results describing these processes.

In the present paper we study the Gross-Neveu model in both S^2 and H^2 spaces. The renormalized effective potential has been obtained for arbitrary curvature in the framework of large N expansion. The curvature-induced phase transition and chiral symmetry restoration have been shown to occur for any value of coupling constant at sufficiantly large curvature in S^2 . It does not take place, unlike previous case, in H^2

The exact relation between critical values of coupling constant and curvature has been derived and the phase diagram has been constructed in the space S^2 .

Gross-Neveu model in an external gravitational field is described by the following action:

$$S = \int d^2x \sqrt{-g}[i\overline{\psi}\gamma^\mu(x)\nabla_\mu\psi + \frac{\lambda}{2N}(\overline{\psi}\psi)^2]. \qquad (1)$$

Introducing the auxilary field

$$\sigma = -\frac{\lambda}{N}(\overline{\psi}\psi) \qquad (2)$$

and using the standard large N expansion, we get the effective potential as follows:

$$V(\sigma) = \frac{\sigma^2}{2\lambda} + \mathrm{Sp}\int_0^\sigma D(x,x,s)ds, \qquad (3)$$

where D is the Euclidean propagator.

Let us begin from the positive curvature case - space S^2. The solution of equation for D is given by the linear combination of hypergeometric function. Taking into account the appropriate boundary conditions and selecting the renormalization condition of the form

$$V''(\sigma)|_{R=0,\sigma=\mu} = \frac{1}{\lambda}, \tag{4}$$

we find the derivative of the renormalized effective potential

$$V'(\sigma) = \frac{\sigma}{\lambda}\left\{1 + \frac{\lambda}{2\pi}\left[\psi\left(1 + i\sqrt{\frac{2\sigma^2}{R}}\right) + \psi\left(1 - i\sqrt{\frac{2\sigma^2}{R}}\right) - 2 - \ln\frac{2\mu^2}{R}\right]\right\}. \tag{5}$$

$\psi(x)$ is the digamma function here.

The asymtotes of effective potential can be calculated for both small and strong curvature. In the small curvature limit our expression coincides with the former results and the chiral symmetry is broken, as it happens in flat spacetime. And vice versa, we can conclude from the other asymtote that this symmetry is restored at the large curvature.It means that there is the curvature induced phase transition in this model.

These results are confirmed completely by numerical analysis,which clearly exhibits the continuous character of the phase transition.

With the notation

$$R_{\mathrm{cr}} = 2\mu^2 \exp\left[2(\gamma + 1) - 2\pi/\lambda\right], \tag{6}$$

where γ is the Euler constant, we obtain from the expression (5) that the point $\sigma = 0$ is a minimum for $R > R_{\mathrm{cr}}$ and a maximum for $R < R_{\mathrm{cr}}$. Thus, for any value of λ there exists a value of R, above which chiral symmetry is restored. In other words, formula (6) defines the curve on the $R - \lambda$ plane, which divides it into two areas with broken and unbroken symmetry.

In H^2 case $V(\sigma)$ looks very similar to expression (5), however phase transition never occurs and chiral symmetry is always broken.

Finally, I wish to express my deep gratitude to E. Elizalde, S. Leseduarte, S. Odintsov for collaboration, to DFG for financial support of my participation in the Workshop and to A. Letwin for support of my scientific activity.

Spinning Particles in Taub-NUT Background

Mihai Visinescu

In the last time the pseudo-classical limit of the Dirac theory of a spin-1/2 particle in curved space-time is described by the supersymmetric extension of the simple (spin-less) relativistic point particle. The spinning space represents the extension of the ordinary space-time with anti-symmetric Grassmann variables $\{\psi^\mu\}$ to describe the spin degrees of freedom [1].

The equations of motion of the pseudo-classical Dirac particles can be derived from the Lagrangian of the one-dimensional supersymmetric σ-model

$$\mathcal{L} = \int_a^b d\tau \left(\frac{1}{2} g_{\mu\nu}(x)\, \dot{x}^\mu \dot{x}^\nu + \frac{i}{2} g_{\mu\nu}(x)\, \psi^\mu \frac{D\psi^\nu}{D\tau} \right) \tag{1}$$

where $g_{\mu\nu}$ is the metric of the curved space. Here and in the following the overdot denotes an ordinary proper-time derivative $d/d\tau$ and the covariant derivative of ψ^μ is defined by

$$\frac{D\psi^\mu}{D\tau} = \dot{\psi}^\mu + \dot{x}^\lambda\, \Gamma^\mu_{\lambda\nu}\, \psi^\nu. \tag{2}$$

In general the symmetries of a spinninig particle model can be divided into two classes. First, there are four independent *generic* symmetries which exist in any theory [1]. The second kind of conserved quantities, called *non-generic*, depend on the explicit form of the metric $g_{\mu\nu}(x)$. In this paper we shall consider in more details the case of spinning particles in Taub-NUT space [2-4]. Much attention has been paid to the Euclidean Taub-NUT metric since in the long-distance limit the relative motion of two monopoles is described approximately by its geodesics. On the other hand the Kaluza-Klein monopole was obtained by embedding the Taub-NUT gravitational instanton into five-dimensional Kaluza-Klein theory.

In a special choice of co-ordinates the Euclidean Taub-NUT metric takes the form:

$$ds^2 = (1 + \frac{4m}{r})[dr^2 + r^2 d\theta^2 + r^2 \sin^2\theta\, d\varphi^2] + \frac{16m^2}{(1+\frac{4m}{r})}[d\chi + \cos\theta\, d\varphi]^2. \tag{3}$$

Let us expand a conserved quantity $\mathcal{J}$ in a power series in the covariant momentum Π_μ:

$$\mathcal{J}(x, \Pi, \psi) = \sum_{n=0}^\infty \frac{1}{n!} \Pi^{\mu_1} \cdots \Pi^{\mu_n}\, \mathcal{J}^{(n)}_{\mu_1\cdots\mu_n}(x, \psi). \tag{4}$$

We have for the coefficients $\mathcal{J}^{(n)}_{\mu_1\cdots\mu_n}$ the following generalized Killing equations:

$$\mathcal{J}^{(n)}_{(\mu_1\cdots\mu_n;\mu_{n+1})} + \frac{\partial \mathcal{J}^{(n)}_{\mu_1\cdots\mu_n}}{\partial \psi^\sigma} \Gamma^\sigma_{\mu_{n+1})\lambda}\, \psi^\lambda = \frac{i}{2}\, \psi^\sigma\, \psi^\lambda\, R_{\sigma\lambda\nu(\mu_{n+1}}\mathcal{J}^{(n+1)\nu}_{\mu_1\cdots\mu_n)}. \tag{5}$$

We present an analysis of the generalized Killing equations for the configuration space of spinning particles. The first generalized Killing equation (n=0) shows that with each Killing vector there is an associated Killing scalar. This equation has been solved in Refs.[2,4]. The second generalized Killing equation (n=1) gives the corrections of the Killing vectors which lead to a modification of the Runge-Lenz vector of the Taub-NUT spinning space [3].

A special attention is paid to the homogeneous parts of the generalized Killing equation (5) [5]. Simple solutions of the homogeneous part of these equations are expressed in terms of the Killing-Yano tensors [4],[6]. These solutions are connected with quantities which are separated conserved.

Acknowledgements

I would like to thank Dr. M. Bordag for well-organized and stimulated workshop. The hospitality and support during my stay in Leipzig is greately appreciated.

References

[1] See e.g. R.H. Rietdijk, "Applications of supersymmetric quantum mechanics", PhD.Thesis, Univ. Amsterdam (1992).

[2] M. Visinescu, *Class.Quant.Grav.* **11** (1994) 1867.

[3] M. Visinescu, *Phys.Lett.* **B339** (1994) 28.

[4] J.W. van Holten, *Phys.Lett.* **B342** (1995) 47.

[5] D. Vaman and M. Visinescu, in preparation.

[6] G.W. Gibbons, R.H. Rietdijk and J.W. van Holten, *Nucl.Phys.* **B404** (1993) 42.

Quantum Optics of Accelerated 2-level Systems on Arbitrary Stationary Trajectories

Jürgen Audretsch

Recently, a new physical picture for the spontaneous excitation of a uniformly accelerated two-level atom has been put forward [1]. Following a quantum optical approach it is based on the distinction of the two competing mechanisms which operate to excite an atom in the quantum vacuum: vacuum fluctuations and radiation reaction. Vacuum fluctuations tend to excite an atom in its ground state and de-excite it in an excited state. On the other hand, radiation reaction leads always to a loss of internal excitation energy. For an inertial atom in the ground state, the two contributions cancel exactly, so that a very sublime balance between vacuum fluctuations and radiation reaction prevents the spontaneous excitation of the atom [2]. If the atom is in the excited state, both contribution add up to the well-known spontaneous emission rate. It has been shown in [1] that uniform acceleration will disturb this balance. Spontaneous transitions from the ground state to the excited state become possible: the Unruh effect [3].

Continuing earlier work [1] and [4] more general states of motion are discussed in considering arbitrary stationary trajectories in Minkowski space [5]. We obtained the following results: The contribution of radiation reaction is the same for all stationary accelerated trajectories as in the inertial case. Accordingly, radiation reaction is a purely local concept that is not sensitive to the actual state of motion of the atom. Next we showed that the contribution of vacuum fluctuations differs for all atoms on accelerated stationary trajectories from that of an inertial atom. Together, these two results demonstrate that a generalized Unruh effect takes place for all stationary trajectories except the inertial ones. In this sense, the spontaneous excitation of an atom in its ground state is the normal case and the non-occurrence of the Unruh effect for inertial atoms is the exception. We furthermore demonstrated that the radiative energy shift ("Lamb shift") of a two-level atom is modified by acceleration for all stationary trajectories. Again only vacuum fluctuations give rise to the shift. Finally, the Lamb shift for a circulating atom (ring accelerator) is calculated by applying the theorem of Ref. [6].

References

[1] J. Audretsch, R. Müller; Phys. Rev. A **50**, 1755 (1994).

[2] J. Dalibard, J. Dupont-Roc, C. Cohen-Tannoudji; J. Physique **43**, 1617 (1982).

[3] W. G. Unruh; Phys. Rev. D **14**, 870 (1976).

[4] J. Audretsch, R. Müller; Phys. Rev. A **52**, 629 (1995).

[5] J. Audretsch, R. Müller, M. Holzmann; Class. Quantum Grav. **12**, 2927 (1995).

[6] J. Audretsch, R. Müller, M. Holzmann; Phys. Lett. A **199**, 151 (1995).

Radiation-Field Quantization for Linear Dielectrics through Green's Function Expansion

Toralf Gruner and **D.-G. Welsch**

The use of instruments in optical experiments needs careful examination with regard to their action on the quantum statistics of radiation. Instruments that respond linearly to radiation may be regarded as spatially structured dielectrics whose presence can be taken into consideration by quantizing the phenomenological Maxwell theory. Since radiation in dielectric matter undergoes dispersion and absorption, the losses must necessarily be included in a quantization scheme. This central problem can be solved by using a novel approach that is based on a Green's function expansion of the radiation field in place of a mode expansion, which fails when the losses are taken into account. The theory is applied to calculate quantum-theoretically consistent input-output relations for dispersive and absorptive multi-slab dielectric plates. Assuming that the polarization field and the electric field are locally related to each other, the phenomenological Maxwell equations governing the propagation of radiation in linear dielectrics, without external sources, can be quantized by supplementing them by an appropriately chosen noise source associated with the losses in the dielectric matter. For the sake of transparency, let us consider a multi-slab dielectric plate with permittivity $\epsilon(x,\omega)$ and restrict attention to radiation that propagates along the x axis and is polarized in z direction. In this case the basic equation for the vector potential (in the Fourier space) reads as [1]

$$\left[\frac{\partial^2}{\partial x^2} + \frac{\omega^2}{c^2}\epsilon(x,\omega)\right] \hat{A}(x,\omega) = \hat{j}_n(x,\omega), \qquad \hat{j}_n(x,\omega) = \frac{\omega}{c^2}\sqrt{\frac{\hbar}{\pi\epsilon_0}}\epsilon_i(x,\omega)\,\hat{f}(x,\omega), \quad (1)$$

where the noise source $\hat{j}(x,\omega)$ is related to a bosonic field $\hat{f}(x,\omega)$ through $\epsilon_i(x,\omega) = \mathrm{Im}\{\epsilon(x,\omega)\}$. By means of the Green function $G(x,x',\omega)$ of the inhomogeneous wave equation in (1), the vector potential can be given by

$$\hat{A}(x) = \int_0^\infty d\omega\, \frac{\omega}{c^2}\sqrt{\frac{\hbar}{\pi\epsilon_0\mathcal{A}}} \int_{-\infty}^\infty dx'\, G(x,x',\omega)\,\sqrt{\epsilon_i(x',\omega)}\,\hat{f}(x',\omega) + \mathrm{H.c.} \qquad (2)$$

($\mathcal{A}$, normalization area), which implies that $[\hat{A}(x),\hat{E}(x')] = -i\hbar(\mathcal{A}\epsilon_0)^{-1}\delta(x-x')$. For an N-slab device, such as an $(N-2)$-slab dielectric plate of thickness l in free space or in (two) dielectrics, Eq. (2) can be rewritten as

$$\hat{A}(x) = \sum_{j=1}^N \chi_j(x) \int_0^\infty d\omega\, \sqrt{\frac{\hbar}{4\pi c\omega\epsilon_0\beta_j(\omega)\mathcal{A}}}\, \frac{\beta_j(\omega)}{n_j(\omega)} \left[e^{i\beta_j(\omega)\omega x/c}\,\hat{a}_{j+}(x,\omega)\right.$$

$$+e^{-i\beta_j(\omega)\omega x/c}\hat{a}_{j-}(x,\omega)\Big] + \text{H.c.,} \qquad (3)$$

where $\chi_j(x)$ is equal to unity inside the jth slab and zero otherwise, and $\sqrt{\epsilon_j(\omega)} = n_j(\omega) = \beta_j(\omega) + i\gamma_j(\omega)$. In Eq. (3) the amplitude operators $\hat{a}_{j+}(x,\omega)$ and $\hat{a}_{j-}(x,\omega)$ associated with the (damped) waves propagating to the right and left, respectively, satisfy quantum Langevin equations of the type

$$\frac{\partial}{\partial x}\hat{a}_{j\pm}(x,\omega) = \mp\gamma_j(\omega)\frac{\omega}{c}\,\hat{a}_{j\pm}(x,\omega) \mp i\sqrt{2\frac{\omega}{c}\gamma_j(\omega)}\,e^{\mp\omega\beta_j(\omega)x/c}\hat{f}(x,\omega). \qquad (4)$$

Equation (4) together with the boundary conditions for the Green function at the surfaces of discontinuity can be used to relate the output operators $\hat{a}_{1-}(-l/2,\omega)$ and $\hat{a}_{N+}(l/2,\omega)$ to the input operators $\hat{a}_{1+}(-l/2,\omega)$ and $\hat{a}_{N-}(l/2,\omega)$ and (bosonic) noise sources $\hat{g}_{\pm}(\omega)$ associated with the losses within an $(N-2)$-slab dielectric plate [2]:

$$\begin{pmatrix} \hat{a}_{1-}(-l/2,\omega) \\ \hat{a}_{N+}(l/2,\omega) \end{pmatrix} = \mathbf{T}(\omega)\begin{pmatrix} \hat{a}_{1+}(-l/2,\omega) \\ \hat{a}_{N-}(l/2,\omega) \end{pmatrix} + \mathbf{A}(\omega)\begin{pmatrix} \hat{g}_{+}(\omega) \\ \hat{g}_{-}(\omega) \end{pmatrix}. \qquad (5)$$

When the plate is embedded in free space, the input operators are the ordinary (x-independent) photon destruction operators $\hat{a}_{1+}(\omega)$, $\hat{a}_{N-}(\omega)$ associated with the incoming modes in free space and the characteristic transformation and absorption matrices $\mathbf{T}(\omega)$ and $\mathbf{A}(\omega)$, respectively, satisfy the conditions

$$|T_{11}(\omega)|^2 + |T_{12}(\omega)|^2 + |A_{11}(\omega)|^2 + |A_{12}(\omega)|^2$$

$$= |T_{21}(\omega)|^2 + |T_{22}(\omega)|^2 + |A_{21}(\omega)|^2 + |A_{22}(\omega)|^2 = 1, \qquad (6)$$

$$T_{11}(\omega)T_{21}^*(\omega) + T_{12}(\omega)T_{22}^*(\omega) + A_{11}(\omega)A_{21}^* + A_{12}(\omega)A_{22}^* = 0. \qquad (7)$$

These conditions ensure the preservation of the bosonic commutation relations, so that the output operators are the (x-independent) photon operators associated with the outgoing modes in free space. Disregarding the losses, $\mathbf{A}(\omega) \approx 0$, the the well-known results of unitary transformation $\mathbf{T}(\omega)$ are recognized.

References

[1] For the three-dimensional case, see T. Gruner, D.-G. Welsch, Preprint FSUJ TPI QO-01/95, to be published in Phys. Rev. A.

[2] T. Gruner, D.-G. Welsch, Preprint FSUJ TPI QO-06/95, quant-ph/9511041.

Localization of Electromagnetic Waves in two Dimensions

Arkadiusz Orlowski, M. Rusek and J. Mostowski

The localization of the electron wave functions is the well-known concept in contemporary condensed matter physics. It originates from investigations of the electron transport in disordered solids, usually semiconductors [1], where the propagation of electrons is altered by the presence of a random potential. This phenomenon is completely based on the interference effects in multiple elastic scattering. Since interference is the common property of all wave phenomena, many generalizations of the Anderson localization to other matter-waves (neutrons) as well as classical waves (electromagnetic and acoustic waves) have been proposed. We focus our attention on electromagnetic waves. There is a variety of experimental investigations in this case, both in the optical and microwave domains. The question as to whether interference effects in strongly-scattering random media can reduce the diffusion constant to zero producing purely localized states depends on dimension of a sample under consideration. Despite some reasonable indications that strong localization could be possible in three-dimensional random dielectric structures the convincing experimental demonstration has been given only for two dimensions [2]. In this case the strongly-scattering medium has been provided by a set of dielectric cylinders randomly placed between two parallel aluminium plates on half the sites of a square lattice.

In the standard approach to localization of electromagnetic waves [3] a monochromatic wave is called localized in a non-dissipative dielectric medium if the squared modulus of the electric field $|\vec{\mathcal{E}}(\vec{r})|^2$ is localized. This definition is based on the analogy between the Helmholtz equation and the time-independent Schrödinger equation. Detailed analysis of the scattering of electromagnetic waves has pointed out that the only conserved quantity is the energy density of the field [4]. We also prefer to say that a monochromatic field is localized, if the time-averaged energy density of the field vanishes far from a certain region of space.

We believe that what really counts for localization is the scattering cross-section and not the geometrical shape and real size of the scatterer. Therefore we will represent the dielectric cylinders located at the points $\vec{\rho}_a$ by two-dimensional dipoles

$$\vec{\mathcal{P}}(\vec{r}) = \sum_{a=1}^{N} \vec{p}_a \, \delta^{(2)}(\vec{\rho} - \vec{\rho}_a). \tag{1}$$

Since the polarization of our system varies only at a certain plane, we have introduced cylindric coordinates $\vec{r} = (\vec{\rho}, z)$ in the above formula. For the sake of simplicity let us now assume that both the free field and the medium are linearly polarized along the z-axis. Therefore our discussion may be restricted to the scalar theory. It is now evident from that the electric field of the wave radiated by the a-th dipole reads as

$$\vec{\mathcal{E}}_a(\vec{r}) = \vec{e}_z \, k_0^2 \, p_a \, 2 \, K_0(-ik_0|\vec{\rho}|) \tag{2}$$

where K_0 denotes the modified Bessel function of the second kind.

The crucial point is how each dipole should be coupled to the electromagnetic field. The standard Lorenz-Lorentz formula [5] is now rather useless, because it is only approximately valid in the macroscopic limit. To provide a realistic and self-consistent description we must assume that the average energy is conserved in the scattering process. Therefore, if we isolate a single dipole then the time-averaged field energy flux integrated over a closed surface surrounding it should vanish for an arbitrary incident wave. It is remarkable that this simple and obvious requirement gives an explicit form of the field-dipole coupling. After simple calculations using the Kirchhoff integral formula [5] we get

$$i\pi\, k_0^2\, p_a = \frac{1}{2}(e^{i\phi} - 1)\mathcal{E}'(\vec{\rho}_a), \tag{3}$$

where the field of the wave incident on the a-th dipole,

$$\mathcal{E}'(\vec{\rho}_a) = \mathcal{E}^{(0)}(\vec{\rho}_a) + \sum_{b \neq a} \mathcal{E}_b(\vec{\rho}_a),, \tag{4}$$

is the sum of the free field and the waves radiated by other dipoles and ϕ_a is some arbitrary real number.

Inserting (3) into (2) and using (4), we finally arrive at the very simple set of linear algebraic equations determining the field acting on each dipole $\mathcal{E}'(\vec{\rho}_a)$ for a given free field $\mathcal{E}^{(0)}(\vec{\rho}_a)$. If we solve it and calculate the dipole moments we are able to find the electromagnetic field everywhere in space using the Maxwell equations. To prove localization in the simple case of fields sources vanishing outside of a certain region of space one has to investigate the electromagnetic field in the free space outside of the sources. According to our definition of localization if an electromagnetic wave is localized then the time-averaged field energy density must vanish far from the sources. It is evident from that in this region the radiated field becomes zero. Therefore the free field must vanish on a sufficiently large closed surface. Thus the Kirchhoff integral formula implies that the free field is zero everywhere inside this surface. It is now evident that if the system of dipoles (1) provides localization of the electromagnetic wave then the system of equations (4) should have a nonzero solution for vanishing free field $\mathcal{E}^{(0)} = 0$. This means that the eigenvalues λ corresponding to the eigenvectors of the system (1)

$$\lambda\, \mathcal{E}'(\vec{\rho}_a) = \mathcal{E}^{(0)}(\vec{\rho}_a), \quad a = 1, \dots N, \tag{5}$$

describing localized waves should be equal to zero. In a general case the eigenvalues λ depend on the positions of dipoles $\vec{\rho}_a$ and the phases ϕ_a describing their coupling to the field. However, if $\vec{\rho}_a$ are given, then for each eigenvector of the system of equations (4) there exists a certain angle ϕ for which the modulus of the corresponding eigenvalue λ takes a minimal value.

As a simple example let us consider a system of 100 dipoles (1) distributed randomly in a square with the density of one dipole per wavelength squared. We have calculated and diagonalized numerically the matrix describing this situation. Then we have chosen a certain eigenmode of the system (4) and checked if the corresponding eigenvalue can approach zero. We obtained the minimal value of its modulus $\propto 10^{-2}$. Therefore the field incident on each dipole $|\mathcal{E}'(\vec{\rho}_a)|^2$ is large compared to the free field $|\mathcal{E}^{(0)}(\vec{\rho}_a)|^2$ calculated at the dipole. Therefore the time-averaged energy density in the medium under consideration can be much greater than the energy density in the surrounding free space. Such a quasi-localization is practically indistinguishable from the perfect one.

Obviously perfect localization is impossible in systems (1) consisting of a finite number of dipoles. However, for a fixed density of the medium under consideration quasi-localization becomes better for an increasing number of dipoles. Indeed, the minimal possible eigenvalue from *all* eigenmodes decreases as a function of the number of dipoles N. We have constructed a model of a system of dielectric cylinders with the frequency-dependent permeability $\epsilon(\omega)$ chosen to provide the best possible localization for all frequencies. Indeed, in such an abstract medium *quasi*localization takes place practically for all sufficiently large densities. Of course in an experiment the dielectric cylinders must have finite diameter to provide scattering cross-section sufficiently large for localization which imposes a limit on the maximal density.

In summary, we have presented a novel approach to localization of electromagnetic waves in two dimensions based on Maxwell equations in integral form. Dielectric medium providing localization is modeled by a system of discrete dipoles. Each dipole corresponds to a single dielectric cylinder. In this treatment the scattering properties of the medium may be eliminated from the considerations and the best possible localization for a given distribution of particles can be studied. It was shown that in a medium consisting of a finite number of dielectric cylinders with specially chosen frequency-dependent permeability perfect light localization is impossible. However quasi-localization becomes better with increasing density of the medium or number of scattering particles.

References

[1] P.W. Anderson, Phys. Rev. **109**, 1492 (1958).

[2] R. Dalichaouch, J.P. Armstrong, S. Schultz, P.M. Platzman, and S.L. McCall, Nature **354**, 53 (1991).

[3] S. John, Physics Today, 32 (May 1991).

[4] B.A. van Tiggelen and E. Kogan, Phys. Rev. A **49**, 708 (1994).

[5] M. Born and E. Wolf, *Principles of Optics* (Pergamon Press, Oxford, 1965).

Spontaneous Emission by Atoms near a Dielectric Waveguide

Wladyslaw Zakowicz and A. Blędowski

Spontaneous emission of photons by an atom placed near the dielectric waveguide is analyzed using global free electromagnetic modes, satisfying necessary continuity conditions at the waveguide boundaries [1]. These modes include modes extended over the whole space (travelling) and modes trapped to the waveguide. To describe angular properties of the spontaneous emission, a parameterization of the travelling modes by the outgoing waves has been proposed. Otherwise, using a more common parameterization of these modes by the incoming waves, one has to take into account the interference of different reflected and transmitted waves accompanying photons incoming from both sides of the waveguide [2], what was overlooked in [3]. Angular characteristics of the travelling photon emission (a), distributions of the trapped photon emission (b), as well as global decay rates (c), have been calculated using the quantum approach and standard perturbation theory. Enhancement and suppression of radiative decay rates in various configurations have been shown. A certain calculational error, in the treatment of the density of trapped states, in an earlier paper by Khosravi and Loudon [4], affecting emission rates for thin waveguides, has been corrected.

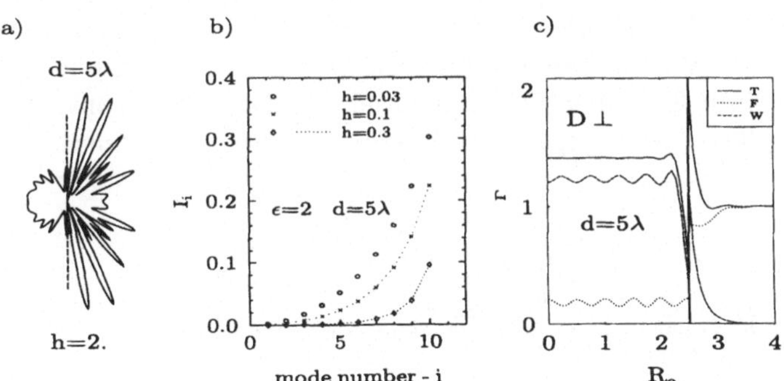

References

[1] W. Zakowicz, A. Blędowski Phys.Rev. **A 52**,1640 (1995).

[2] W. Zakowicz Phys.Rev. **A 52**,882 (1995). R.J. Glauber and M. Lewenstein, Phys.Rev. **A 43**,467 (1991).

[3] H. Khosravi and R. Loudon, Proc. R.Soc.Lond.A **436**,373(1992).

The Vacuum of $(QED)_{0+1}$ and Functional Integral Measure with Boundary Conditions

Jürgen Löffelholz

Gauge theories like QCD are expected to have θ-vacuum structure [1]. Unfortunately it is not clear whether starting from some lattice approximation - and only this we control - the heuristic picture survives the continuum limit (role of instantons, topological term, boundary conditions). Also for chiral symmetry breaking by a Fermion mass arguments may fail. We looked for a quantum mechanical model - simulating such properties - we are able to formulate with mathematical rigour and hence give proper physical meaning. Our model called $(QED)_{0+1}$ can be defined by the Hamiltonian $H = ((p - eA)^2 + E^2)/2 + M(1 - \cos\varphi)$, where $(\varphi, A) \in S \times R$ are coordinates on a cylinder and p, E canonical momenta. Let $\mathcal{A}$ denote the C^*-algebra generated by Weyl operators $e^{i(n\varphi + \beta p)} \otimes e^{i(aA + bE)}$, $n \in Z$. After trivializing the center we obtain a unique irreducible representation π so that H exists and has a ground state [2]. Since $Q = \varphi - E/e$ commutes with H the GNS-Hilbert space is the nonseparable sum of sectors $\mathcal{H}(\theta)$ which are invariant under the dynamics and characterized by values $e^{i\theta}, \theta \in S$, for $U = e^{iQ}$. Clearly, U^n yield large gauge transformations. If $M = 0$ all the vectors $\Omega_\theta \simeq \lim_{T \nearrow +\infty} e^{-TH}(e^{ie\theta A}\Omega)$, $\theta \in S$, get the same energy whereas for $M > 0$ the true vacuum Ω belongs to the sector with $\theta = 0$. We construct the associated functional measure $d\mu_{[-T,T]}$ governing trajectories $t \in [-T, T] \to (\varphi(t), A(t))$ and, vice versa, recover the quantum mechanics exploiting OS-positivity [3]. We learn that in the case $M > 0$ starting with local boundary conditions in the limit $T \nearrow +\infty$ for normalized correlation functions of gauge invariant variables the phase factor $\exp ie\theta(A(T) - A(-T))$ becomes irrelevant.

References

[1] C. Callan, R. Dashen, D. Gross, *The structure of the gauge theory vacuum* Phys.Letters **B 63**, 334-340 (1976)

[2] J. Löffelholz, G. Morchio, F. Strocchi, *The $(QED)_{0+1}$ model and a possible dynamical solution of the strong CP problem* Preprint University of Pisa IFUP-TH 7 (1995)

[3] K. Osterwalder, R. Schrader, *Axioms for Euclidean Green's functions* Comm. in Math.Phys. **31**, 83-112 (1973)

Part VI

Appendix

Program of the Workshop

Monday, Sept. 18   [Casimir Effect]

Morning session   (chairman G. Barton)

K.A. Milton (Univ. of Oklahoma) *The Casimir Effect in Dielectric Media (Spherical Geometry)*

V. Ritus (Lebedev Phys. Inst., Moscow) *Boson and Fermion Radiations of Accelerated Mirror in 1+1-Space-Time and Their Connection With Radiations of Electric And Scalar Charges in 3+1-Space-Time*

H. Osborn (Univ. of Cambridge) *Conformal Invariance and Its Implications for Quantum Field Theories with Plane Boundaries in General Dimensions*

A. Actor (Penn State Univ.) *Hard, Semi-Hard and Soft Boundary Conditions*

Afternoon Session   (chairman K.A. Milton)

G. Barton (Univ. of Sussex) *On quantum Radiation from Mirrors Moving Sideways*

C. Eberlein (Univ. of Illinois) *Sonoluminescence as Quantum Vacuum Radiation*

C. Villarreal (Univ. Nacional Autónoma de México) *Particle Emission, Squeezed States, and Uniformly Moving Boundaries*

V. Dodonov (Lebedev Physics Inst., Moscow) *Photons Generation and Squeezing in a Cavity with Vibrating Walls*

Evening Session   (chairman M. Bordag)

D. Robaschik *Fluctuations of the Casimir Pressure at Finite Temperature*

I. Brevik (Univ. Trondheim) *Casimir Theory for the Piecewise Uniform String*

V. Mostepanenko (Technological Inst., St.Petersburg) *Casimir Force Between the Real Boundaries: The Countributions of Mechanical and Electrical Imperfections*

Tuesday, Sept. 19   [QED and QCD in Ext. Fields]

Morning session   (chairman I. Buchbinder)

M. Reuter (DESY-Theorie) *Scale-Dependent Effective Actions and the Structure of the QCD Vacuum*

V. Skalozub (Dniepropetrovsk State University) *Effective Lagrangians of Arbitrary Inhomogeneous Electromagnetic Fields in a Dense Fermionic Medium*

A. Wipf (Univ. Jena) *Gauge Theories in a Bag*

A. Abrikosov Jr. (Moscow) *Instantons in the Bag Model*

Afternoon Session   (chairman H. Osborn)

V. Skarzhinsky (Lebedev Phys. Inst., Moscow) *QED in the External Aharonov-Bohm Field*

E. Moreira Jnr. (Queen Mary and Westfield College, London) *Quantum Fields in a Conical Background*

E. Ferrer (Univ. of New York at Fredonia) *Magnetic Response of Chern-Simons Gauge Theory at Finite Density*

N. Khusnutdinov (Kazan State Ped. Univ.) *Self - Interaction Force for Charged Particle in the Space - Time of Supermassive Cosmic String*

Evening Session   (chairman A. Wipf)

G. Bhattacharya (Saha Inst. of Nucl. Physics, Calcutta) *Exact Evaluation of Euler-Heisenberg Effective Action in some Special External Field Configurations*

A. Nikishov (Lebedev Phys. Inst., Moscow) *Energy-Stress Tensor of Particles Produced by an External Field*

Y. Sitenko (Inst. for Theoret. Physics, Kiev) *Nonlocality, Self-Adjointness of the Hamiltonian and Vacuum Polarization in Spaces with Nontrivial Topology*

S. Falkenberg (Univ. Jena) *Hydrogen Atom in a String Background*

Wednesday, Sept. 20   [Ground State in Ext. Fields]

Morning session   (chairman A. Actor)

S. Dowker (Univ. Manchester) *Functional Determinants on Möbius Corners*

E. Elizalde (Univ. of Barcelona) *Applications of Zeta Function Regularization in QFT*

M. Bordag, K. Kirsten (Leipzig Univ.) *Ground State Energy in Smooth Background Fields*

A. Rebhan (DESY, Gruppe Theorie) *Quantum Field Theory at Finite Temperature and Cosmological Perturbations*

Afternoon Session   (chairman J. Audretsch)

P. Henning (GSI Darmstadt) *Finite Temperature Field Theory: Physical Effects of Nontrivial Spectral Functions*

K. Kirsten (Leipzig Univ.) *On Bose-Einstein Condensation in External Fields*

J. Latorre (Univ. of Barcelona) *Speed of Light in Non-Trivial Vacua*

V. Zhukovsky (Moscow State Univ.) *Physical Effects in Quantum Field Theory, Influenced by External Fields, Finite Temperature and Matter Density*

Evening Session   (chairman E. Elizalde)

I. Avramidi, (Univ. Greifswald) *Effective Potential in Yang-Mills Theory and the Stability of the Chromomagnetic Vaccuum*

R. Schimming (Univ. Greifswald) *Comparison of Algorithms for the Calculation of the Heat Kernel Coefficients*

J. Lindig (Leipzig Univ.) *Local Energy Density in Smooth Background Fields*

S. Voropaev (Vernadsky Inst., Moscow) *Contact Interactions of Free Anyons*

Thursday, Sept. 21   [Quantum Fields in Black Hole Background]

Morning session   (chairman G. Esposito)

V. Frolov (Univ. of Alberta) *Entropy of Black Holes*

D. Hochberg (Univ. Auton. de Madrid) *Thermodynamical Features of Black Holes Dressed with Quantum Fields*

A. Bytsenko (State Techn. Univ., St.Petersburg), S. Zerbini, G. Cognola, L. Vanzo *Finite Temperature Effects in Black Hole Background*

S. Krasnikov (Astronomical Observ. Pulkovo) *On the Quantum Instability of the Time Machine*

Afternoon Session   (chairman V. Mostepanenko)

A. Popov (Pedag. Univ., Kazan) *Selfconsistent Semiclassical Solution with a Throat in the Theory of Gravity*

V. de la Incera (Univ. of New York at Fredonia) *Global Symmetries and Mass Eigenvalues of the Bosonic String in Electromagnetic Background*

S. Sushkov (State Pedag. Inst., Kazan) *Casimir Effect in the Bounded Throat of the Wormhole and a Model of the Elementary Charged Particle*

Evening Session   (chairman M. Reuter)   [Quantum Optics]

J. Audretsch (Univ. Konstanz) *Quantum Optics of Accelerated 2-Level-Systems on Arbitrary Stationary Trajectories*

T. Gruner (Univ. Jena) *Radiation-Field Quantization for Arbitrary Linear Dielectrics Through Green's Function Expansion*

A. Orlowski (Inst. of Physics, Warsaw) *Analytical Approach to Localization of Electromagnetic Waves*

W. Zakowicz (Inst. of Physics, Warsaw) *Spontaneous Emissions Near the Dielectric Slab*

R. Saibatalov (EF University, St.Petersburg) *Electromagnetic Field in Causal and Acausal Rotating Universe*

Friday, Sept. 22   [Topics in (Quantum) Gravity]

Morning session   (chairman V. Frolov)

I. Buchbinder (State Pedag. Inst., Tomsk) *Quantum Theory of Bosonic String in the Background Fields of Massive Modes*

G. Esposito (INFN, Napoli) *Non-Local Properties in Euclidean Quantum Gravity*

D. Galtsov (Moscow State University) *Geroch Group for Dilaton-Axion Gravity*

A. Aminova (Kazan State Univ.)   *Quantizations of Kähler Spaces Addmiting H-Projective Mappings*

Afternoon Session   (chairman D. Galtsov)

S. Odintsov (Univ. of Barcelona) *Renormalization Group and Effective Action in Renormalizable Quantum Gravity with Matter*

A. Grib, <u>E.A. Poberii</u> (EF University, St.Petersburg) *Vacuum Instability of Minimally Coupled Scalar Field in Curved Background: On the Problem of Minimal and Conformal Couplings in General Relativity.*

Y. Shil'nov (Kharkov State University) *Phase Structure of 2D Gross-Neveu Model in Spacetimes of Constant Curvature*

M. Visinescu (Inst. of Atomic Physics, Bucharest) *Spinning Particles in Taub-Nut Background*

J. Löffelholz (Leipzig) *Vacuum Structure of the $(QCD)_{0+1}$ - Model, Euclidean Greens Functions with Boundary Conditions, and Uniqueness of Their Infinite Volume Limit*

List of Participants

Abrikosov Jr., Alexei — Institute for Theoretical & Experimental Physics, 117259 Moscow, Russia,
e-mail: persik@vitep5.itep.ru

Actor, Alfred — Department of Physics, The Pennsylvania State University, Fogelsville, PA 18051, USA,
e-mail: aaa2@psuvm.psu.edu

Audretsch, Jürgen — Univertsität Konstanz, Fakultät für Physik, Postfach 5560 M673, 78434 Konstanz, Germany,
e-mail: audre@spock.physik.uni-konstanz.de

Aminova, Asya — Dept. of General Relativity and Gravitation, Kazan State University, Lenin Str. 18, 420008 Kazan, Russia,
e-mail: aminova@phys.ksu.ras.ru

Avramidi, Ivan — Fachbereich Mathematik, Universität Greifswald, Fr.-L.-Jahn-Str. 15a, 17487 Greifswald, Germany,
e-mail: avramidi@math-inf.uni-greifswald.d400.de

Barton, Gabriel — Physics & Astronomy Division, University of Sussex, Brighton BN1 9QH, England,
e-mail: barton@sussex.ac.uk

Bhattacharya, Gautam — Saha Inst. of Nucl. Physics, 1/AF Bidhannagar, 700064 Calcutta, India,
e-mail: gautam@tnp.saha.ernet.in

Bordag, Michael — Institut für Theoretische Physik, Universität Leipzig, Augustusplatz 10, 04109 Leipzig, Germany,
e-mail: bordag@qft.physik.uni-leipzig.d400.de

Brevik, Iver — Applied Mechanics, The Norwegian Institute of Technology, 7034 Trondheim, Norway,
e-mail: brevik@unit.no

Buchbinder, Ioseph — Department of Theoretical Physics, Tomsk State Pedagogical Institute, 634041 Tomsk, Russia,
e-mail: josephb@tspi.tomsk.su

Bytsenko, Andrei — Dept. of Theoretical Physics, State Techn. Univ., 195251 St.Petersburg, Russia,
e-mail: abyts@tuexph.spb.su

Cognola, Guido — Dept. of Physics, Univ. of Trento, via Sommarive, 38053 Povo/Trento, Italy,
e-mail: cognola@science.unitn.it

Dodonov, Victor — Lebedev Physics Institute, Leninsky Pr. 53, 117924 Moscow, Russia,
e-mail: dodonov@sci.fian.msk.su

Dowker, Stuart — Department of Theoretical Physics, The University of Manchester, Manchester M130PL, England,
e-mail: dowker@a3.ph.man.ac.uk

Duffy, Gavin — Univ. College Dublin, Belfiled, Dublin 4, Ireland,
e-mail: duffy@ollamh.ucd.ie

Eberlein, Claudia — TCM, Cavendish Lab, Madingley Rd, Cambridge CB3 0HE, England,
e-mail: cce20@phy.cam.ac.uk

Elizalde, Emilio — Center for Advanced Study CEAB, CSIC, 17300 Blanes, and Department ECM and IFAE, Faculty of Physics, University of Barcelona, Diagonal 647, 08028 Barcelona, Spain,
e-mail: eli@zeta.ecm.ub.es

Erdmenger, Johanna — Univ. of Cambridge Silver Street, Cambridge CB3 9EW, England,
e-mail: erdmenger@amtp.cam.ac.uk

Esposito, Giampiero — Instituto Nazionale di Fisica Nucleare, Sezione di Napoli, Mostra d'Oltremare Padiglione 20, 80125 Napoli, Italy, Dipartimento di Scienze Fisiche, Mostra d'Oltremare Padiglione 19, 80125 Napoli, Italy,
e-mail: esposito@napoli.infn.it

Falkenberg, Sven — Friedrich-Schiller-Universität Jena, Theoretisch-Physikalisches Institut, M.-Wien-Platz 1, 07743 Jena, Germany,
e-mail: falkengb@tph100.physik.uni-leipzig.de

Ferrer, Efraim J. — Physics Dept., SUNY, Fredonia, NY 14063, USA,
e-mail: ferrer@fredonia.edu

Frolov, Valeri — CIAR Cosmology Program; Theoretical Physics Institute, University of Alberta, Edmonton T6G 2J1, Canada and Lebedev Physics Institute, Leninsky Pr. 53, 117924 Moscow, Russia,
e-mail: frolov@phys.ualberta.ca

Galtsov, Dmitri — Department of Theoretical Physics, Moscow State University, 119899 Moscow, Russia,
e-mail: galtsov@grg.phys.msu.su

Geyer, Bodo — Institut für Theoretische Physik, Universität Leipzig, Augustusplatz 10, 04109 Leipzig, Germany,
e-mail: geyer@ntz.uni-leipzig.de

Gruner, Torsten — Friedrich-Schiller-Universität Jena, Theoretisch-Physikalisches Institut, Max-Wien-Platz 1, 07743 Jena, Germany,
e-mail: gruner@hpfs1.physik.uni-jena.de

Henning, Peter — Theoretische Physik, Gesellschaft für Schwerionenforschung (GSI), P.O. Box 110552, 64220 Darmstadt, Germany,
e-mail henning@gsi.de

Hochberg, David — Departamento de Física Teórica, C-XI, Universidad Autónoma de Madrid, Cantoblanco, 28049 Madrid, Spain,
e-mail hochberg@ccuam3.sdi.uam.es

Incera, Vivian de la — Physics Dept., SUNY, Fredonia, NY 14063, USA,
e-mail: ferrer@fredonia.edu

Kaiser, Hans-Jürgen	DESY-IFH Zeuthen, 15735 Zeuthen, Germany, e-mail: hkaiser@ifh.de
Khusnutdinov, Nail	Department of Theoretical Physics, Kazan State Pedagogical University, Mezhlauk str.1, 420021 Kazan, Russia, e-mail: nail@dtp.ksu.ras.ru
Kirsten, Klaus	Institut für Theoretische Physik, Universität Leipzig, Augustusplatz 10, 04109 Leipzig, Germany, e-mail: kirsten@tph100.physik.uni-leipzig.de
Krasnikov, Sergei	Astronomical Observ. of RAN, Pulkovo, 196140 St.Petersburg, Russia, e-mail: redish@pulkovo.spb.su
Krtous, Pavel	Theoretical Physics Institute, University of Alberta, Edmonton T6G 2J1, Canada, e-mail: krtous@phys.ualberta.ca
Latorre, Jose I.	Departament E.C.M, Facultat de Física, Univ. of Barcelona, Diagonal 647, 08028 Barcelona, Spain, e-mail: latorre@sophia.ecm.ub.es
Lindig, Joachim	Institut für Theoretische Physik, Universität Leipzig, Augustusplatz 10, 04109 Leipzig, Germany, e-mail: lindig@tph100.physik.uni-leipzig.de
Löffelholz, Jürgen	Frankenheimer Weg 28, 04205 Leipzig, Germany, e-mail: loeffelholz@qft.physik.uni-leipzig.de
Luig, Klaus	Fachbereich Physik, Universität Dortmund, Otto-Hahn-Str. 4, 44221 Dortmund, Germany, e-mail: luig@het.physik.uni-dortmund.de
Milton, Kimball A.	Department of Physics and Astronomy, The University of Oklahoma, Norman OK 73019, USA, e-mail: milton@phyast.nhn.uoknor.edu
Moreira Jnr., Edisom S.	Physics Department, Queen Mary and Westfield College, Mile End Rd., London E1 4NS, UK, e-mail: moreira@qmw.ac.uk
Mostepanenko, Vladimir	St.Petersburg State Technological Institute and Friedmann Laboratory for Theoretical Physics, 7-Krasnoarmeyskaya str. 6/8, 198052 St.Petersburg, Russia, e-mail: mostep@tu.spb.ru
Müller, Uwe	DESY-IFH Zeuthen, 15735 Zeuthen, Germany, e-mail: umueller@convex.ifh.de
Nikishov, Anatoly	Lebedev Physical Institute, Moscow, Leninsky pr. 53, 117924 Moscow, Russia, e-mail: nikishov@lpi.ac.ru
Odintsov, Sergei	Dept. of Mathematics and Physics, Pedagogical Institute, 634041 Tomsk, Russia, and Dept. ECM, Faculty of Physics,

University of Barcelona, Diagonal 647, 08028 Barcelona, Spain,
e-mail: sergei@ecm.ub.es

Orlowski, Arkadiusz | Instytut Fizyki, Polska Akademia Nauk, Aleja Lotników 32/46, 02–668 Warszawa, Poland,
e-mail: orlow@beta2.ifpan.edu.pl

Osborn, Hugh | DAMTP, Silver St., Cambridge CB3 9EW, England
e-mail: osborn@damtp.cam.ac.uk

Plunien, Günter | Institut für Theoretische Physik, TU Dresden, 01062 Dresden, Germany,
e-mail: plunien@ptprs9.phy.tu-dresden.de

Poberii, Eugene | UEF University, Griboedov k. 30/32, 191023 St.Petersburg, Russia
e-mail: pober@friedman.usr.lgu.spb.su

Pollifrone, Giuseppe | Dipartimento di Fisica, Univ. di Roma "La Sapienza", P. le Aldo Moro 2, 00185 Roma, Italy,
e-mail: pollifrone@roma1.infn.it

Popov, Arkadiy | Department of Geometry, Kazan Pedagogical University, Mezhlauk str. 1, 420021 Kazan, Russia,
e-mail: popov@kspu.ksu.ras.ru

Rebhan, Anton K. | Deutsches Elektronen-Synchrotron (DESY), Notkestraße 85, 22603 Hamburg, Germany,
e-mail: rebhana@x4u2.desy.de

Reuter, Martin | Deutsches Elektronen-Synchrotron (DESY), Notkestraße 85, 22603 Hamburg, Germany,
e-mail: t00reu@dsyibm.desy.de

Ritus, Vladimir | Lebedev Physics Institute, Leninsky Pr. 53, 117924 Moscow, Russia,
e-mail: ritus@lpi.ac.ru

Robaschik, Dieter | Daumierstr. 13, 04157 Leipzig, Germany,
e-mail: robaschi@qft.physik.uni-leipzig.de

Saibatalov, Rustam | UEF University, Griboedov k. 30/32, 191023 St.Petersburg, Russia
e-mail: rustam@friedman.usr.lgu.spb.su

Scharnhorst, Klaus | Department of Physics, University of Wales, Singleton Park, Swansea SA2 8PP, UK,
e-mail: scharnhorst@swansea.ac.uk

Schimming, Rainer | Fachbereich Mathematik, Universität Greifswald, Jahnstr. 15a, 17489 Greifswald, Germany,
e-mail: schimming@math-inf.uni-greifswald.d400.de

Shil'nov, Yurii | Dept. of Theoretical Physics, Faculty of Physics, Kharkov State University, Svobody Sq., 4, 310077 Kharkov, Ukraine,
e-mail: shilnov@office.kharkov.ua

Sitenko, Yurii	Bogolyubov Institute for Theoretical Physics, National Academy of Ukraine, 252143 Kiev, Ukraine, e-mail: yusitenko@gluk.apc.org
Skalozub, Vladimir	Dnepropetrovsk State University, 320625 Dnepropetrovsk, Ukraine, e-mail: skalozub@uni.dnepropetrovsk.ua
Skarzhinsky, Vladimir	Lebedev Physics Institute, Leninsky Pr. 53, 117924 Moscow, Russia, e-mail: skarzh@npad.fian.msk.su
Sushkov, Sergey	Department of Geometry, Kazan Pedagogical University, Mezhlauk str. 1, 420021 Kazan, Russia, e-mail: sushkov@kspu.ksu.ras.ru
Vanzo, Luciano	Dept. of Physics, Univ. of Trento, via Sommarive, 38053 Povo/Trento, Italy, e-mail: vanzo@alpha.science.unitn.it
Villarreal, Carlos	Instituto de Física, Universidad Nacional Autónoma de México, Apdo. Postal 20-364, México 1000, D.F., México, e-mail: villarreal@ifunam.ifisicacu.unam.mx
Visinescu, Mihai	Department of Theoretical Physics, Institute of Atomic Physics, P.O.Box MG-6, Magurele, Bucharest, Romania, e-mail address: mvisin@roifa.ifa.ro
Voropaev, Sergey	Vernadsky Institute, Lab. of Theoretical and Mathematical Physics, Kossygin Street 19, 117918 Moscow, Russia, e-mail: voropaev@tph100.physik.uni-leipzig.de
Wipf, Andreas	Friedrich-Schiller-Universität Jena, Theoretisch-Physikalisches-Institut, Max-Wien-Pl. 1, 07743 Jena, Germany, e-mail: aww@hpfs1.physik.uni-jena.d400.de
Yanovski, Aleksander	Kliment Ohridski University, James Boucher Boul. 5, 1126 Sofia, Bulgaria, e-mail: yanovski@fmi.uni-sofia.bg
Zakowicz, Wladyslaw	Institute of Physics, Polish Academy of Sciences, Al.Lotników 32, 02-668 Warsaw, Poland, e-mail: zakov@ifpan.edu.pl
Zerbini, Sergio	Dept. of Physics, Univ. of Trento, via Sommarive, 38053 Povo/Trento, Italy, e-mail: zerbini@alpha.science.unitn.it
Zimmerschied, Frank	Fachbereich Physik, Universität Kaiserslautern, Erwin-Schrödinger Str., 67663 Kaiserslautern, Germany, e-mail: zimmers@physik.uni-kl.de
Zhukovsky, Vladimir	Department of Theoretical Physics, Faculty of Physics, Moscow State University, 119899 Moscow, Russia, e-mail: grats@grg1.phys.msu.su

TEUBNER-TEXTE zur Physik

Preisänderungen vorbehalten.

B. G. Teubner Stuttgart · Leipzig